INTERNATIONAL CENTRE FOR MECHANICAL SCIENCES

COURSES AND LECTURES - No. 286

HYDRODYNAMICS OF LAKES

EDITED BY

K. HUTTER
FEDERAL INSTITUTE OF TECHNOLOGY
ZURICH

SPRINGER-VERLAG WIEN GMBH

This volume contains 95 illustrations.

Originally published by Springer Verlag Wien-New York in 1984

ISBN 978-3-211-81812-1 **ISBN 978-3-7091-2634-9 (eBook)**
DOI 10.1007/978-3-7091-2634-9

PREFACE

With the growing recognition of the significance of our resources and with the increase in contamination of our environment by industrial and municipal waste, our communities have become more and more sensitive to the rôle which lakes play in the natural environment. While the biology and chemistry of lakes have been part of limnological research ever since lakes have been scientifically studied, the investigation of their *physical dynamics*, i.e. the response of lakes to meteorological forcings, is relatively recent.

The purpose of this monograph (an outgrowth of a 1983 - summer course held at the International Centre for Mechanical Sciences (CISM) in Udine, Italy) is to summarize current understanding of large scale water circulation in lakes and to outline the methodologies by which this understanding has been aquired. The intention is to provide the necessary background for the analysis of transport of water and particulate matter, to present the theoretical and computational methods of solutions, and to establish a thorough interrelationship between theoretically deduced results and observations. Apart from presenting the fundamentals of continuum physics that apply to large scale motion of a fluid in a basin on the rotating Earth, the book contains presentations of basin wide wave dynamics, non-linear wave action, precise studies of the vertical current structure, detailed investigations of barotropic and baroclinic current forecast modelling as well as an account of the history of physical limnology.

A quick glance through the table of contents will show that the book will be useful not only to physical limnologists, but equally also to oceanographers, meteorologists, fluid dynamicists, physicists and applied mathematicians with an interest in environmental or geophysical fluid dynamics. Because its content is fairly coherent, it may also serve as a basic text for a general course in lake hydrodynamics.

The lectures were given by six renowned scientists, and thus, repetitions and certain inhomogeneities in the text were not able to be avoided, even though an attempt was made to obtain a coherent manuscript. The final manuscript was typed in Zurich. It is my pleasant duty to express my sincere thanks to the authorities of CISM for the organisation of the course and to Mr. F. Langenegger for typing the final version of the camera ready manuscript. Last but not least I thank Prof. D. Vischer, director of the Laboratory of Hydraulics, Hydrology and Glaciology ETH, Zurich for his permission to pursue this project.

Zurich, October 1983 K. Hutter

PREFACE

[illegible] the pro[illegible] of [illegible] water [illegible] resources [illegible] with the increase in [illegible] our [illegible] quality [illegible] which takes place in the natural environment [illegible] lakes have been [illegible] relatively recent.

The purpose of this [illegible] the [illegible] has been [illegible] for the [illegible] present [illegible] established [illegible] biology.

A [illegible] through [illegible] not [illegible] meteorology [illegible] mathematicians with [illegible] fluid dynamics [illegible] hydrodynamics.

The [illegible] twenty six [illegible] in the [illegible] to obtain a coherent [illegible] the [illegible] during the final [illegible] this volume.

Zürich, October 19[illegible] [illegible]

CONTENTS

Hydrodynamics of Lakes: CISM Lectures
edited by K. Hutter, 1984
Springer Verlag Wien-New York

FUNDAMENTAL EQUATIONS AND APPROXIMATIONS

Kolumban Hutter
Laboratory of Hydraulics, Hydrology
and Glaciology,
Gloriastrasse 37/39, ETH,
CH-8092 Zurich, Switzerland

Air photograph of Lake of Lugano, photographed
from the South to the North (Photo SWISSAIR).

1. INTRODUCTION

The fundamental physical principles governing the motion of lake waters are the conservation laws of mass, momentum and energy. When diffusion processes of active or passive tracer substances are also considered, these laws must be complemented by transport equations of tracer mass. All these statements have the form of *balance laws* and in each of them flux terms arise, for which, in order to arrive at *field equations*, phenomenological postulates must be established. Hydrodynamics of lakes can be described by a *Navier-Stokes-Fourier-Fick fluid* or its simplifications. Its field equations are partial differential equations for the velocity vector $\underline{v}$, the pressure p, the temperature T and, possibly, the mass concentrations c_α ($\alpha = 1, \ldots, N$) of N different tracers (i.e. a suspended sediment, phosphate, nitrate, salinity, etc.). Boundary conditions for $\underline{v}$, p, T and c_α must also be established; in view of the fact that surfaces may deform and that evaporation may occur, these are not alltogether trivial. In fact the equations of motion of the free or of internal surfaces of density discontinuity - these are the so-called *kinematic surface equations* - serve as further field equations with the surface displacements as unknown boundary variables. Additional boundary conditions have to be formulated at the lake bottom and along the shore. The latter play a more significant role in physical limnology than in oceanography because for many phenomana the boundedness of the lake domain will affect the details of the processes while oceans may for the same processes be regarded as infinite or semi-infinite. This, for instance, implies that by and large wave spectra in the ocean are continuous, while they are often quantized in lakes.

The most general set of field equations is seldom used, and if so, one is very soon confronted with numerical techniques. Simplifications are introduced by either restricting attention to special processes (for which certain equations and variables drop out) or by using scaling arguments which may render certain terms in the governing equations unimportant. For instance, one may disregard diffusion mechanisms of dissolved matter. The velocity $\underline{v}$, the pressure p and the temperature T are then the only field variables and balance of mass, momentum and energy the corresponding field equations. Or density (and consequently also temperature) is assumed to remain constant for each fluid particle. This adiabatic assumption is valid whenever thermal diffusion is negligible as compared to convection.

Most significant, however, are the *Boussinesq-*, the *hydrostatic pressure-* and the *incompressibility* assumptions. With these, density variations enter the expression of the buoyancy force only, and pressure can be eliminated as a field variable. The role of the pressure gradient is then played by gradients of the free surface and of the internal "discontinuity" surfaces of the density. Two distinct classes of motion can then be distinguished and, depending upon whether the rotation of the Earth is significant, these can be complemented by a third. Very roughly, these may be classified as follows:

(i) Gradients in the free surface elevation drive the system, but inhomogeneities in the density field are unimportant. Processes are in this case gravity driven (surface deformation is significant) and are called *barotropic* (because density differences are unimportant).

(ii) Inhomogeneities in the density field are significant and the processes are essentially governed by buoyancy forces. They are again gravity driven and are classified as *baroclinic*. Since amplitudes of particle displacements are large where density gradients are large which occurs inside the lake, such motions are often also referred to as *internal*.

Except for spin-up and quasi-steady circulation processes time scales of barotropic and baroclinic oscillations are hours and days, respectively; they are sufficiently separated which explains why very often free surface deformations are ignored when internal oscillations are studied. Such "*rigid*" assumptions are also helpful in analyses of

(iii) rotation dominated processes. These only exist because of the rotation of the Earth and their existence requires a non-uniform bottom profile*). Time scales are generally an order of magnitude larger than for (internal) gravity-dominated processes.

To emphasize the significance of the non-uniformity in topography, these motions are called *topographic*. They may arise for both, barotropic and baroclinic processes but gravity plays a secondary role. It is for this reason that in wave analyses gravity dominated waves are called waves of the *first class* while those due to the rotation of the Earth are of *second class*.

To a large extent hydrodynamics of lakes deals with the identification of these separate forms of motions, with their generation by external (wind) forces and with their interaction. In what follows an attempt is

---

*) *More precisely, either the Coriolis parameter or the topography or both must vary with position. The Coriolis parameter is the quantity arising in the governing equations, which is responsible for the effects of the rotation of the Earth. It varies with geographical latitude, but the size of lakes is usually too small that such planetary effects would come to bear.*

made to present the governing equations and the motivations which lie behind the approximations that allow to single out the various simplified processes. As for notation, we shall use symbolic and Cartesian tensor notation interchangingly. Vector and tensor operators will be denoted by grad, curl, div etc. when they are applied to vectors and tensors in physical 3-space. When coordinates are restricted to the horizontal plane, gradient, divergence and curl will symbolically be written as ∇c, $\nabla \cdot \underline{a}$ and $\nabla \wedge \underline{a}$. In other words, the Nabla-operator will refer to two dimensions.

2. FIELD EQUATIONS AND BOUNDARY CONDITIONS

Fundamental to all continuum physics is the balance law for a physical quantity. Let g be its specific value (per unit volume), p its specific production, s its specific supply from external sources and $\underline{\phi}\cdot\underline{n}$ its flux through an areal increment with unit normal vector $\underline{n}$. The balance law for g within an arbitrary body part P with boundary ∂P is then the statement

$$\frac{d}{dt}\int_P g\,dv = \int_P p\,dv + \int_P s\,dv - \int_{\partial P} \underline{\phi}\cdot\underline{n}\,da\,; \qquad (2.1)$$

it simply sais that the time rate of change of g within P is balanced by its production, its supply and its flux through the boundary ∂P *). If g and $\underline{\phi}$ are differentiable within the body, (2.1) implies the local form

$$\frac{\partial g}{\partial t} + \operatorname{div}(g\underline{v} + \underline{\phi}) - s - p = 0, \qquad (2.2)$$

valid at any point of the body ($\underline{v}$ is the velocity vector). On the other hand, on a surface of discontinuity whose points travel with the velocity $\underline{u}$ and at which g, $\underline{\phi}$, s and p may experience finite jumps, (2.1) implies

$$[\![g(\underline{v} - \underline{u}) \cdot \underline{n}]\!] + [\![\underline{\phi} \cdot \underline{n}]\!] = \pi_g\,. \qquad (2.3)$$

Here, the bracket $[\![f]\!] = f^+ - f^-$ denotes the difference of the quantity enclosed as the surface is approached from the two sides. Further, $\underline{n}$ is the unit surface vector pointing into the positive side, and π_g is the surface production of g.

The laws (2.1) - (2.3) lie at the heart of ensuing developments as the physical laws of conservation of mass, momentum and energy can all be

*) *For more details, consult a text on continuum mechanics, such as Müller (1973), Chadwick (1976), Wang & Truesdell (1973), Becker & Bürger (1976).*

brouth into the form (2.2), and associated dynamic boundary conditions do have the form (2.3). Below, this will not be explained in all detail, but the reader can easily corroborate the identifications.

2.1 The basic field equations

In what follows, all fields A(t) will be separated into a large-scale-organized component $\bar{A}(t)$ and a randomly fluctuating component $A'(t)$, $A(t) = \bar{A}(t) + A'(t)$. As is customary in turbulence theory, field equations are only established for the averaged quantities $\bar{A}$. This requires that typical time scales, over which $\bar{A}$ and A' vary, are sufficiently different, so that $\bar{A'} = 0$. This implies that field equations for averaged quantities can be deduced from those for the total fields.

It is not our intention to derive these equations from basic principles. For that the reader may consult the pertinent literature*). Here it may suffice to state that the basic laws for the averaged quantities have the same structure as those for the total fields. These are the conservation laws of mass, momentum and energy and the balance laws of tracer mass and have the local forms

$$\begin{aligned}
&\text{mass:} && \dot{\rho} + \rho\,\mathrm{div}(\underline{v}) = 0, \\
&\text{tracer mass:} && \rho\,\dot{c}_\alpha = -\,\mathrm{div}(\underline{j}_\alpha + \underline{j}^R) + \pi_\alpha\,, \quad (\alpha = 1,2,\ldots,N), \\
&\text{momentum:} && \rho\,[\dot{\underline{v}} + 2\,\underline{\Omega} \wedge \underline{v}] = -\,\mathrm{grad}\,p + \mathrm{div}\,(\underline{t}^E + \underline{t}^R) + \rho\,\underline{g}\,, \\
&\text{energy:} && \rho\,c_p\,\dot{T} = -\,\mathrm{div}(\underline{q} + \underline{q}^R) + \mathrm{tr}(\underline{t}^E\,\underline{D}) + p^R + \rho\,r,
\end{aligned} \qquad (2.4a)^{**)}$$

in which overbars, identifying averaged fields, have been omitted, and where superimposed dots denote material time derivatives,

$$\psi^{\cdot} = \frac{\partial\psi}{\partial t} + \mathrm{grad}\,\psi \cdot \underline{v}\,. \qquad (2.5)$$

---

*) *Turbulence is studied, among others, by Batchelor (1952), Bradshaw (1976), Hinze (1959), Monin & Yaglon (1971) and Spalding (1982).*

**) *Balance of angular momentum simply implies that $\underline{t}^E$ and $\underline{t}^R$ must be symmetric. It does not form a generic equation.*

The jump conditions of mass momentum and energy, on the other hand read

$$\text{mass:} \qquad [\![\rho(\underline{v}-\underline{u})\cdot\underline{n}]\!] = 0,$$

$$\text{tracer mass:} \quad [\![\rho\, c_\alpha(\underline{v}-\underline{u})\cdot\underline{n}]\!] + [\![(\underline{j}_\alpha + \underline{j}_\alpha^R)\cdot\underline{n}]\!] = 0, \quad (\alpha = 1,2,\ldots,N),$$

$$\text{momentum:} \qquad [\![\rho\underline{v}(\underline{v}-\underline{u})\cdot\underline{n}]\!] - [\![(\underline{t}^E + \underline{t}^R - p\,\underline{1})\cdot\underline{n}]\!] = 0, \tag{2.4b}$$

$$\text{energy:} \qquad [\![\rho(c_p T + \tfrac{v^2}{2})(\underline{v}-\underline{u})\cdot\underline{n}]\!] + [\![(\underline{q}+\underline{q}^R)\cdot\underline{n}]\!] = 0.$$

Furthermore, $\wedge$ is the exterior vector product, and the various symbols have the following meanings:

ρ = mass density,
$\underline{v}$ = velocity vector,
$c_\alpha = \rho_\alpha/\rho$ mass concentration of tracer α,
$\underline{j}$ = diffusive molecular flux of tracer α,
$\underline{j}_\alpha^R = \overline{-\rho\, c'_\alpha\, \underline{v}'}$ diffusive turbulent mass flux of tracer α,
π_α = specific production rate of tracer mass α,
$\underline{\Omega}$ = angular velocity of the Earth,
p = pressure,
$\underline{t}^E$ = molecular viscous stress tensor,
$\underline{t}^R = \overline{-\rho\, \underline{v}' \otimes \underline{v}'}$ turbulent Reynolds stress tensor,
$\underline{g}$ = gravity vector,
T = temperature,
c_p = specific heat,
$\underline{q}$ = molecular heat flux vector,
$q^R = \overline{-\rho\, c_p T' v'}$ turbulent heat flux vector,
$p^R = \overline{\mathrm{tr}(\underline{t}^{E'}\, \underline{D}')}$ turbulent energy production,
r = specitic radiation,
$\underline{D} = 1/2(\mathrm{grad}\,\underline{v} + \mathrm{grad}^T\,\underline{v})$ stretching tensor, where T denotes the transpose.

We have omitted above the balance law of *entropy* (second law for thermodynamics). This law also has the form (2.1) with $g = \rho\eta$, $\underline{\phi} = \underline{q}/T$, $p = \rho\gamma$, where η is the specific entropy and γ its production. It is the expression of the second law that $\gamma \geq 0$.

To arrive at partial differential equations for $\underline{v}$, p, T and c_α phenomenological relations for the molecular and turbulent fluxes (the terms with the divergence operator) and for the turbulent energy production must be established. It is known that the molecular fluxes are one to three orders of magnitude smaller than the corresponding turbulent quantities. One may, therefore, ignore ($\underline{j}_\alpha$, $\underline{t}^E$, q) in comparison to ($\underline{j}^R$, $\underline{t}^R$, $\underline{q}^R$). The latter quantities are expressed in terms of the mean fields. The finding of such relationships is one of the most challenging problems of theoretical turbulence. The modern trend is to seek further balance laws for the turbulent fluxes. One such higher order model is known as the k - ε model, see Spalding (1982), but its use has been limited in problems of physical limnology. Here we shall take a more traditional approach and shall be satisfied with relations of the form

$$\psi = \hat{\psi}(c_\alpha, T, \text{grad } c_\alpha, \underline{D}, \text{grad } T). \qquad (2.6)$$

In particular we postulate

$$\begin{aligned}
\underline{j}_\alpha^R &= -\sum_{\beta=1}^{N} \underline{D}_{\alpha\beta}^R \text{ grad } c_\beta, & j_{\alpha k}^R &= -\sum_{\beta=1}^{N} (D_{k\ell}^R)_{\alpha\beta}\, c^\beta_{,\ell}\,, \\
\underline{t}^R &= \underline{M}^R \underline{D}\,, & t_{k\ell}^R &= M^R{}_{k\ell mn}\, D_{mn}, \qquad (2.7)\\
\underline{q}^R &= -\underline{\kappa}^R \text{ grad } T, & q_k^R &= -\kappa_{k\ell}^R\, T_{,\ell}\,,
\end{aligned}$$

and shall also set $p^R = 0$. The diffusive mass flux $\underline{j}_\alpha^R$ of tracer α is related to the concentration gradients, grad c_β with diffusivity tensors $\underline{D}_{\alpha\beta}^R$ which may also depend on scalar combinations of the remaining independent variables of (2.6). In the common case cross coupling between various tracers is weak, so that $\underline{j}_\alpha^R = \underline{D}_{\alpha\alpha}^R$ grad c_α (no sum over α) is a sufficient approximation to $(2.7)_1$. This relationship is known as *Fick's law*. Similarly, the Reynolds stress tensor $\underline{t}^R$ is linearly related to the stretching tensor $\underline{D}$ with a fourth rank tensor $\underline{M}^R$ whose elements have the dimension of viscosity. The number of independent coefficients which $\underline{M}$ may have depends on the kind of anisotropy one imposes upon the turbulence structure. For an isotropic incompressible fluid there is only one coefficient left and the emerging fluid is the familiar *Navier Stokes fluid*. For an orthotropic structure in which vertical turbulent diffusion differs from the horizontal diffusion, more details will be given below. Finally, the turbulent heat flux is related to the temperature gradient with the tensor of heat conduction $\underline{\kappa}^R$. This relationship is reminiscent of *Fourier's law of heat conduction*.

Besides (2.7) an equation of state for the density ρ must hold. In principle this equation has the form

$$\rho = \rho(T, c_1, \ldots, c_N, p), \qquad (2.8)$$

but for lakes with moderate depths the pressure dependence can be ignored. Furthermore, the mineral components have usually a negligible effect on the density, or can once and for all be worked into (2.8) so that $\rho = \rho(T)$ is sufficient. There are well known formulae*) for the temperature dependence of ρ, but for computational purposes it suffices to relate the *density anomaly* $\sigma = (\rho - \rho_*)/\rho_*$ to the *potential temperature* $\theta = (T - T_*)$, where $\rho_* = \rho(T_*)$ and $T_* = 4^\circ C$; the relation is

$$\sigma = -\varepsilon\,\theta^2, \qquad \varepsilon = 6.8 \cdot 10^{-6} \cdot C^{-2}. \tag{2.9}$$

Finally, we remark that equations (2.3) are referred to a steady rotating frame which is fixed on the Earth. The acceleration is given in this case by

$$\dot{\underline{v}} + 2\,\underline{\Omega} \wedge \underline{v} + \underline{\Omega} \wedge (\underline{\Omega} \wedge \underline{x}), \tag{2.10}$$

in which the second term is the Coriolis acceleration and the third the centrifugal acceleration. The first of these is contained in $(2.7)_3$, the second is small and can be thought of being ignored or absorbed into the gravity acceleration. It is also customary in limnology and oceanography to work with a Cartesian coordinate system, in which the (xy)-plane is tangential to the globe, with the positive x-coordinate pointing towards East and the y-coordinate pointing towards North. The z-axis is then radial as shown in Figure 1. In this coordinate system the Coriolis acceleration

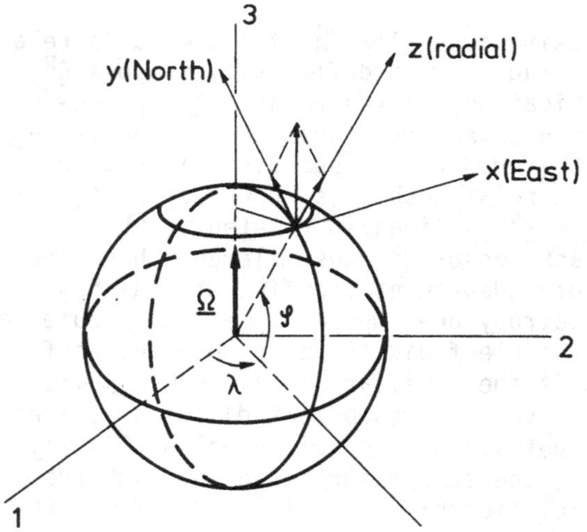

Figure 1

View of the globe with a stationary coordinate system, 1, 2, 3 and an Earth fixed coordinate system x, y, z. $\underline{\Omega}$ is the angular velocity of the Earth, λ and φ measure the position of the (x,y,z)-system relative to the 1,2,3 system. In the (x,y,z)-system the vector $\underline{\Omega}$ has two components in the y- and z-directions, as indicated.

*) *See i.e. Bowman & Schoonover (1967), Tilton & Taylor (1937), Ambühl & Bührer (1974) and the Handbook of Chemistry.*

has the components

$$2\underline{\Omega} \Lambda \underline{v} = [\tilde{f} w - f v, f u, -\tilde{f} u] \tag{2.11}$$

where

$$f = 2 |\underline{\Omega}| \sin \varphi, \quad \tilde{f} = 2 |\underline{\Omega}| \cos \varphi. \tag{2.12}$$

Here, u, v, w are the velocity components in the x-, y- and z-directions, φ is the latitude angle and f and $\tilde{f}$ are Coriolis parameters. With $|\underline{\Omega}| = 1.46 \cdot 10^{-4}\,s^{-1}$, f and $\tilde{f}$ are $O(10^{-4})$. In general, since $\varphi = \varphi(y)$, both quantities vary with position; however the horizontal dimension of most lakes is so small that f and $\tilde{f}$ may be regarded as constants, except at the equator where $f \cong 0$. This case will be excluded in these lectures*).

2.2 Boundary conditions

These must be established at the lake *bottom* and at the *free surface* and are of two different kinds, *dynamic* and *kinematic*.

Consider the *kinematic condition* first. Let $S(\underline{x},t) \equiv 0$ be the equation of the surface S. Because this equation is an identity, valid for all times, the time derivative following the surface, DS/Dt, must also vanish. Thus

$$\frac{DS}{Dt} \equiv \frac{\partial S}{\partial t} + \text{grad}\, S \cdot \underline{u} = 0, \tag{2.13}$$

where $\underline{u}$ is the surface propagation velocity. Therefore

$$\begin{aligned} \frac{\partial S}{\partial t} + \text{grad}\, S \cdot \underline{v}^{\pm} &= \text{grad}\, S \cdot (\underline{v} - \underline{u})^{\pm} \\ &= \| \text{grad}\, S \| \, (\underline{v} - \underline{u})^{\pm} \cdot \underline{n} \end{aligned} \tag{2.14}$$

$\underline{v}^{\pm}$ are the particle velocities on the positive and negative side of S and

$$\underline{n} \equiv \frac{\text{grad}\, S}{\| \text{grad}\, S \|} . \tag{2.15}$$

The important kinematic statement is Equation (2.13), but Equation (2.14) is more useful for practical applications. Important is that there are two forms of it referred to the (+) and (-) part of S. Further, the unit normal vector $\underline{n}$ is defined with the aid of Equation (2.15), and earlier it was assumed that $\underline{n}$ points into the positive part P^+; in other words,

---

*) *df/dy is denoted by β. These so-called β-effects are important in global atmospheric and oceanic processes, but are of no importance for lake dynamics.*

Equation (2.15) defines P^+ amd P^-. The left hand side of (2.14) may also be written as $(dS/dt)^{\pm}$, and $(\underline{v}-\underline{u})^{\pm}\cdot\underline{n}$ on the right can be interpreted as diffusive volume fluxes. We thus define

$$a_{\perp} \equiv (\underline{v}-\underline{u})^{-}\cdot\underline{n} = \frac{\rho^+}{\rho^-}(\underline{v}-\underline{u})^{+}\cdot\underline{n}, \tag{2.16}$$

in which $\rho^{\pm}$ are the mass densities of the body to the left and right of S, respectively. Hence, Equation (2.14) may more conveniently also be written as

$$\left(\frac{dS}{dt}\right)^{-} = \| \operatorname{grad} S \| \, a_{\perp}, \tag{2.17}$$

in which the (-) sign will be identified with the water side.

Boundary surfaces are stationary or moving surfaces of singularity. *Dynamic boundary conditions* emerge, therefore, by applying the jump condition (2.3) to the balance laws of mass, tracer mass, momentum and energy and by recognizing that in none of these cases there is a surface production term $\pi_g = 0$. At the *lake bottom* one usually assumes no-slip or sliding over the basal surface. In the first case the bottom is material, in the second it is not, but drainage vanishes. In both cases $[\![g(\underline{v}-\underline{u})\cdot\underline{n}]\!] = 0$, and (2.3) reduces to $[\![\underline{\phi}\cdot\underline{n}]\!] = 0$, or when applied to the balance laws of tracer mass, momentum and energy*)

$$[\![\underline{j}_\alpha\cdot\underline{n}]\!] = 0, \quad [\![\underline{t}\cdot\underline{n}]\!] = \underline{0}, \quad [\![\underline{q}\cdot\underline{n}]\!] = 0. \tag{2.18}$$

Accordingly, the tracer flux normal to the bottom is continuous and so are the corresponding components of the stresses and the heat flux. In the boundary value problem, heat and tracer mass flux are usually prescribed as the geothermal heat flow and as the tracer flow into the sediment.

The *free surface* is another boundary and it need not necessarily be material. For many problems, certainly all purely mechanical problems, it is, however, regarded as material and then the conditions (2.18) apply again. In more familiar terms these read

$$\underline{j}_\alpha\cdot\underline{n} = J_\alpha^{atm}, \quad \underline{t}\cdot\underline{n} = \underline{t}_{(n)}^{atm}, \quad \underline{q}\cdot\underline{n} = Q^{atm}, \tag{2.19}$$ *)

in which $\underline{J}_\alpha^{atm}$, $\underline{t}_{(n)}^{atm}$, Q^{atm} are the atmospheric tracer mass flow of constituent α, the wind stress and atmospheric pressure (collected as a vector) and the atmospheric heat flow, respectively.

---

*) *The stress tensor in this formula is the difference* $\underline{t}^R - p\,\underline{1}$ *of the Reynolds stress tensor* ($\underline{t}^E = \underline{0}$) *and the pressure tensor.*

In energy budget calculations and models of the seasonal development of the thermocline, however, evaporation may have to be included in the description. The free surface as a surface of phase change is then no longer material, and jump conditions apply in the form (2.3). One property of a surface of phase change is that the temperature is continuous, $[\![T]\!] = 0$. Thermodynamic arguments then show that*)

$$[\![q\cdot n]\!] = \rho^- L a_\perp , \tag{2.20}$$

which follows from an entropy balance and in which L is the heat of evaporation. The jump conditions (2.3) of tracer mass, momentum and entropy now yield

$$\begin{aligned}
&[\![\underline{j}_\alpha\cdot\underline{n}]\!] + [\![c_\alpha]\!]\,\rho^- a_\perp = 0, \Rightarrow [\![\underline{j}_\alpha\cdot\underline{n}]\!] \simeq 0, \quad (\alpha = 1,\dots,N)\\
&[\![\underline{t}\cdot\underline{n}]\!] - [\![\underline{v}]\!]\,\rho^- a_\perp = 0, \quad \Rightarrow [\![\underline{t}\cdot\underline{n}]\!] \simeq 0, \\
&[\![\underline{q}\cdot\underline{n}]\!] + L\rho^- a_\perp = 0, \quad \Rightarrow [\![\underline{q}\cdot\underline{n}]\!] + \rho^- L a_\perp = 0.
\end{aligned} \tag{2.21}$$

All three contain a term, which is proportional to $\rho^- a_\perp$, the mass flux through the surface, but, clearly in the jump conditions of tracer mass and of momentum these terms are so small in comparison to the first terms that they can safely be ignored. In the entropy jump, the term should, however, be kept. It introduces a new variable, $a_\perp$, for the determination of which the continuity requirement of the temperature, $[\![T]\!] = 0$, serves, as complementary equation.

This last statement needs clarification. $[\![T]\!] = 0$ means that at a free evaporating surface the temperature of the water particles equals that of the vapor and air particles just across the surface. The thermal diffusion process in the atmospheric boundary layer can, of course be incorporated. The usual procedure is to relate Q^{atm} with T_s^{atm} and T_∞^{atm} according to

$$Q^{atm} = h\cdot(T_s^{atm} - T_\infty^{atm}),$$

Here, T_s^{atm} is the atmospheric temperature at the surface, T_∞^{atm} is the

---

*) *Balance of entropy yields the jump condition*

$$[\![\frac{\underline{q}\cdot\underline{n}}{T}]\!] + [\![\rho\eta(\underline{v}-\underline{u})\cdot n]\!] = 0,$$

where η *is the specific entropy. For* $[\![T]\!] = 0$ *and with* $L \equiv T[\![\eta]\!]$ *Equation (2.20) is immediate. However, Equation (2.20) is also equivalent to the energy jump condition, which at a surface of phase change can be brought into a form* $[\![\underline{q}\cdot\underline{n}]\!] = \dots\dots$ *. Only one of these equations is needed and that for entropy is simpler. For more details see Müller (1973), Hutter (1983).*

atmospheric temperature outside the boundary layer and h is the *heat transfer coefficient* which itself depends on the flow state of the atmosphere.

As a special case of a kinematic condition, consider the free surface $S \equiv \zeta(x,y,t) - z = 0$. Equation (2.17) then reads

$$\frac{\partial \zeta}{\partial t} + \nabla\zeta \cdot (u,v) - w = a, \quad \text{on} \quad z = \zeta(x,y,t)$$
$$\text{with} \qquad a = \sqrt{1 + \|\dot{\nabla}\zeta\|^2}\, a_\perp . \tag{2.22}$$

Similarly, if $S \equiv -H(x,y) - z = 0$ is the equation of the bottom, (2.17) with $a_\perp = 0$ reduces to

$$\nabla H \cdot (u,v) + w = 0, \quad \text{on} \quad z = -H(x,y), \tag{2.23}$$

which is nothing else than the tangency condition of a sliding flow.

With regard to dynamic boundary conditions at the bottom [Equations (2.18)] the common assumption is to require that the geothermal heat flux vanishes and that the α-th tracer mass flux into the sediment is given by $J_\alpha^\perp$. Hence, since $\underset{\sim}{n} = (-\nabla H, -1)/\sqrt{1 + \|\nabla H\|^2}$, one obtains from the first and third of (2.18)

$$\begin{aligned} &\nabla H \cdot (q_x, q_y) + q_z = 0, \\ &\nabla H \cdot (j_x^\alpha, j_y^\alpha) + j_z^\alpha = -\sqrt{1 + \|\nabla H\|^2}\, J_\alpha^\perp \end{aligned} \tag{2.24}$$

for each α, valid on $z = -H(x,y)$. The term on the right hand side of Equation $(2.24)_2$ excluding the minus sign is the tracer mass flux into the sediment through a unit area in the horizontal plane. Sometimes it is not the geothermal heat flux that is prescribed but the temperature. The first of Equation (2.24) is then replaced by $T = T_b$ where T_b is known. The dynamic condition $[\![\underset{\sim}{q} \cdot \underset{\sim}{n}]\!] = 0$ still holds in this case but it determines the geothermal heat flux from the sediment as an after the fact information. An analogous situation exists for the jump condition of traction, which may be written as

$$-p\underset{\sim}{n} + \underset{\sim}{t}^R \underset{\sim}{n} = \underset{\sim}{t}_b, \tag{2.25}$$

determining the bottom traction $\underset{\sim}{t}_b$ (usually called bottom stress). Equation (2.25) is useful to evaluate basal traction but does not serve as boundary condition. *No slip* or *viscous sliding* is the mechanical boundary condition that is usually applied. If $\hat{\underset{\sim}{t}}_{(n)}$ denotes the tangential traction, one commonly assumes $\hat{\underset{\sim}{t}}_{(n)} = k\underset{\sim}{u}$ where k is a slip coefficient. With $k = 0$, $\hat{\underset{\sim}{t}}_{(n)} = \underset{\sim}{0}$ i.e. no shear traction and one has perfect slip, with $k \to \infty$, $\underset{\sim}{u} = \underset{\sim}{0}$, the no slip condition is obtained. With $\hat{\underset{\sim}{t}}_{(n)} = \underset{\sim}{t}^R\underset{\sim}{n} - (\underset{\sim}{n} \cdot \underset{\sim}{t}^R \underset{\sim}{n})\underset{\sim}{n}$ the viscous sliding law can be written as

$$\underline{t}^R \underline{n} - (\underline{n} \cdot \underline{t}^R \underline{n})\, \underline{n} = k\, \underline{u}\ , \tag{2.26}$$

which automatically satisfies the tangency condition $\underline{u} \cdot \underline{n} = 0$.

3. BASIC APPROXIMATIONS

In this section the popular approximations of physical limnology and oceanography will be discussed.

3.1 Hydrostatic equations of motion and Boussinesq assumption

At the heart of the hydrostatic equations lies in the *shallow water approximation*. It may, perhaps, best be introduced by sketching, how equations are conveniently non-dimensionalized. To this end, let L and H be typical horizontal and vertical length scales and U a typical horizontal velocity. In the shallow water approximation horizontal distances are scaled with L, vertical distances with H, horizontal velocities with U, vertical velocities with $W = H/L \cdot U$ and time with the Coriolis parameter f. The choice for the scale W is crucial, for it sets the ratio of the velocity scale W/U equal to the aspect ratio $\mathbb{A} = H/L$ which is generally a small number. Introducing non-dimensional variables (denoted by asterisks) according to

$$\begin{aligned} t^* = f\,t, \quad (x^*, y^*) = \frac{1}{L}(x, y), \quad z^* = \frac{z}{H} \\ (u^*, v^*) = \frac{1}{U}(u, v), \quad w^* = \frac{w}{W} \end{aligned} \tag{3.1}$$

and complementing these by appropriate scalings for the remaining variables it can be shown (see e.g. Simons, 1980; LeBlond & Mysak, 1980) that

(i) in the vertical momentum equation the dimensionless acceleration and turbulent friction terms are multiplied by $\mathbb{A}^2$, while the pressure gradient and the buoyancy force are not,

(ii) in the horizontal momentum equation the non-dimensional version of $\tilde{f}$ is multiplied by $\mathbb{A}$, while f is not.

In the scaled equations the dimensionless variables are of order unity and the significance of a particular coefficient arises as a scaling factor such as $\mathbb{A}$, and $\mathbb{A} < 0.1$ is small. In the limit as $\mathbb{A} \to 0$ (the shallow water approximation), of the rotation dependent parameters in the horizontal momentum equations only f survives, and the vertical momentum balance

(2.4), reduces to the statement

$$\frac{\partial p}{\partial z} = -\rho g, \tag{3.2}$$

which, upon integration, becomes

$$p(z) = \underbrace{p^{atm} + \rho_* g(\zeta - z)}_{p_e} + \underbrace{\rho_* g \int_z^{\zeta} \sigma(z')dz'}_{p_i}; \tag{3.3}$$

p^{atm} is the atmospheric pressure, $z = \zeta(x,y,t)$ the equation of the free surface and ρ_* the density of water at 4°C. The first two terms on the right of (3.3) are the external or barotropic pressure function p_e which depends on the atmospheric pressure and the free surface displacement, the integral term is the internal, or baroclinic pressure function p_i depending on the internal density distribution.

The shallow water approximation implies further simplifications, in particular in the turbulent fluxes (2.7). Most significant is, however, that the pressure can be eliminated with the aid of (3.3), and the momentum equations reduce to only two horizontal component equations. These remaining equations are called the *hydrostatic equations of motion*.

The approximation has also its limitations. Vertical velocities must be small in comparison to horizontal velocities. This excludes upwelling and downwelling zones from being properly modelled and precludes proper analysis of flow instabilities.

In the *Boussinesq assumption* one supposes that density differences are small in the sense that density variations are important in the Archimedian buoyancy force, but not, when density arises as a factor of a rate term. In other words, in the balance of mass (2.4) one may ignore $\dot{\rho}$ as compared to div $\underline{v}$, and in the remaining equations (2.4) one may replace ρ by ρ_*, a constant, in all terms on the left, but must keep ρ as depending on temperature in the external gravity force. One implication of this reduction [which can also be based on rigorous scaling arguments (see Greenspan, 1968)] is, that the balance of mass reduces to the statement div $\underline{v} = 0$, the classical incompressibility assumption*). The other is that as far as ρ is concerned the equations are linearized.

In summary, the shallow water approximation and the Boussinesq assumption reduce the mechanical field equations to

*) *Our incompressibility assumption is $\partial\rho/\partial p = 0$ and is more general than div $\underline{v} = 0$. We require no volume change under isothermal (and isohaline) processes, but permit these when temperature (and salinity) changes arise. Thus $\dot{\rho}$ does not vanish for all processes.*

$$\nabla \cdot \underline{v} + \frac{\partial w}{\partial z} = 0 \tag{3.4}$$

$$\frac{d}{dt}[\underline{v}] + f(\underline{k} \wedge \underline{v}) = -\frac{\nabla}{\rho_*}(p_e + p_i) - \nabla \cdot \underline{\Gamma} - \frac{\partial \underline{\gamma}}{\partial z}, \tag{3.5}$$

in which $\underline{v}$ is a two-vector with the components (u,v); $\underline{k}$ is a unit vector in the z-direction [implying $\underline{k} \wedge \underline{v} = (-v,u)$] and

$$\frac{d}{dt}[\underline{\phi}] = \frac{\partial \underline{\phi}}{\partial t} + \nabla \underline{\phi} \cdot \underline{v} + \frac{\partial \underline{\phi}}{\partial z} w = \frac{\partial \underline{\phi}}{\partial t} + \nabla \cdot (\underline{\phi} \otimes \underline{v}) + \frac{\partial}{\partial z}(\underline{\phi}\, w) \tag{3.6}$$

is the material time derivative operator. p_e and p_i are defined in (3.3), and $\underline{\Gamma} = \underline{\Gamma}^T$ and $\underline{\gamma}$ are a two-tensor and a two-vector which together form part of the Reynolds stress deviator $\underline{t}^R$ as indicated in (3.7):

$$\underline{t}^R = \left[\begin{array}{cc|c} \multicolumn{2}{c|}{-\rho^* \underline{\Gamma}} & -\rho^* \underline{\gamma} \\ \hline t_{13}^R & t_{23}^R & t_{33}^R \end{array}\right]. \tag{3.7}$$

If the turbulent closure conditions (2.7) are used and if it is assumed that the turbulent diffusion in the vertical direction differs from that in the horizontal direction (but all horizontal directions are equivalent), it can be shown (see Hutter & Trösch, 1975) that in the limit as $\mathbb{A} \to 0$ one has

$$\Gamma_{xx} = -a_1 \frac{\partial u}{\partial x} + \frac{a_2}{2}\frac{\partial v}{\partial y}, \quad \Gamma_{xy} = -a_2\left(\frac{\partial u}{\partial y} + \frac{\partial v}{\partial x}\right), \quad \Gamma_{yy} = +\frac{a_2}{2}\frac{\partial u}{\partial x} - a_1 \frac{\partial v}{\partial y},$$
$$\gamma_x = -\nu \frac{\partial u}{\partial z}, \quad \gamma_y = -\nu \frac{\partial v}{\partial z}, \tag{3.8}$$

in which a_1 and a_2 are horizontal turbulent viscosities and ν is the vertical turbulent viscosity and all three may depend on objective scalar combinations of $\underline{D}$, ρ, grad T. We should mention that the representations (3.8) do not agree with common turbulence postulates in the literature. Section 3.5 will elaborate more one this.

The forms of the tracer balance laws and of the energy equation, which are consistent with the shallow water and the Boussinesq assumptions and conform with orthotropic turbulence, are

$$\frac{d}{dt}[c_\alpha] = \nabla \cdot \left[\sum_{p=1}^{N} D_{\alpha\beta}^H \nabla c_\beta\right] + \frac{\partial}{\partial z}\left[\sum_{\beta=1}^{N} D_{\alpha\beta}^V \frac{\partial c_\beta}{\partial z}\right] + \pi_\alpha \tag{3.9}$$

$$(\alpha = 1,2,\ldots,N),$$

$$\frac{d}{dt}[T] = \nabla\cdot(K^H \nabla T) + \frac{\partial}{\partial z}(K^V \frac{\partial T}{\partial z}) + r(T), \tag{3.10}$$

in which $D^H_{\alpha\beta}$, $D^V_{\alpha\beta}$ and $K^H = \kappa^H/(\rho_0 c_p)$, $K^V = \kappa^V/(\rho_0 c_p)$ are horizontal and vertical tracer and thermal diffusivities. For further information on these, see section 3.5.

Finally, it should be mentioned that with the scalings (3.1) the non-dimensional convective acceleration terms are multiplied with the dimensionless Rossby-number $R = U/fL$. Convective accelerations are, therefore, only significant if $R = O(1)$. Section 3.2 deals with a simplification for which this is the case.

3.2 Adiabaticity

For processes whose time scales are short as compared to a typical relaxation time of (turbulent) diffusive processes, the flux terms ($\underline{j}^R$, $\underline{t}^R$, g^R) are negligible in comparison to the convective terms arising in the respective equations. For conservative tracers ($\pi_\alpha = 0$) and vanishing radiative heat ($r = 0$), Equations $(2.4)_2$ and $(2.4)_4$ then reduce to

$$\frac{d\,c_\alpha}{dt} = 0, \quad \alpha = 1,2,\dots,N, \quad \frac{dT}{dt} = 0. \tag{3.11}$$

Because incompressibility also requires that $\partial\rho/\partial p = 0$, (3.11) imply therefore

$$\frac{d\rho}{dt} = 0, \tag{3.12}$$

which is the adiabaticity statement. For an equation of state $\rho = \rho(T)$, the second of (3.11) already implies (3.12). It should be pointed out in this case that (3.12) is simply the reduced energy equation $dT/dt = (dT/d\rho)(d\rho/dt) = 0$, in which thermal diffusion and radiation are discarded. Sometimes, equation (3.12) is said to be a consequence of the incompressibility assumption, but this is a misconception because one independent equation ($\dot\rho + \rho \operatorname{div} \underline{v} = 0$) cannot be transformed into two equations ($\dot\rho = 0$ and $\operatorname{div} \underline{v} = 0$); so (3.12) is a genuine consequence of (3.11).

The field equations for adiabatic processes are therefore

$$\begin{aligned} &\operatorname{div} \underline{v} = 0, \\ &\dot\rho = 0, \\ &\rho(\dot{\underline{v}} + 2\underline{\Omega} \wedge \underline{v}) = -\operatorname{grad} p + \rho \underline{g} \end{aligned} \tag{3.13}$$

and these can be further reduced by invoking the shallow water and Boussinesq assumptions. If this is done they become

$$\nabla \cdot \underline{v} + \frac{\partial w}{\partial t} = 0,$$

$$\frac{\partial \rho}{\partial t} + \nabla \rho \cdot \underline{v} + \frac{\partial \rho}{\partial z} w = 0, \tag{3.14}$$

$$\frac{\partial \underline{v}}{\partial t} + \nabla \underline{v} \cdot \underline{v} + \frac{\partial \underline{v}}{\partial z} w + f \underline{k} \wedge \underline{v} = -\nabla(p_e + p_i)/\rho_*$$

where p_e and p_i are given in (3.3).

3.3 Linear approximations

These approximations are introduced as small perturbations of a basic simple solution of the nonlinear equations. Such a basic state is *hydrostatic equilibrium*

$$\underline{v} = \underline{0}, \quad \zeta = 0,$$
$$p_0(z) = p^{atm} + \int_z^0 \rho_0(z')\, g\, dz', \tag{3.15}$$

where $\rho_0(z)$ is the known vertical density distribution determined from vertical temperature and electrical resistivity measurements. An important quantity characterizing the state of rest and perturbations about it is the Brunt-Väisälä frequency, defined by

$$N^2 = -g \frac{d\rho_0/dz}{\rho_0}. \tag{3.16}$$

$N^2 > 0$ characterizes a stable state of rest.

Departures from the state of rest (3.15) are described by writing $p = p_0 + p'$, $\rho = \rho_0 + \rho'$ and $\underline{v} = \underline{v}'$, inserting these into the governing equations and linearizing the emerging equations in the primed quantities. This yields

$$\frac{\partial \rho'}{\partial t} + w \frac{d\rho_0}{dz} = 0,$$

$$\operatorname{div} \underline{v} = 0, \tag{3.17}$$

$$\rho_0 \frac{\partial \underline{v}}{\partial t} + 2 \rho_0 \underline{\Omega} \wedge \underline{v} + \operatorname{grad} p' + \rho' \underline{g} = 0,$$

appropriate for adiabatic processes. Notice that since $\rho_0 = \rho_0(z)$, the Boussinesq assumption is not incorporated in (3.17). This would require to replace on the left of $(3.17)_3$ ρ_0 by ρ_* and to approximate the Brunt-

Väisälä frequency by $N^2 = -g(d\rho_0/dz)/\rho_*$. This last simplification hardly affects the value of N.

Combining the Boussinesq and the small perturbation assumptions transforms (3.17) into

$$\frac{\partial\rho'}{\partial t} + w\frac{\partial\rho_0}{\partial z} = 0, \qquad \nabla\cdot\underline{v} + \frac{\partial w}{\partial z} = 0,$$
$$\rho_*\frac{\partial\underline{v}}{\partial t} + f\rho_*\underline{k}\wedge\underline{v} + \nabla p' = 0, \qquad \rho_*\frac{\partial w}{\partial t} + \frac{\partial p'}{\partial z} + \rho' g = 0, \tag{3.18}$$

in which $\underline{v}$ is the two-vector of the horizontal velocity and in the shallow water approximation p' could, with the aid of (3.3) and (3.15), be written as

$$\frac{1}{\rho_*}p' = g\zeta + \int_z^0 \frac{\rho-\rho_0}{\rho_*}\,g\,dz'; \tag{3.19}$$

however, often it is preferable to keep p' as an independent variable. Often one writes $(\rho'/\rho_*)\underline{g} = \underline{g}'$.

Linearizing field equations does not suffice to arrive at linear initial boundary value problems. At the free surface $z = \zeta(x,y,t)$ we apply $(2.19)_2$ and (2.22) and obtain

$$\left.\begin{aligned} &p_0(\zeta) + p'(x,y,\zeta,t) = p^{atm}, \\ &\frac{\partial\zeta}{\partial t} + \nabla\zeta\cdot\underline{v} - w = 0, \end{aligned}\right\} \text{ at } z = \zeta, \tag{3.20}$$

in which p_0 is the hydrostatic pressure (3.15) and p' the dynamic perturbation. The nonlinearity has entered (3.20) in the convective terms of the kinematic condition but also because ζ appears as an independent variable in p', $\underline{v}$ and w. Assuming small amplitudes Taylor series expansion about $z=0$ may be used to transform (3.20) to the linearized conditions

$$\begin{aligned} &-\rho_0 g\zeta + p'(x,y,0,t) = 0, \\ &\frac{\partial\zeta}{\partial t} = w(x,y,0,t), \end{aligned} \qquad \text{at } z = 0, \tag{3.21}$$

which could also be written as a single equation in p', namely

$$\left(\frac{\partial^2}{\partial t^2} + N^2\right)p' + g\frac{\partial p'}{\partial z} = 0, \qquad \text{at } z = 0. \tag{3.22}$$

In the derivation of (3.21), (3.15) was used, and (3.22) follows from (3.21) and a combination of $(3.17)_1$, the third component of $(3.17)_3$ and (3.16). The first term in the bracket is the non-hydrostatic term.

3.4 Vertically integrated transport equations

The linear adiabatic equations lie at one end of the range of approximations; much of the theory of large scale circulation in lakes has, however, been developed on the basis of models in which the vertical structure of the current is ignored. The procedure is in this case, to average the equations by integrating them over the water depth. The basic unknown in this case is not the local velocity field, but the integrated volume transport defined by

$$\underline{V} = (U,V) \equiv \int_{-H}^{\zeta} \underline{v}\, dz = \int_{-H}^{\zeta} (u,v)\, dz \,. \tag{3.23}$$

A measure of the horizontal current is then obtained by averaging this volume transport over depth, $\bar{\underline{v}} = \underline{V}/D$, where $D = H + \zeta$. Alternatively one may solve for the vertical current distribution in a second step. This separation of the full dynamical problem into the global horizontal transport and the vertical structure of it, is known as the Ekman problem and has an extensive literature*). The separation is exact under very limited conditions only and results obtained with it should be regarded as qualitative. For instance, the vertical structure of the temperature and its evolution in time cannot be properly modelled this way. The domain of application is therefore primarily the barotropic case.

In the shallow water Boussinesq approximation, Equations (3.4), (3.5), must be integrated over depth, subject to the boundary conditions at the bottom and free surface, respectively.

Integrating the continuity equation

$$\int_{-H}^{\zeta} \left(\nabla \cdot \underline{v} + \frac{\partial w}{\partial z}\right) dz = 0 \tag{3.24}$$

yields, upon using Leibnitz's rule,

$$\nabla \cdot \int_{-H}^{\zeta} \underline{v}\, dz - \underline{v}\,(\zeta)\cdot\nabla\zeta - \underline{v}\,(-H)\cdot\nabla H + w(\zeta) - w(-H) = 0 \tag{3.25}$$

or, when the boundary conditions (2.22) and (2.23) for a material free surface and the definition (3.23) are used

*) *Among these see i.e. Ekman (1905, 1923), Fieldstad (1930), Hidaka (1933), Welander (1957), Platzman (1963), Jelesnianski (1967, 1970), Heaps (1972), Heaps & Jones (1975), Thomas (1975), Lai & Rao (1976), Witten & Thomas (1976), etc.*

$$\frac{\partial \zeta}{\partial t} + \nabla \cdot \underline{V} = 0. \tag{3.26}$$

This has the form of the *kinematic wave equation*. In a similar fashion the momentum equation (3.5) is integrated,

$$\int_{-H}^{\zeta} \left(\frac{\partial \underline{v}}{\partial t} + f\,\underline{k} \wedge \underline{v} + \frac{1}{\rho_*} \nabla (p_e + p_i) + \nabla \cdot (\underline{\Gamma} + \underline{v} \otimes \underline{v}) + \frac{\partial}{\partial z} (\underline{\gamma} + \underline{v}\, w) \right) dz = 0$$

and yields, upon interchanging differentiations and using Leibnitz's rule in appropriate terms

$$\begin{aligned} \frac{\partial \underline{V}}{\partial t} + f\,\underline{k} \wedge \underline{V} + \frac{\zeta + H}{\rho_*} \nabla p_e &= -\frac{1}{\rho_*} \int_{-H}^{\zeta} \nabla p_i \, dz - \nabla \cdot \int_{-H}^{\zeta} (\underline{v} \otimes \underline{v} + \underline{\Gamma}) \, dz \\ &\quad - (\underline{\gamma} - \underline{\Gamma} \nabla \zeta)_{z=\zeta} + (\underline{\gamma} + \underline{\Gamma} \nabla H)_{z=-H} \\ &\quad + \underline{v}(\zeta) \underbrace{\left[\frac{\partial \zeta}{\partial t} + \underline{v} \cdot \nabla \zeta - w \right]_{z=\zeta}}_{0} + \underline{v}(-H) \underbrace{\left[\underline{v} \cdot \nabla H + w \right]_{z=-H}}_{0} . \end{aligned} \tag{3.27}$$

In view of the boundary conditions (2.22) and (2.23), the last two terms on the right hand side vanish. On the other hand, since at the free surface

$$\underline{n} = \frac{(-\nabla \zeta, 1)}{\sqrt{1 + \| \nabla \zeta \|^2}}$$

and because $\underline{t}^R/\rho_*$ is given by (3.7), one has

$$\frac{1}{\rho_*} \underline{t}^R \cdot \underline{n} = \frac{\underline{\Gamma} \nabla \zeta - \underline{\gamma}}{\sqrt{1 + \| \nabla \zeta \|^2}} ;$$

this must equal $\underline{t}^{atm} \cdot \underline{n}/\rho_*$. Thus

$$\underline{\Gamma} \nabla \zeta - \underline{\gamma} = \frac{1}{\rho_*} \underline{t}^{wind}$$

and similarly

$$\underline{\Gamma} \nabla H + \underline{\gamma} = \frac{-1}{\rho_*} \underline{t}^{bottom}$$

Therefore, (3.27) reduces to

$$\frac{\partial \underline{V}}{\partial t} + f\,\underline{k} \wedge \underline{V} = -\frac{\zeta + H}{\rho_*} \nabla p_e + \underline{F}, \tag{3.28}$$

with

$$\underline{F} = -\frac{1}{\rho_*} \int_{-H}^{\zeta} \nabla p_i \, dz - \nabla \cdot \int_{-H}^{\zeta} (\underline{v} \otimes \underline{v} + \underline{\Gamma})\, dz + \frac{\underline{t}^{\text{wind}} - \underline{t}^{\text{bottom}}}{\rho_*} .$$

The force $\underline{F}$ consists of the baroclinic pressure, the advective acceleration terms $\underline{v} \otimes \underline{v}$, the horizontal turbulent Reynolds stresses $\underline{\Gamma}$ and the contribution from wind stress and bottom friction. Equation $(3.28)_1$ is not a closed system and can only be solved if $\underline{F}$ is expressed in terms of $\underline{V}$ and the wind forces. It is also evident that there is never an exact field equation in terms of $\underline{V}$, for $\int_{-H}^{\zeta} \underline{v} \otimes \underline{v}\, dz$ can never exactly be expressed in terms of $\underline{V}$, unless of course, the current is uniformly distributed with depth, in which case we may write

$$\int_{-H}^{\zeta} \underline{v} \otimes \underline{v} \, dz = \frac{1}{\zeta + H} \underline{V} \otimes \underline{V} . \tag{3.29}$$

This implies that (3.28) is only reasonable for barotropic processes. If horizontal turbulent diffusion is also ignored ($\underline{\Gamma} = \underline{0}$) and if

$$\underline{t}^{\text{bottom}} = + \rho^* C \, |\underline{V}| \, \underline{V}, \tag{3.30}$$

where $C \simeq 2 \cdot 10^{-3}$ is a dimensionless drag coefficient [see i.e. Csanady (1978), Simons (1980), Ramming & Kowalik (1980), Oman (1982)], an approximation of (3.28) is

$$\frac{\partial \underline{V}}{\partial t} + f\,\underline{k} \wedge \underline{V} = -\nabla \cdot \left(\frac{\underline{V} \otimes \underline{V}}{\zeta + H}\right) - \rho_* C \, |\underline{V}| \, \underline{V} - \frac{\zeta + H}{\rho_*} \nabla p_e + \frac{\underline{t}^{\text{wind}}}{\rho_*} . \tag{3.31}$$

Equations (3.26) and (3.31) together form a system of nonlinear partial differential equations for ζ and $\underline{V}$ which must be solved in a domain defined by the lake shore. If $\underline{N} = (N_x, N_y)$ is the unit normal vector along the shore-line and no-through flux is the appropriate boundary condition, the boundary condition imposed upon (3.26) and (3.31) is

$$\underline{V} \cdot \underline{N} = 0 \tag{3.32}$$

along the shore.

A criterion to judge whether (3.31) is an appropriate approximation follows from the fact that (3.31) is not suitable when vertical velocity profiles deviate too much from uniformity. This must be expressible as a balance between *horizontal advection* and *vertical turbulent diffusion* (Hutter et al., 1982). If U and L are a typical horizontal velocity and length, then $T_{adv} = L/U$ will be a time characteristic of advective transports. Similarly, if H_0 is the depth of the lake and ν a typical turbulent vertical viscosity, then $T_{diff} = H_0^2/4\nu$ will be a time characteristic of vertical diffusion. The dimensionless ratio

$$\Delta = \frac{T_{diff}}{T_{adv}} = \frac{H_0^2\, U}{4\nu L} \tag{3.33}$$

is a measure of how significant turbulent diffusion is as compared to advection. For $\Delta \ll 1$, (3.31) is an appropriate approximation, otherwise it is not, *and* the full three-dimensional equations must be solved. On the other hand, if $\Delta \ll 1$ and the Rossby number $R = UL/f \ll 1$, then (3.31) applies with the advective terms ignored. This reduced form of the integrated momentum balance is the basis of many classical problems of barotropic circulation dynamics. Finally we mention that integrating equations over layers is efficiently used in many numerical models of baroclinic (and barotropic) processes. By increasing the number of layers the density stratification and the non-linear convective acceleration terms may be better and better approximated.

3.5 Remarks on turbulent closure conditions

Let us return to section 3.1 in which the quasistatic equations (3.4) to (3.10) were derived. The turbulent closure conditions were specialized there for orthotropic turbulence. In other words, it was assumed that turbulent properties in the vertical and horizontal directions differ from each other while all horizontal directions are treated as equivalent (horizontal isotropy). Invoking also the shallow water approximation resulted in the expressions (3.8) for $\underline{\Gamma}$ and $\underline{\gamma}$, while the turbulent mass and heat flux vectors are given by [see Equations (3.9) and (3.10)]:

$$\begin{aligned} \left(j^R_{\alpha x}, j^R_{\alpha y}\right) &= - \sum_{\beta=1}^{N} D^H_{\alpha\beta}\, \nabla c_\beta\,, & j^R_{\alpha z} &= - \sum_{\beta=1}^{N} D^V_{\alpha\beta} \frac{\partial c_\beta}{\partial z} \\ \frac{1}{\rho_* c_p}\left(q^R_x, q^R_y\right) &= - K^H \nabla T, & \frac{1}{\rho_* c_p} q^R_z &= - K^V \frac{\partial T}{\partial z}. \end{aligned} \tag{3.34}$$

In the case of one single tracer only, the summation in $(3.34)_1$ and the Greek indices may be dropped. Expressions (3.34) are those arising in the oceanographic and limnological literature, however (3.8) does not conform with the traditional closure condition for the turbulent Reynolds-stresses. Most writers replace*)

*) *See i.e. Ekman (1905, 1923), Welander (1957), Phillips (1969), Krauss (1973), Simons (1980), Bennett (1977), Ramming & Kowalik (1980).*

$$(\nabla\cdot\underline{\Gamma})_x \quad \text{by} \quad \nabla\cdot(-A\,\nabla u)$$
$$(\nabla\cdot\underline{\Gamma})_y \quad \text{by} \quad \nabla\cdot(-A\,\nabla v)\,, \tag{3.35}$$

where $(u,v) = \underline{v}$ is the horizontal velocity vector and A a typical kinematic viscosity. Clearly, there is no choice of coefficients a_1 and a_2 in (3.8) which would cosistently yield (3.35). But (3.8) is *objective*, i.e. invariant under changes of coordinate systems, while the right hand sides of (3.35) are not. This should be borne in mind, and in fact it would be more reasonable to work with (3.8).

In certain numerical applications the closure conditions (3.35) give rise to numerical instabilities. Bennett (1977), therefore, suggested to replace

$$(\nabla\cdot\underline{\Gamma})_x \quad \text{by} \quad \frac{\partial}{\partial x}(-A_1\;\nabla\cdot\underline{v})$$
$$(\nabla\cdot\underline{\Gamma})_y \quad \text{by} \quad \frac{\partial}{\partial y}(-A_1\;\nabla\cdot\underline{v})\,. \tag{3.36}$$

This is not objective either but numerical stability properties of finite difference approximations of the governing equations tend to be improved with (3.36). I regard it as important that numerical analysts start to use the objective laws (3.8).

A, A_1, ν, D^H, D^V, K^H and K^V have all the same dimension $[m^2/s]$, numerical values of which have been determined by many authors. Of these, the values for A and A_1 seem to be poorly known, but a physically realistic value of A (or A_1) is $1\,m^2/s$ (see Csanady, 1978). Furthermore, since turbulence is the dominant diffusion mechanism, thermal diffusion is no different from tracer mass diffusion, implying that

$$D^H \simeq K^H\,, \quad D^V \simeq K^V\,. \tag{3.37}$$

This, of course requires that cross diffusivities among different tracer components vanish. Experimentally determined values for the diffusivities differ in the hypo- and epilimnion and are generally scattered substantially. Table 1 is a slight extension of values collected by Oman (1982).

Theoretically, the diffusivities may vary with the state of motion, in particular with ρ, grad ρ and $\underline{D}$, where $\underline{D}$ is the stretching tensor [compare the text after (2.7)]. One objective functional relation for ν is $\nu = \nu(\rho, \|\text{grad}\,\rho\|, D_{(2)})$, where $D_{(2)} = D_{ij}D_{ij}$ is the second stretching tensor invariant. With these, a dimensionless number may be formed, namely

$$|Ri| = \frac{\dfrac{g}{\rho}\,\|\,\text{grad}\;\rho\,\|}{2\,D_{(2)}}\,. \tag{3.38}$$

Table 1

Horizontal diffusivities D^H (or K^H) in $[m^2/s]$.

Author	Year	Epilimnion	Hypolimnion
Okubo	(1971	0.5 - 1.0	
Kullenberg	(1972)	0.5	
Murthy	(1975)	1.0	0.1
Shuter et al.	(1978)	1	
Imboden & Emerson	(1978)	-	0.001 - 0.1

Vertical diffusivities ν, D^V (or K^V) in $[m^2/s]$.
(Some of the expressions show a dependence on the Richardson number Ri (see text) or wind speed W)

Author	Year	ν(Epilimnion)	ν(Hypolimnion)	K^V(Epilimnion)	K^V(Hypolimnion)
Jacobsen	(1913)		0.00019-0.00038		0.00002 - 0.00006
Thorade	(1914)	$1.02 \cdot 10^{-4}\ W^3$ W in [m/s]			
Brennecke	(1921)	0.0160			
Durst	(1924)	0.025-0.1500			
Sverdrup	(1926)	0-0.1			
Thorade	(1928)	0.0075-0.1720			
Fjeldstad	(1929)	0.001-0.040			
Defant	(1932)			0.0320	
Fjeldstad	(1933)			0.0002-0.0016	
Suda	(1936)	0.0680-0.7500		0.0030-0.0080	
Suda	(1936)	0.0150-0.1460		0.0007-0.0090	
Munk & Anderson	(1948)	$\nu_0/(1+\frac{10}{3}Ri)^{1/2}$		$K_0(1+10\ Ri)^{3/2}$	
Wüst	(1955)	0.0007-0.0050			
Mamayev	(1959)	$\nu_0\ \exp(-m\ Ri)$		$K_0\ \exp(-n\ Ri)$	
Nan'niti	(1964)	0.0100			
Neumann & Pierson	(1964)	$0.1825 \cdot 10^{-4} W^{\frac{5}{2}}$			
Hunkins	(1966)	0.0024			
Assaf	(1971)	0.0150-0.0225		0.00	
Horn	(1971)			0.024-0.031	
Csanady	(1972)	0.0065-0.0160			
Hoeber	(1972)			0.0340-0.500	
Jones	(1973)	0.0020-0.0700			
Prümm	(1974)			0.0265-0.0480	
Ostapoff & Worthem	(1974			0.0067-0.0085	
Williams & Gibson	(1974)		0.0012-0.0025		0.000052-0.0027
Arsen'yev et al.	(1975)	0.0060-0.0750			
Imboden & Emerson	(1978)				0.000005 -0.00001
Baker	(1980)	0-0.0025			

In a stratified fluid the dominant term in grad ρ is $d\rho_0/dz$ where ρ_0 is the density of the equilibrium state. Similarly, since $|w| \ll |u|, |v|$, and because $|\partial u/\partial x| \ll |\partial u/\partial z|$, $|\partial u/\partial y| \ll |\partial u/\partial z|$ etc., one may show that $D_{(2)} \simeq 1/2 \, ||\partial \underline{v}/\partial z||^2$. Hence, an approximate value for Ri is

$$\mathrm{Ri} = \frac{-\dfrac{g}{\rho_0}\dfrac{d\rho_0}{dz}}{||\partial \underline{v}/\partial z||^2} \tag{3.39}$$

which is the *classical Richardson number*. It measures the stability of the stratification. Most semi-theoretical formulas for ν are based on functional relationships of the form $\nu = \nu(\mathrm{Ri})$ and a popular choice is (Bennett, 1977)

$$\nu = \frac{\nu_0 + \nu_1}{\sqrt{1 + \frac{10}{3}\mathrm{Ri}}} \tag{3.40}$$

in which $\nu_0 = 100\,[s]\, \underline{t}^{wind}/\rho_\star$ depends explicitly on the wind stress and $\nu_1 = 0.0004\ m^2/s$. More on this will certainly be said when baroclinic numerical models are closely analysed.

4. VORTICITY AND POTENTIAL VORTICITY EQUATIONS

In this section, a vector and a scalar identy will be discussed which follow from momentum balance.

4.1 The vorticity equation

Taking the curl of the momentum equation $(2.4)_3$, defining the vorticity $\underline{\omega}$ of the velocity field $\underline{v}$ by

$$\underline{\omega} = \mathrm{curl}\ \underline{v} \tag{4.1}$$

and observing the vector identity

$$\underline{\omega} \wedge \underline{v} = \underline{v}\ \mathrm{grad}\ \underline{v} - \mathrm{grad}\left(\frac{v^2}{2}\right), \tag{4.2}$$

the vorticity equation

$$\frac{\partial \underline{\omega}}{\partial t} + \mathrm{curl}\left((2\underline{\Omega} + \underline{\omega}) \wedge \underline{v}\right) = \frac{\mathrm{grad}\,\rho \wedge \mathrm{grad}\,p}{\rho^2} + \mathrm{curl}\left(\frac{\mathrm{div}\ \underline{t}^R}{\rho}\right) \tag{4.3}$$

is obtained. In (4.3) the assumption was made that the external force field is conservative so that its curl vanishes. To obtain the final form of the vorticity equation we apply the vector identity

$$\mathrm{curl}\,(\underline{a} \wedge \underline{b}) = \mathrm{grad}\,\underline{a}\cdot\underline{b} - \mathrm{grad}\,\underline{b}\cdot\underline{a} + \mathrm{div}\,\underline{b}\ \underline{a} - \mathrm{div}\,\underline{a}\ \underline{b}, \tag{4.4}$$

valid for any two vectors $\underline{a}$ and $\underline{b}$ and apply it to $\underline{a} = 2\underline{\Omega} + \underline{\omega}$ and $\underline{b} = \underline{v}$ and

obtain
$$\operatorname{curl}\left((2\underline{\Omega}+\underline{\omega})\wedge\underline{v}\right) = \operatorname{grad}(2\underline{\Omega}+\underline{\omega})\,\underline{v} - (\operatorname{grad}\underline{v})(2\underline{\Omega}+\underline{\omega}) + \operatorname{div}\underline{v}\,(2\underline{\Omega}+\underline{\omega}), \tag{4.5}$$

since $\operatorname{div}(2\underline{\Omega}+\underline{\omega}) = 0$. Substituting this into (4.3) and writing $(2\underline{\Omega}+\underline{\omega}) = \underline{\omega}_a$ yields

$$\frac{d\underline{\omega}_a}{dt} = \operatorname{grad}\underline{v}\cdot\underline{\omega}_a - \underline{\omega}_a \operatorname{div}\underline{v} + \frac{\operatorname{grad}\rho \wedge \operatorname{grad} p}{\rho^2} + \operatorname{curl}\left(\frac{\operatorname{div}\underline{t}^R}{\rho}\right). \tag{4.6}$$

The material rate of change of the *absolute vorticity* $\underline{\omega}_a$ is made up of the four terms on the right hand side of (4.6). The last two terms are the production of vorticity due to the baroclinic vector and due to the curl of the turbulent friction. They vanish when $\operatorname{grad}\rho$ and $\operatorname{grad} p$ are parallel (barotropicity) and when the internal friction is ignored. The first and the second term describe the production of vorticity due to vortex tilting and stretching. For adiabatic processes these are the only surviving terms which govern the evolution of the total or relative vorticity.

4.2 The potential vorticity

More useful than the concept of vorticity is the concept of *potential vorticity*, which was introduced by Ertel (1942). In the presentation below, we follow essentially Pedlosky (1982).

To introduce the potential vorticity, replace in (4.6) $\operatorname{div}\underline{v}$ by $-\dot{\rho}/\rho$ to write it in the form

$$\frac{d}{dt}\left(\frac{\underline{\omega}_a}{\rho}\right) = \frac{1}{\rho}\operatorname{grad}\underline{v}\cdot\underline{\omega}_a + \frac{\operatorname{grad}\rho\wedge\operatorname{grad}p}{\rho^3} + \frac{1}{\rho}\operatorname{curl}\left(\frac{\operatorname{div}\underline{t}^R}{\rho}\right). \tag{4.7}$$

Consider a scalar quantity λ which satisfies the balance statement

$$\frac{d\lambda}{dt} = \Psi_\lambda\,, \tag{4.8}$$

where Ψ_λ may incorporate flux, supply and production terms. Let $\operatorname{grad}\lambda$ be the gradient field of λ and form the inner product of (4.7) with $\operatorname{grad}\lambda$. This yields

$$\operatorname{grad}\lambda\cdot\frac{d}{dt}\left(\frac{\underline{\omega}_a}{\rho}\right) = \left(\frac{1}{\rho}\operatorname{grad}\underline{v}\cdot\underline{\omega}_a\right)\cdot\operatorname{grad}\lambda + \operatorname{grad}\lambda\cdot\frac{\operatorname{grad}\rho\wedge\operatorname{grad}p}{\rho^3} + \frac{\operatorname{grad}\lambda}{\rho}\cdot\operatorname{curl}\left(\frac{\operatorname{div}\underline{t}^R}{\rho}\right). \tag{4.9}$$

If one adds the identity

$$\frac{\underline{\omega}_a}{\rho}\cdot\frac{d}{dt}(\operatorname{grad}\lambda) = \frac{\underline{\omega}_a}{\rho}\cdot\operatorname{grad}\left(\frac{d\lambda}{dt}\right) - \operatorname{grad}\lambda\cdot\left(\operatorname{grad}\underline{v}\cdot\frac{\underline{\omega}_a}{\rho}\right) \tag{4.10}$$

and uses (4.8), Equations (4.9) and (4.10) combine to yield

$$\frac{d}{dt}\left(\frac{\underline{\omega}_a \cdot \operatorname{grad}\lambda}{\rho}\right) = \frac{\underline{\omega}_a \cdot \operatorname{grad}\Psi_\lambda}{\rho} + \operatorname{grad}\lambda \cdot \frac{\operatorname{grad}\rho \wedge \operatorname{grad}p}{\rho^3} + \frac{\operatorname{grad}\lambda}{\rho} \cdot \operatorname{curl}\left(\frac{\operatorname{div}\underline{t}^R}{\rho}\right). \tag{4.11}$$

The quantity

$$\pi_\lambda = \frac{2\underline{\Omega} + \underline{\omega}}{\rho} \cdot \operatorname{grad}\lambda \tag{4.12}$$

is called the *potential vorticity*. It follows that if

(i) λ is conserved for each fluid particle i.e., $\Psi_\lambda = 0$, or Ψ_λ = constant,

(ii) the turbulent friction is negligible,

(iii) the fluid is barotropic, $\operatorname{grad}\rho \wedge \operatorname{grad}p = \underline{0}$ or λ can be considered a function of ρ and/or p only,

then the potential vorticity for each particle remains constant.

As an example, consider adiabatic processes and $\lambda = \rho$. Then $\dot{\rho} = 0$ ($\Psi_\rho = 0$) $\underline{t}^R \simeq \underline{0}$; thus π_ρ is conserved and

$$\frac{d}{dt}\left(\frac{2\underline{\Omega} + \underline{\omega}}{\rho} \cdot \operatorname{grad}\rho\right) = 0 \tag{4.13}$$

must hold. A second, important special case is obtained for adiabatic barotropic processes. For these the horizontal velocity components are independent of z, so that (3.26) implies

$$\frac{d(\zeta + H)}{dt} + (H + \zeta)\,\nabla\cdot\underline{v} = 0\,. \tag{4.14}$$

On the other hand, integrating the continuity equation $\operatorname{div}\underline{v} = 0$ from $-H$ to z and applying the bottom boundary condition yields

$$w(z) = -(z + H)\,\nabla\cdot\underline{v} - \nabla H\cdot\underline{v}. \tag{4.15}$$

Eliminating $\nabla\cdot\underline{v}$ between (4.14) and (4.15) and observing that $w(z) = dz/dt$, we obtain

$$\frac{dz}{dt} + \nabla H\cdot\underline{v} - \frac{z + H}{\zeta + H}\,\frac{d(\zeta + H)}{dt} = 0$$

$$\frac{d(z + H)/dt}{z + H} - \frac{d(\zeta + H)/dt}{\zeta + H} = 0$$

$$\frac{d}{dt}\left(\frac{z + H}{\zeta + H}\right) = 0. \tag{4.16}$$

It follows that $\lambda = (z + H)/(\zeta + H)$ is conserved on particle paths. The barotropic potential vorticity (ρ as a constant is irrelevant)

$$\pi_S = (2\underline{\Omega} + \underline{\omega}) \cdot \operatorname{grad} \lambda$$

must, therefore, also be conserved along particle paths. In the shallow water approximation the dominant component of grad λ is $\hat{\underline{k}}/(\zeta + H)$, so

$$\pi_S = \frac{f + \omega z}{\zeta + H} \tag{4.17}$$

satisfies the evolution equation

$$\frac{d}{dt}\left(\frac{f + \omega z}{\zeta + H}\right) = 0, \tag{4.18}$$

an equation which will be of importance in the study of topographic barotropic waves.

To see this significance, consider the rigid lid approximation of (4.14) ($\partial\zeta/\partial t \simeq 0$). In this case the linearized version of (4.14) is identically satisfied by the barotropic stream function ψ in terms of which one has

$$\frac{\partial\psi}{\partial y} = Hu\,, \quad \frac{\partial\psi}{\partial x} = -Hv \tag{4.19}$$

and the linearized form of (4.18) reduces to

$$\nabla \cdot \left(\frac{\nabla \frac{\partial\psi}{\partial t}}{H}\right) - J\left(\psi, \frac{f}{H}\right) = 0. \tag{4.20}$$

in which

$$J(a,b) = \frac{\partial a}{\partial x}\frac{\partial b}{\partial y} - \frac{\partial b}{\partial x}\frac{\partial a}{\partial y}\,.$$

A thorough study of (4.20) is given by Prof. Mysak in his lectures.

5. TWO-LAYER SHALLOW WATER EQUATIONS

A full three-dimensional treatment of the general equations requires numerical solution of the governing equations. One way of approximation is to subdivide the lake domain into horizontal layers; this discretization corresponds to a finite difference representation of the governing equations in the vertical direction. A description of this procedure is given by Dr. Kielmann in his lectures.

In this section we present a two layer model. The basic idea to introduce it, is the form of the mean stratification that develops during the summer months and exists for time periods which are large in comparison to characteristic processes of internal wave dynamics. Typically, there is an upper layer of several meters depth in which the temperature is high and the mean density low, and the deep portion of the basin with temperature below 4 - 6 °C and mean density large. The transition zone is generally narrow and vertical temperature gradients are large (see Fig. 2). The upper layer is called *epilimnion*, the lower layer *hypolimnion* and the transition zone is the *metalimnion*. The location of largest mean temperature gradient is often referred to as the *thermocline*. For many problems, the true mean stratification is approximated by a two layer

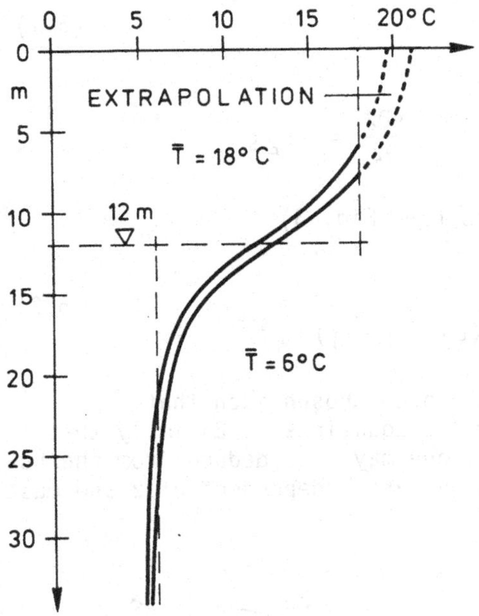

Figure 2

Upper and lower bound temperature profiles as measured in Lake of Zurich during August / September 1978. The dotted lines are extrapolations. Also shown are the two layer approximations with density discontinuity at 12 m depth and upper and lower layer temperatures 18 °C and 6 °C, respectively.

model with constant density in the epilimnion, a larger density in the hypolimnion and a density discontinuity surface between the two at the position of the thermocline. Motions of this two-layer system manifest themselves as perturbations about the stable state of rest with no deformation of the free surface and the interfacial surface.

The discontinuous density model is an approximation and has the shortcoming that it allows only two modes of oscillations, a barotropic mode and a first baroclinic mode, while the continuous density model has a countably infinite number of the baroclinic modes. The situation is akin to the oscillation of a bar which has been discretized by two concentrated mass points. As long as a detailed vertical velocity distribution is not sought the limitation to the two layer model is not serious, however, since most of the energy is usually contained in the first two modes and interface oscillations can well be identified with the motions of the isotherm corresponding to the thermocline depth.

In ensuing developments we will ignore turbulent diffusion and will invoke the shallow water approximation implying the hydrostatic pressure relation. Denoting the upper and lower layers by subscripts "1" and "2", respectively, the continuity and momentum equations reduce to

$$
\begin{aligned}
&\nabla \cdot \underline{v}_1 + \frac{\partial w_1}{\partial z} = 0\,, \\
&\frac{d\underline{v}_1}{dt} + f\,\underline{k} \wedge \underline{v}_1 + \frac{1}{\rho_1}\nabla p_1 = 0, \qquad \frac{\partial p_1}{\partial z} = -\rho_1 g\,, \\
&\nabla \cdot \underline{v}_2 + \frac{\partial w_2}{\partial t} = 0\,, \\
&\frac{d\underline{v}_2}{dt} + f\,\underline{k} \wedge \underline{v}_2 + \frac{1}{\rho_2}\nabla p_2 = 0, \qquad \frac{\partial p_2}{\partial z} = -\rho_2 g\,.
\end{aligned} \tag{5.1}
$$

Integrating the pressure equations yields (see Fig. 3)

$$
\begin{aligned}
p_1 &= \rho_1 g(\zeta_1 - z) + p^{atm}, \\
p_2 &= \rho_2 g(\zeta_2 - z - H_1) - \rho_1 g(\zeta_2 - \zeta_1 - H_1) + p^{atm}\,,
\end{aligned} \tag{5.2}
$$

in which the constants of integration have been chosen such that $p_1(z=\zeta_1) = p^{atm}$ and $p_1 = p_2$ at $z = -H_1 + \zeta_2$. Equations (5.2) imply that ∇p_1 and ∇p_2 are functions of x and y only; one may thus deduce from the horizontal momentum equations that $\underline{v}_1$ and $\underline{v}_2$ are independent of z and must obey the equations

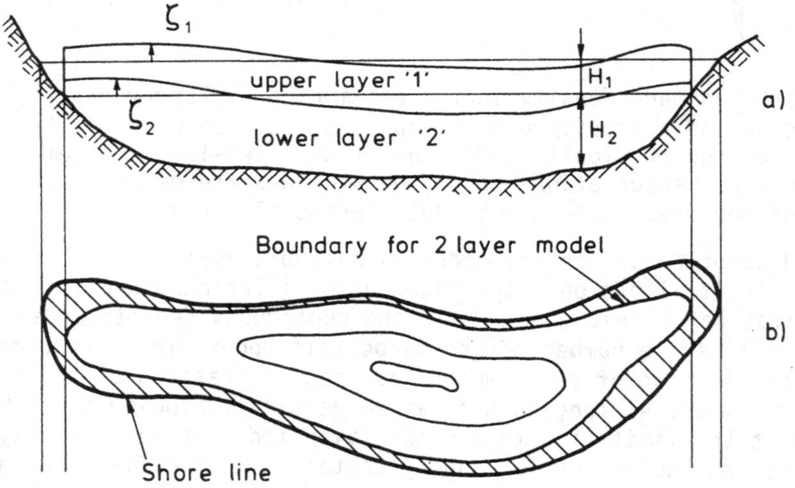

Figure 3 Two layer model with upper layer depth H_1 and lower layer depth H_2. The mathematical model uses a vertical wall as boundary at the depth contour of H_1. The dashed portion in b) is ignored in the model calculations.

$$\frac{d\underline{v}_1}{dt} + f\,\underline{k} \wedge \underline{v}_1 + g\,\nabla\zeta_1 = 0,$$
$$\frac{d\underline{v}_2}{dt} + f\,\underline{k} \wedge \underline{v}_2 + g'\nabla(\zeta_2 - \zeta_1) + g\nabla\zeta_1 = 0, \tag{5.3}$$

where $g' = g(\rho_2 - \rho_1)/\rho_2$ is the reduced gravity constant for the two layer model.

A second set of equations can be deduced by vertically integrating the continuity equations and using the kinematic boundary conditions*)

$$\frac{\partial\zeta_1}{\partial t} + \nabla\zeta_1 \cdot \underline{v}_1 - w_1 = 0, \quad \text{at} \quad z = \zeta_1(x,y,t),$$
$$\frac{\partial\zeta_2}{\partial t} + \nabla\zeta_2 \cdot \underline{v}_1 - w_1 = 0, \quad \text{at} \quad z = -H_1 + \zeta_2(x,y,t). \tag{5.4}$$
$$\nabla H_2 \cdot \underline{v}_2 - w_2 = 0, \qquad \text{at} \quad z = -(H_1 + H_2).$$

This integration is from $z = -(H_1 + H_2)$ to $z = -H_1 + \zeta_2$ and from $z = -H_1 + \zeta_2$ to $z = \zeta_1$, respectively, and yields

$$\frac{\partial(\zeta_1 - \zeta_2)}{\partial t} + \nabla\cdot\Big((\zeta_1 - \zeta_2 + H_1)\underline{v}_1\Big) = 0,$$
$$\frac{\partial\zeta_2}{\partial t} + \nabla\cdot\Big((\zeta_2 + H_2)\,\underline{v}_2\Big) = 0\,. \tag{5.5}$$

Equations (5.3) and (5.5) form a nonlinear system of partial differential equations for the unknowns v_1, v_2, ζ_1 and ζ_2 and must be solved in a domain of the (x,y)-plane common to both layer equations. Because these are obtained by integrating over depth, (5.3) and (5.5) can only make sense in lake domains where the total depth is larger than H_1. In other words, (5.3) and (5.5) are subject to the no-flow through condition,

$$\underline{v}_1 \cdot \underline{N} = 0, \quad \underline{v}_2 \cdot \underline{N} = 0 \tag{5.6}$$

along a closed boundary which is defined by $z = -D < -H_1$ and does not, in general, coincide with the lake shore (see Fig. 3b). This may seriously affect the validity of the two layer model whenever a stratified lake

---

*) *The second of equations (5.4) can equally be written in the form*

$$\frac{\partial\zeta_2}{\partial t} + \nabla\zeta_2 \cdot \underline{v}_2 - w_2 = 0, \quad \text{at} \quad z = -H_1 + \zeta_2(x,y,t)\,.$$

When written in terms of the upper layer velocities the material particles just above the interface are followed, when lower layer velocities are used one follows particles just below the interface.

possesses large shallow portions for which the depth is comparable to or smaller than H_1. This should be borne in mind.

Another word of caution is necessary. Application of the shallow water approximation, strictly, implies that $|\zeta_\alpha| \ll \min(H_1, H_2)$, $\alpha = 1,2$. Because amplitudes of thermocline oscillations may, on occasion be comparable to layer thicknesses, the approximating equation (5.3), (5.5) may be inappropriate. When $|\zeta| \ll \min(H_1, H_2)$ holds, then it is also permissible to linearize (5.5). If, furthermore, layer-Rossby numbers are small (5.3) may also be linearized. In this linear limit the field equations are

$$\begin{aligned} &\frac{\partial \underline{v}_1}{\partial t} + f\,\underline{k} \wedge \underline{v}_1 + g\nabla\zeta_1 = 0, \\ &\frac{\partial \underline{v}_2}{\partial t} + f\,\underline{k} \wedge \underline{v}_2 + g'\nabla(\zeta_2 - \zeta_1) + g\nabla\zeta_1 = 0, \\ &\frac{\partial(\zeta_1 - \zeta_2)}{\partial t} + \nabla\cdot(H_1\,\underline{v}_1) = 0, \\ &\frac{\partial \zeta_2}{\partial t} + \nabla\cdot(H_2\,\underline{v}_2) = 0. \end{aligned} \tag{5.7}$$

This system of equations contains the barotropic and baroclinic gravity waves (first class) and the topographic waves (second class). In small lakes, the latter can be filtered out by ignoring f; by introducing the rigid lid approximation, on the other hand, the barotropic gravity waves are filtered out. These will be explained in the sections which deal with linear and nonlinear waves.

REFERENCES

Arsen'yev, S.A., Dobroklonsky, S.V., Mamedov, R.M. and Shelkovnikov, N.K., 1975. Direct measurements of small scale marine turbulence characteristics from a stationary platform in the open sea. Izs., Atmos. Ocean Phys., 11, pp. 845-850.

Assaf, G., Gerard, R. and Gordon, A.L., 1971. Some mechanics of ocean mixing revealed in aerial photographs. J. Geophys. Res. 76, pp. 6550-6572.

Baker, J.R., 1980. Currents as a Function of Depth in the Water: Solution of a Linear Model. M.S. Thesis, University of Minnesota, Duluth.

Batchelor, G.K., 1953. The Theory of Homogeneous Turbulence. Cambridge Monograph on Mechanics and Applied Mathematics. Cambridge University Press.

Becker, E. and Bürger, W., 1976. Kontinuumsmechanik. B.G. Teubner, Stuttgart.

Bowman, H.A, and Schoonover, R.M., 1976. Procedure for high precision density determinations by hydrostatic weighing. J. Res. Natl. Bureaux of Standards, 7/c, 3, 179.

Bradshaw, P. 1976. Turbulence. Topics in Applied Physics. Springer Verlag, Berlin - Heidelberg - New York.

Brennecke, W., 1921. Die ozeanographischen Arbeiten der Deutschen Antarktischen Expedition 1911-1912. Arch. dtsch. Seewarte, 39, 206.

Bührer, H. and Ambühl, H.A., 1975. Die Einleitung von gereinigtem Abwasser in Seen. Schw. Z. Hydrologie, 37(2), pp. 347-369.

Chadwick, P., 1976. Continuum Mechanics. Concise Theory and Problems. George Allen and Unwin, London.

Csanady, G.T., 1972. Frictional currents in the mixed layer at the sea surface. J. Phys. Oceanogr., 2, pp. 498-508.

Csanady, G.T. 1978. Water circulation and dispersal mechanisms. In: Lakes: Chemistry, Geology, Physics (Ed. A. Lerman). Springer Verlag, New York - Heidelberg - Berlin.

Defant, A., 1932. Die Gezeiten und inneren Gezeitenwellen des Atlantischen Ozenas. Wiss. Erg. Deut. Atlantische Expedition Meteor, 1925 - 1927, 7, 318.

Durst, C.S., 1924. The relationship between wind and current. Quart. J. Roy. Met. Soc. 50, 113.

Ekman, V.W., 1905. On the influence of the earth's rotation on ocean currents. Ark. Mat. Astron. Fys., 2(11), 52 p.

Ekman, V.W., 1923. Ueber Horizontalzirkulation bei winderzeugten Meeresströmungen. Ark. Mat. Astron. Fys., 17(26), 74 p.

Ertel, H. 1942. Ein neuer hydrodynamischer Wirbelsatz. Meteorolog. Z., 59, pp. 277-281.

Fjeldstad, J.R., 1930. Ein Problem aus der Windstromtheorie. Z. Angew. Math. & Mech., 10, pp. 121-137.

Fjeldstad, J.R., 1933. Wärmeleitung in Meer. Geofysiske Publikasjoner, Vol. 10, No. 7, 20 pp. Oslo.

Greenspan, H.P., 1968. The Theory of Rotating Fluids. Cambridge University Press.

Heaps, N. 1972. On the numerical solution of the three dimensional hydrodynamical equations for tides and storm surges. Mem. Soc. R. Sci. Liege Ser., 6(2), pp. 143-180.

Heaps, N. and Jones J.E., 1975. Storm surge computations for the Irish Sea using a three dimensional numerical model. Mem. Soc. R. Sci. Liege Ser., 6(7), pp. 289-333.

Hidaka, K., 1933. Non-stationary ocean currents. Mem. Imp. Mar. Obs. Kobe, 5, pp. 141-266.

Hinze, J.O., 1959. Turbulence. McGraw Hill, New York.

Hoeber, H., 1972. Eddy thermal conductivity in the upper 12 m of the tropical Atlantic. J. Phys. Oceanogr., 2, pp. 303-304.

Horn, W., 1971. Die zeitliche Veränderlichkeit der Temperatur der ozeanischen Deckschicht im Gebiet der grossen meteorbank. Meteor. Forsch. Ergebn. (A), No. 9, S. 42 - 51, Berlin-Stuttgart.

Hunkins, K., 1966. Ekman drift currents in the arctic oceans. Deap Sea Res., 13, pp. 607-620.

Hutter, K., 1983. Theoretical Glaciology. Reidel Publishing Comp. Dordrecht - Boston.

Hutter, K. and Trösch, J., 1975. Ueber die hydrodynamischen und thermodynamischen Grundlagen der Seezirkulation. Mitteilung No. 20 der Versuchsanstalt für Wasserbau, Hydrologie und Glaziologie, ETH, Zürich.

Hutter, K., Oman, G. and Ramming, H.G., 1982. Wind-bedingte Strömungen des homogenen Zürichsees. Mitteilung No. 61 der Versuchsanstalt für Wasserbau, Hydrologie und Glaziologie, ETH, Zürich.

Imboden, D.M. and Emerson, S., 1978. Natural radon and phosphorus as limnologic tracers: Horizontal and vertical diffusion in Greifensee. Limnology & Oceanography, Vol. 23, pp. 77-90.

Jacobsen, J.P., 1931. Beiträge zur Hydrographie der Dänischen Gewässer. Komm. f. Havunders Medd. Ser. Hydr., 2(2), 94.

Jelesnianski, C.P., 1967. Numerical computations of storm surges with bottom stress. Mon. Weather Rev., 95, pp. 740 - 756.

Jelesnianski, C.P., 1970. Bottom-stress time-history in linearized equations of motion for storm surges. Mon. Weather Rev., 98, pp. 462-478.

Jones, J.H., 1973. Vertical mixing in the equatorial undercurrent. J. Phys. Oceanogr., 3, pp. 286-296.

Kullenberg, G., 1972. Apparent horizontal diffusion in stratified shear flow. Tellus, 24, pp. 17-28.

Krauss, W., 1973. Dynamics of the homogeneous and the quasi-homogeneous ocean. Bornträger, Berlin.

Lai, R.Y.S. and Rao, D.B., 1976. Wided drift currents in a deep sea with variable eddy viscosity. Arch. Meteor. Geophys. Bioklimat., A 25, pp. 131-140.

LeBlond, P.H. and Mysak, L.A., 1978. Waves in the Ocean. Elsevier Oceanography Series. Elsevier Scientific Publishers Comp., Amsterdam, Oxford, New York.

Mamayev, O.I., 1958. The influence of stratification on vertical turbulent mixing in the sea. Izv. Geophys. Ser., 1, pp. 870-875.

Monin, A.S. and Yalgon, A.M., 1971. Statistical Fluid Dynamics. MIT Press, Cambridge, Mass.

Müller, I., 1973. Rationale Thermodynamik. Die Grundlagen der Materialtheorie. Bertelsmann Universitätsverlag..

Munk, W.H. and Anderson, E.R., 1948. Notes on a theory of the thermocline. J. Mar. Res., 7, pp. 277-295.

Murthy, C.R., 1975. Horizontal diffusion characteristics in Lake Ontario. J. Phys. Oceanogr., 6, pp. 76 - 84.

Nan'niti, T., 1964. Some observed results of oceanic turbulence. In: Studies on Oceanography. K. Yoshida Editor, University of Washington Press.

Neumann, G. and Pierson, W.J., 1964. Principals of Physical Oceanography. Prentice-Hall.

Okubo. A., 1971. Oceanic diffusion diagrams, Deap Sea Res., 18, pp. 789 - 802.

Oman, G., 1982. Das Verhalten des geschichteten Zürichsees unter äusseren Windlasten. Mittelung No. 60 der Versuchsanstalt für Wasserbau. Hydrologie und Glaziologie, ETH, Zürich.

Ostapoff, F. and Worthem, S., 1974. The intradiurnal temperature variations in the upper ocean layer. J. Phys. Oceanogr., 4, pp. 601-612.

Pedlosky, J., 1982. Geophysical Fluid Dynamics. Springer Verlag, Berlin - Heidelberg - New York.

Phillips, O.M., 1969. The dynamics of the upper ocean. Cambridge Monographs on Mechanics and Applied Mathematics. Cambridge University Press.

Platzman, G.W., 1963. The dynamic prediction of Wind tides on Lake Erie. Meteorol. Monogr., 4(26), 44 p.

Prümm, O., 1974. Height dependence of diurnal variations of wind velocity and water temperatures near the air-sea interface of the tropical Atlantic. Boundary-Layer Meteorol., 6, pp. 341-347.

Ramming, H.G. and Kowalik, Z., 1980. Numerical Modelling of Marine Hydrodynamics. Application to Dynamic Physical Processes. Elsevier Oceanography Series. Elsevier Scientific Publishing Comp. Amsterdam, Oxford, New York.

Shuter, V., Oman, G., Stortz, K. and Sydor, M., 1978. Turbidity dispersion in Lake Superior through use of landsat data. J. Great Lakes Res., 4, pp. 359-360.

Simons, T.J., 1980. Circulation models of lakes and inland seas. Can. Bull. Fisheries and Aquatic Sci., No. 203.

Spalding, D.B., 1982. Turbulence Models, a Lecture Course. Imperial College of Science and Tehcnology.

Suda, K., 1936. On the dissipation of energy in the density current. Geophys. Mag., 10, pp. 131-243.

Sverdrup, H.U., 1926. Dynamics of tides on the North Siberian Shelf. Results of the Maud Expedition. Geofyisks Publikasjoners, 4, 75.

Thomas, J.H., 1975. A theory of wind-driven currents in shallow water with variable eddy viscosty. J. Phys. Oceanogr., 5, pp. 136-142.

Thorade, H., 1914. Die Geschwindigkeit von Triftströmungen und die Ekman'sche Theorie. Ann. d. Hydrogr. u. Math. Meteor., 42, pp. 379-391.

Thorade, H., 1928. Gezeitenuntersuchungen in der Deutschen Bucht der Nordsee. Arch. Dtsch. Seewarte, 46, pp. 1-85.

Tilton, L.W. and Taylor, J.K., 1937. Accurate representation of the refractivity and density of distilled water as a function of temperature. J. Res. Nat. Bureaux of Standards, 18, 205.

Wang, C.C. and Truesdell, C.A., 1973. Introduction to Elasticity. Noordhoff International Publishing, Leyden.

Welander, P, 1957. Wind action on a shallow sea: Some generalizations of Ekman's theory. Tellus, 9, pp. 45-52.

Williams, R.B. and Gibson, C.H., 1974. Direct measurements of turbulence in the Pacific Equatorial undercurrent. J. Phys. Oceanogr., 4, pp. 104-108.

Witten, A,J. and Thomas J.H., 1976. Steady wind-driven currents in a large lake with depth-dependent eddy viscosity. J. Phys. Oceanogr., 6, pp. 85-92.

Wüst, G., 1975. Strömungsgeschwindigkeiten in Tiefen- und Bodenwasser des Atlantischen Ozeans. Deap Sea Res. Papers in Marine Biology and Ocenaography, pp. 373-392.

Hydrodynamics of Lakes: CISM Lectures
edited by K. Hutter, 1984
Springer Verlag Wien-New York

LINEAR GRAVITY WAVES, KELVIN WAVES AND POINCARE WAVES, THEORETICAL MODELLING AND OBSERVATIONS

Kolumban Hutter
Laboratory of Hydraulics, Hydrology and Glaciology,

Gloriastrasse 37/39, ETH,
CH-8092 Zurich, Switzerland

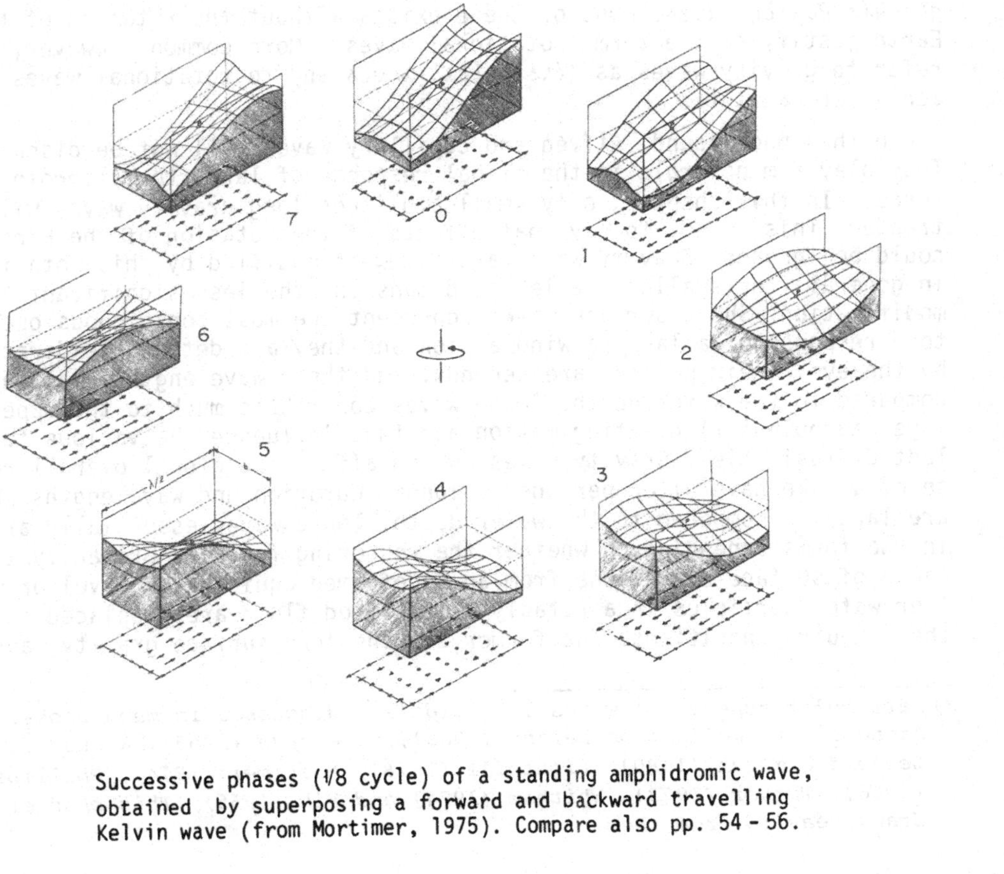

Successive phases (1/8 cycle) of a standing amphidromic wave, obtained by superposing a forward and backward travelling Kelvin wave (from Mortimer, 1975). Compare also pp. 54 - 56.

1. PRELIMINARY REMARKS

Waves in lakes are primarily generated by external meteorological forces. The latter are complex in their spatial and temporal structure and thus impose a large spectrum of the typical physical scales. Their explanation and relation to the primary cause is one of the principal goals of physical limnology.

Lake water, as a compressible electrically conducting fluid on the rotating Earth with free surface allows the propagation of *sound waves* (owing to its compressibility), *Alfvēn waves* (due to its electrical conductivity), short wavelength *capillary waves* (in which surface tension effects come to bear), *gravity waves* and *rotational waves**). Gravity waves arise through the restoring action of gravity on water particles displaced from equilibrium levels, such as a free surface or an internal isopycnal surface. The rotation of the Earth introduces the Coriolis force, acting at right angles to the velocity vector which leads to the existence of *inertial* or *gyroscopic* waves and also gives rise to *planetary* and/or *topographic Rossby waves*. None of these exists without the rotation of the Earth justifying the term "rotational waves". More common, however, is to refer to gravity waves as *first class waves* and to rotational waves as *second class waves*.

In this book sound, Alfvēn and capillary waves will not be discussed. They play a minor role in the global response of lakes to meteorological forces. In this chapter, only small amplitude long gravity waves will be treated. This is not to say that effects of the rotation of the Earth could be ignored. Gravity waves are somewhat modified by this rotation: in general, the smaller the lake's dimension, the less significant this modification. Short surface waves represent the most conspicuous oscillatory response of a lake to wind action and they are detectable directly by the eye. Their periods are seconds, and their wavelengths small as compared to the water depth. Those waves contribute much to the upper layer turbulent fluctuating motion and thus influence the various turbulent diffusivities. Gravity waves, which affect the global overall response of a lake have often periods of longer duration and wavelengths which are large as compared to the water depth. These waves essentially arise in two forms depending on whether the restoring force is primarily the cause of surface elevations from an undeformed equilibrium level or whether water particles in a (stably) stratified fluid are displaced from their neutral positions. The former are the long surface gravity waves;

---

*) *Particular aspects of waves in fluids are discussed in many books. Among others we mention Defant (1961), Greenspan (1968), Krauss (1966), LeBlond & Mysak (1980), Lighthill (1978), Mortimer (1974), Phillips (1966), Turner (1973), Whitham (1974) and Yih (1965), This chapter draws heavily from them.*

density inhomogeneities are irrelevant for their existence and modes are referred to as external or *barotropic*. The latter are *internal* gravity waves, inhomogeneities in the density distribution are crucial for their existence and modes are called *baroclinic*. A characteristic speed of barotropic gravity waves is $c = \sqrt{g\,h}$, where g is the gravity acceleration and h the water depth; thus it is typically from 10 to 50 m/s. Corresponding phase speeds of internal waves are about one to two orders of magnitude smaller. Periods of oscillations in an elongated lake of 10 to 100 km length, say, are therefore between minutes and hours for barotropic oscillations, but tens of hours or days for baroclinic processes. This separation of time scales is also manifest in the significance of the Coriolis effects which modify the corresponding gravity waves which now become Kelvin- and Poincaré waves, respectively. The parameter, which measures this significance is the *external* or *internal Rossby radius* defined by $R = c/f$, in which c is the phase speed and f the Coriolis parameter. Typically, R = (500 km, 5 km) for external and internal waves, respectively. If L is a characteristic horizontal distance (i.e. width of the lake) it may be shown that effects of the rotation of the Earth can be ignored provided that $L/R \ll 1$. Recalling the substantial difference of the phase speeds for the external and internal oscillations, this implies that for most lakes Coriolis effects on the structure of surface gravity waves may be negligible, whereas their neglect may be questionable or unjustified when internal oscillations are analysed. Another consequence of the spectral gap between external and internal oscillations is that equations for the description of one class of waves can (and should) be simplified such that processes of the other class are filtered out. This is done either by using the vertically integrated transport equations (section 3.4 of Hutter) and yields the barotropic response, or by employing a "rigid lid" assumption by which surface gravity waves are eliminated.

Because of the boundedness of the lake domain free gravity waves combine and form *standing waves*. These are called *seiches* and represent the eigenmodes of the particular basin under consideration. Observationally, a substantial amount of the total energy is concentrated at these quantized (eigen)frequencies. The exact distribution depends on the wind forcing but the few lowest modes are generally the most likely excited ones. As is the case for free waves, seiches exist as barotropic surface seiches - involving motions of the whole water masses and attaining their maximum amplitude at or near the free surface - or baroclinic internal seiches associated with the density stratification and attaining their maximum amplitude at or near the thermocline. In an elongated basin the motion is primarily longitudinal. This property may constructively be used in the development of *channel models* for which the transverse structure of the motion is ignored. Such simplifications must be introduced with care because rotational effects necessarily entail transversality effects under basically longitudinal motions. However, because many lakes are narrow and elongated the channel approximation is a competitive alternative in which transversal and rotational features can also be incorporated.

In what follows, reference will frequently be made to the chapter "Funfamental equations and approximations". Equation numbers of that chapter will carry the prefix H1, and the chapter will be referred to as "H1"

2. PLANE WAVES IN A ROTATING STRATIFIED FLUID

In ensuing developments which parallel those of LeBlond & Mysak (1980) plane harmonic waves will be studied with the representation

$$\underline{v} = \underline{V} \exp[i(\underline{k}.\underline{x} - \omega t)] \tag{2.1}$$

Here, $\underline{V}$ is the amplitude vector of $\underline{v}$, $\underline{k}$ the wavenumber vector*), indicating the direction of propagation and ω the frequency. With $\underline{k}$ and ω two speeds can be formed, namely

$$\underline{c} = \frac{\omega}{k^2}\underline{k} \quad \text{and} \quad \underline{c}_g = \frac{\partial\omega}{\partial\underline{k}} = \text{grad}_k\, \omega . \tag{2.2}$$

The first is the *phase velocity* and points into the direction of $\underline{k}$, the second, $\underline{c}_g$, is the *group velocity*; it is the speed with which energy propagates, see i.e. Hayes (1974).

If $\underline{v}$ in (2.1) is the velocity vector and incompressibility is assumed, it follows from the continuity equation, div $\underline{v} = 0$, that $\underline{k}\cdot\underline{V} = 0$. In other words, all motion is transverse to the direction of propagation. Because for all waves discussed in this book the above incompressibility assumption is used, all waves are transverse in this sense. This should be borne in mind.

Governing linearized equations for a rotating stratified fluid are Equations (H1, 3.17), which, upon invoking the Boussinesq assumption, have the form

$$\begin{aligned} \rho_* \{\frac{\partial\underline{v}}{\partial t} + 2\underline{\Omega}\wedge\underline{v}\} &= -\,\text{grad}\, p' + \rho'\underline{g}\,, \\ \frac{\partial\rho'}{\partial t} - \frac{\rho_* N^2}{g} w &= 0\,, \end{aligned} \tag{2.3}$$

in which N is the Brunt-Väisälä frequency defined in (H1, 3.16). To simplify calculations we shall assume that $\underline{\Omega}$ and N are constant. Eliminating

*) *In this chapter $\underline{k}$ is not the unit vector in the z-direction. Unit vectors will carry a hat. So $\hat{\underline{k}}$ is the unit vector in the direction of $\underline{k}$.*

ρ' and the perturbation pressure p' (by taking the curl of $(2.3)_1$) and substituting (2.1) into the resulting equation yields

$$\omega^2 \underline{k} \wedge \underline{V} - 2i\,(\underline{\Omega}\cdot\underline{k})\omega\,\underline{V} + \frac{N^2\,W}{g}\,\underline{k} \wedge \underline{g} = 0, \tag{2.4}$$

in which $\underline{V} = (U,V,W)$. This is a homogeneous, linear system for the components of $\underline{V}$, which possesses a nontrivial solution if its determinant vanishes. The deduction of this characteristic equation is simplified if it is recognized that $\underline{V}$ has only two independent components which are orthogonal to $\underline{k}$. With reference to Figure 1 one may thus choose

$$\underline{V} = U_i\,\hat{\underline{i}} + U_j\,\hat{\underline{j}}, \quad W = -U_i\,\sin\Theta', \quad \underline{g} = -g(\sin\Theta\,\hat{\underline{k}} + \sin\Theta'\,\hat{\underline{i}}) \tag{2.5}$$

in which $(\hat{\underline{i}}, \hat{\underline{j}}, \hat{\underline{k}})$ is a unit orthogonal triad with positive parity whose $\hat{\underline{k}}$-vector is parallel to $\underline{k}$ and $\hat{\underline{i}}$ lies in the vertical plane formed by $\underline{k}$ and $\underline{g}$, and $\hat{\underline{j}}$ is horizontal. Substituting (2.5) into (2.4) shows that (2.4) has the component equations

$$\begin{bmatrix} 2i(\underline{\Omega}\cdot\underline{k})\,\omega & \omega^2 k \\ (\omega^2 - N^2\cos^2\Theta)\,k & -2i(\underline{\Omega}\cdot\underline{k})\,\omega \end{bmatrix} \begin{bmatrix} U_i \\ U_j \end{bmatrix} = 0 \tag{2.6}$$

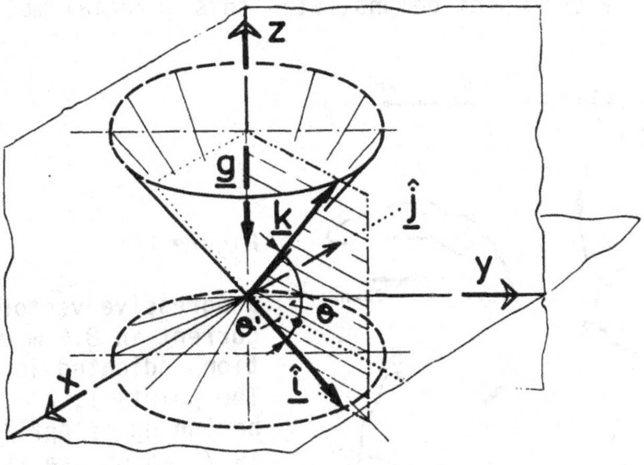

Figure 1 For a wave with wavenumber vector $\underline{k}$ (lying on the mantle of the upper cone), energy propagates in the $\underline{i}$-direction (which is perpendicular to $\underline{k}$ and lies on the mantle of the lower cone). $\underline{g}$, $\underline{k}$ and $\underline{i}$ are all in the same vertical plane to which $\hat{\underline{j}}$ is orthogonal (and therefore in the (x,y)-plane). The angle between $\underline{k}$ and the horizontal plane is Θ, its complement to $\pi/2$ is Θ'.

and permits nontrivial solutions, if the *dispersion relation*

$$\omega^2 [k^2 (\omega^2 - N^2 \cos^2 \Theta) - 4(\underline{\Omega}\cdot\underline{k})^2] = 0 \qquad (2.7)$$

is satisfied. Excluding the steady state $\omega = 0$, the dispersion relation of linear gyroscopic internal waves reads

$$\omega^2 = N^2 \cos^2 \Theta + 4\,\Omega^2 \cos^2 \alpha, \qquad (2.8)$$

where α is the angle between $\underline{\Omega}$ and $\underline{k}$. Evidently, stratification and rotation effects are additive. Using the rules (2.2) it is a straightforward matter to calculate the phase and group velocities; this will be left as an exercise to the reader. More qualitative insight is gained from (2.6) in the limits $N = 0$ and $\underline{\Omega} = \underline{0}$. When there is no rotation the only nontrivial solution of (2.6) is $U_j = 0$, U_i = arbitrary. The fluid particles oscillate in the direction of $\hat{i}$, and the waves are *linearly polarized*. On the other hand, in a homogeneous rotating fluid ($N = 0$) Equation (2.6) implies $|U_i/U_j| = 1$, the waves are *circularly polarized*. Combination of both effects, $\underline{\Omega}$ and N, yields an *elliptically polarized* wave in which fluid particles travel around ellipses; *the direction of rotation is opposite to the direction of* $\underline{\Omega}$. If $\underline{k}$ is nearly vertical, then particles rotate nearly in horizontal planes and loops are traversed in the clockwise direction (on the northern hemisphere). Figure 2 shows a progressive vector diagram of the horizontal current at the indicated station in Lake Geneva which demonstrates this inertial motion, see Bauer et al. (1981).

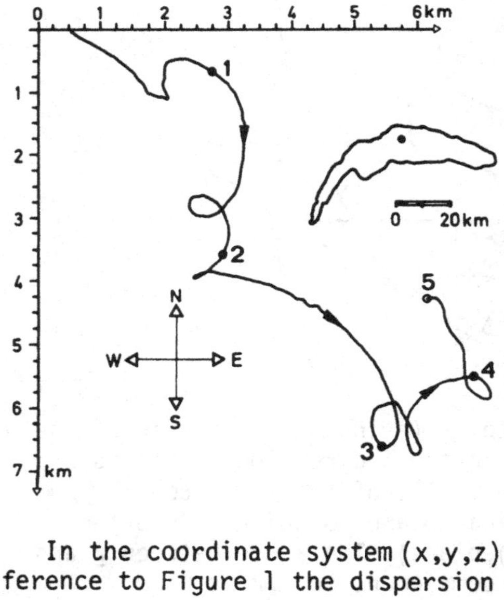

Figure 2

Progressive vector diagram for the current in 8.5 m depth at the station indicated in the insert map. The points 1, 2, ..., 5 mark the beginning of April 2, 3, ..., 6 in 1977. Loops are traversed in the clockwise direction. From Bauer et al. (1981), with changes.

In the coordinate system (x,y,z) one has $\underline{\Omega} = (0, \Omega\cos\phi, \Omega\sin\phi)$. With reference to Figure 1 the dispersion relation (2.8) may then be written as

$$k_1^2(N^2-\omega^2)+k_2^2(N^2-\omega^2+\tilde{f}^2)+2\,k_2\,k_3\,f\,\tilde{f}=k_3^2(\omega^2-f^2),$$

where f and $\tilde{f}$ are defined in (H1, 2.12). Ignoring the term $\tilde{f}$ has been shown in H1 to be part of the shallow water approximation. In this case the last equation reduces to

$$\frac{k_1^2+k_2^2}{k_3^2}=\frac{\omega^2-f^2}{N^2-\omega^2}\,. \tag{2.9}$$

This simplified dispersion relation implies that ω must lie between f and N. Because the usual stratified case is $N \gg f$ one has $f<\omega<N$; f forms a lower bound for inertial gyroscopic waves to exist and N is a corresponding upper bound. For constant f all waves which are not influenced by bottom topography are thus *superinertial*. The existence of an upper bound N for ω, on the other hand explains, why the metalimnion in a stably stratified lake may serve as a wave guide to internal waves. Figure 3 shows a N^2-profile, obtained from density measurements in Lake of Zurich in summer 1978. Often the N^2-distribution is as shown schematically in the insert map. For the selected ω, waves can only propagate in the indicated band of the metalimnion.

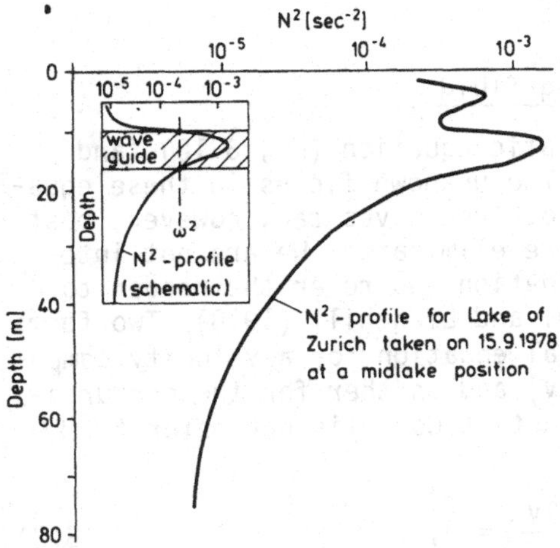

Figure 3

Square Brunt-Väisälä frequency N^2 as determined from measurements at a midlake position in Lake of Zurich, taken on 15 September 1978. The insert map shows its schematic distribution. For the selected ω, waves can only exist in the indicated band.

Finally it should be stressed that there are also waves with subinertial frequencies $\omega < f$. These, however incorporate variations of f with latitude or are guided by topography, see Prof. Mysak's lectures.

Kelvin waves are unidirectional waves along an idealized vertical shore. Their amplitude falls off exponentially as one moves away from the shore and on the northern (southern) hemisphere the shore is to the right (left) when one is looking into the propagation direction. Kelvin waves can have $\omega \gtrless f$. Topographic Rossby waves, on the other hand have $\omega < f$. They exist because of variations in f and/or topography and are discussed by Prof, Mysak in his lecture.

3. CHARACTERISTIC LONG WAVELENGTH WAVE SOLUTIONS

The results of section 2 apply to waves with arbitrary wavelength. Here, we discuss wave propagation in a Boussinesq fluid which obeys the field equations (H1, 3.18). In these, the shallow water approximation has been invoked, except in the vertical momentum balance, which is not reduced to the hydrostatic balance. The additional vertical acceleration term permits a delimitation of the validity of the hydrostatic approximation. Indeed, a combination of $(H1, 3.18)_{1,4}$ yields

$$\left(\frac{\partial^2}{\partial t^2} + N^2\right) w = - \frac{1}{\rho_*} \frac{\partial^2 p'}{\partial z\, \partial t}, \tag{3.1}$$

of which the term $\partial^2 w/\partial t^2$ would be missing, if the hydrostatic pressure assumption had been used. Consequently, the hydrostatic limit equations are valid, if $(f^2 <)\omega^2 < N^2$, where ω is the wave frequency. In ensuing developments this will be assumed to hold.

3.1 Free waves in a stratified Boussinesq fluid

We now focus attention on the hydrostatic equation (H1, 3.18) (and ignore the vertical acceleration term). The unknown fields in these equations are ρ', u, v, w and p'. Propagation of free waves can, however, best be discussed, if all variables but one are eliminated. We are not interested here in the details of this elimination and refer the reader to LeBlond & Mysak (1980) or Phillips (1969) and Lighthill (1978). Two forms are popular, namely a partial differential equation for a velocity component (the *north-south velocity component* v) and another for the perturbation pressure p', respectively. For a constant Coriolis parameter f ($df/dy = 0$) these equations have the forms

$$\nabla^2 v + L \frac{\partial}{\partial z}\left(N^{-2} \frac{\partial v}{\partial z}\right) = 0, \tag{3.2}$$

$$\nabla^2 p' + L \frac{\partial}{\partial z}\left(N^{-2} \frac{\partial p'}{\partial z}\right) = 0, \tag{3.3}$$

in which $L \equiv \partial^2/\partial t^2 + f^2$. Both equations are second order in time; (3.2) is obtained by taking the x-derivative of the vorticity equation and eliminating the remaining variables, (3.3) is essentially the divergence of the momentum equation.

Physically more appealing is, however, a differential equation for w. Straightforward manipulation with Equations (H1, 3.18) yields

$$N^2 \nabla^2 w + L \frac{\partial^2 w}{\partial z^2} = 0. \qquad (3.4)$$

This differential equation is formally different from (3.2) or (3.3).

Solutions are sought by *separation of variables* in the form

$$\begin{aligned} v(x,y,z,t) &= Y_n(z)\, v_n(x,y,t), \\ w(x,y,z,t) &= Z_n(z)\, w_n(x,y,t), \qquad (3.5) \\ p'(x,y,z,t) &= \Pi_n(z)\, p_n(x,y,t). \end{aligned}$$

In other words, the vertical dependence is separated from the horizontal variation. n is an integer whose significance will become apparent in a moment. Substituting $(3.5)_2$ into (3.4), one obtains

$$\frac{d^2 Z_n}{dz^2} + \frac{N^2}{g\,h_n} Z_n = 0 \qquad (3.6a)$$

and

$$L\, w_n - g\, h_n \nabla^2 w_n = 0 \qquad (3.6b)$$

in which $g\,h_n$ is a constant separation parameter and h_n has the dimension of a length; it is called the *equivalent depth*. What has been achieved? The original three dimensional field equation (3.4), which must be solved within the lake domain, has been split into two separate problems; the first is an ordinary differential equation, for the vertical distribution Z_n of the vertical velocity component whose domain of solution is $-H \le z \le 0$. With appropriate boundary conditions this is a Sturm-Liouville eigenvalue problem (see Courant & Hilbert, 1962) with a countable set of positive eigenvalues h_n ($n = 1, 2, \ldots$). Because h_n as the separation constant of the problem cannot depend on any spatial variable, and the eigenvalue of (3.6a) depends on the depth H, it follows that the separation of variables solutions (3.6a,b) are strictly correct only for basins with constant depth.

Once (3.6a) is solved, one may solve (3.6b). For plane waves substitutions of (2.1)*) into (3.6b) reveals the dispersion relation

---

*) *Because a horizontal problem is solved, one has* $v = V_0 \exp(i\,(\underline{k}_h \cdot \underline{x}_h - \omega t))$ *where* $\underline{x}_h$ *is the horizontal position vector.*

$$\omega_n^2 = f^2 + g\, h_n\, k_h^2 , \tag{3.7a}$$

in which k_h is the horizontal component of the wavenumber vector $\underset{\sim}{k}$. Accordingly, the wave frequencies are superinertial and quantized according to the eigenvalues of the *Sturm-Liouville problem* (3.6a).

Phase and group velocity take on the form

$$c = \frac{(g\, h_n\, k_h^2 + f^2)^{1/2}}{k_h^2}\, \underset{\sim}{k}_h , \quad \underset{\sim}{c}_g = \frac{g\, h_n}{\omega_n}\, \underset{\sim}{k}_h ; \tag{3.7b}$$

they are parallel to each other and obey the inequality $|\underset{\sim}{c}_g| < \sqrt{g\, h_n} < |\underset{\sim}{c}|$. These waves were first analysed by Poincaré (1910) and are, therefore called *Poincaré waves*.

In exactly the same fashion the pressure equation may be separated. The horizontal problem is then found to be identical to (3.6b), but the vertical eigenvalue problem for Y_n or Π_n differs from (3.6a) and has the form

$$\frac{d}{dz}\left(N^{-2} \frac{dY}{dz} \right) + \frac{Y}{g\, h_n} = 0, \tag{3.8}$$

in which Y stands for Y_n and Π_n, respectively. The boundary conditions require that $w = 0$ at $z = -H$ and $p' = 0$ at $z = \zeta$, where ζ is the surface elevation. When expressed in terms of the pressure, and when the separations (3.5) are used, they read

$$\frac{d\,\Pi_n}{dz} = 0, \quad \text{at} \quad z = -H \quad \text{and} \quad \frac{d\,\Pi_n}{dz} + \frac{N^2}{g}\,\Pi_n = 0, \quad \text{at} \quad z = 0. \tag{3.9}$$

The general form of the boundary conditions in terms of v or w are very complicated, but when the rotation of the Earth is ignored ($f = 0$), using (H1, 3.18) - (H1, 3.22) one may deduce that

$$w = 0, \quad \text{at} \quad z = -H \quad \text{and} \quad \frac{\partial^3 w}{\partial t^2\, \partial z} - g\,\nabla^2 w = 0, \quad \text{at} \quad z = 0. \tag{3.10}$$

Unfortunately, when trying to separate variables, the second of these cannot be expressed in terms of Z_n alone. In other words, for plane harmonic waves the frequency would enter $(3.10)_2$. This demonstrates, why the pressure equation is advantageous: the differential equation *and* the boundary conditions for Π_n depend on the stratification alone and may be solved once and for all, and in a second step the dispersion relation (3.7) may then be attacked.

Finally we mention that the above differential equations and boundary conditions would be much more complex if β-effects had been included and the Boussinesq assumption would not have been invoked. This is why i.e. equations in LeBlond & Mysak (1980), Lighthill (1978) and Phillips (1969) are different.

3.2 Vertical mode structure for constant depth

Even though $N = \text{constant}$ is a poor approximation of realistic N-distributions, the solution of (3.8) and (3.9) with this simplification gives good qualitative insight into the mode structure. With

$$\varepsilon = \frac{N^2 H}{g} \quad \text{and} \quad \lambda_n^{\,2} = \frac{N^2 H^2}{g\, h_n} \tag{3.11}$$

the vertical distribution of the perturbation pressure reads

$$\Pi_n = A_n [\cos \lambda_n z' - \varepsilon \lambda_n^{-1} \sin \lambda_n z'] \,, \quad z' = z H \,, \tag{3.12}$$

where λ_n satisfies the eigenvalue equation $\tan \lambda_n = \varepsilon \lambda_n^{-1}$. Because $\varepsilon = O(10^{-3})$, approximate satisfaction of the characteristic equation yields $\lambda_0^2 = \varepsilon + O(\varepsilon^2)$, and $\lambda_n^2 = n^2 \pi^2 = n^2 \pi^2 + O(\varepsilon)$, $n = 1,2,\ldots$, corresponding to

$$h_0 = H(1+O(\varepsilon)), \quad h_n = \frac{\varepsilon H}{n^2 \pi^2}(1+O(\varepsilon)),$$
$$\Pi_0 = A_0(1+O(\varepsilon)), \quad \Pi_n = A_n [\cos \frac{n \pi z}{H} + O(\varepsilon)] . \qquad n = 1,2,3,\ldots \tag{3.13}$$

For the lowest barotropic eigenfunction the equivalent depth nearly equals H, and the perturbation pressure is nearly independent of z. For all other baroclinic modes the equivalent depth is much smaller, $h_0 \gg h_1 > h_2 \ldots$, and the pressure distribution $\Pi_n(z)$ in $-H \le z \le 0$ changes its sign n-times. Phase and group velocities follow from (3.7). When $f = 0$

$$\underline{c} = \underline{c}_g = \sqrt{g\, h_n}\,, \quad n = 0,1,2,\ldots \tag{3.14}$$

and the waves are non-dispersive. For barotropic waves the speeds are from 10 to 30 m/s, for baroclinic waves they are $O(\varepsilon^{1/2})$ smaller and typically from 0.3 to 1.0 m/s.

Because according to (3.1) w is proportional to $\omega\, \partial p'/\partial z$, it follows from (3.13) that to lowest order $w_0 \equiv 0$ and $w_n = B_n \sin(n \pi z/H)$ $n = 1,2,\ldots$. Had we solved the eigenvalue problem for the vertical velocity component, the barotropic mode would have been "lost" and only the baroclinic modes could have been found. Furthermore, to within the same order of approximation in ε the sinusoidal relation for w_n indicates that the solution for the *baroclinic* eigenfunction w_n could have been obtained by solving the Sturm-Liouville problem

$$\frac{d^2 Z_n}{dz^2} + \frac{N^2 - \omega^2}{\omega^2 - f^2} k_h^2 Z_n = 0, \qquad -H \le z \le 0$$
$$w_n = 0, \quad \text{at} \quad z = 0, -H. \tag{3.15}$$

(The boundary condition $w_n = 0$ at the upper surface corresponds to the *rigid lid assumption*).

Equation $(3.15)_1$ follows from (3.6a) when $g h_n$ is eliminated with the aid of (3.7). In the numerator of the second term we have also replaced N^2 by $N^2 - \omega^2$, a correction which arises in a non-Boussinesq fluid; it destroys the property that the eigenvalue problem (3.15) is independent of ω, but fascilitates discussions, see Figure 4. For $f < \omega < N$ (3.15) has *oscillatory solutions* and Z_n has (n-1) nodes. The vertical distribution of a unidirectional horizontal velocity component is given by $d Z_n/dz$ and also shown in Figure 4. For the first baroclinic mode the horizontal current is in one direction in the epilimnion and in the opposite direction in the hypolimnion. At the higher modes the horizontal current is to and fro according to the number of nodal points. When $\omega > N$, solutions are exponentially decaying. This is outside the indicated wave guide (dashed in Figure 4). There are many more particulars which can be discussed for continuously stratified fluids. For these we refer to the pertinent

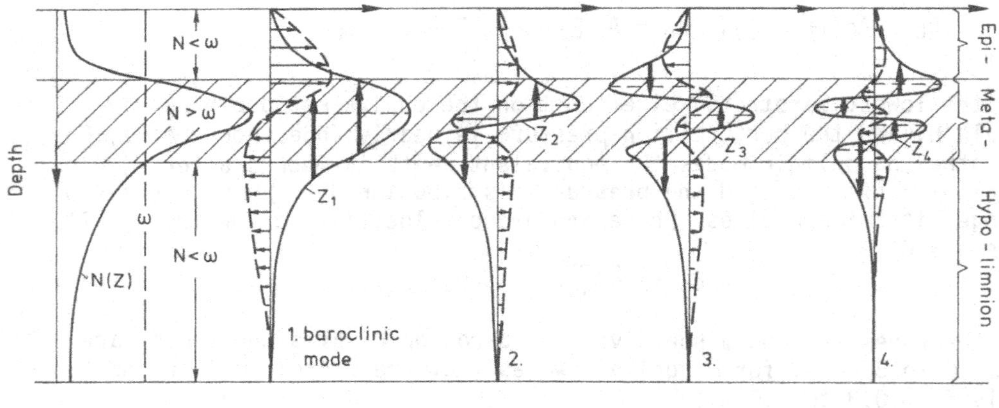

Figure 4 Typical vertical distribution of the Brunt-Väisälä frequency N (left) and the four lowest baroclinic modes (qualitative). Solid curves show the distribution of the vertical velocity component, dashed curves indicate the distribution of the longitudinal velocity component (when $f = 0$). Most energy is usually concentrated in the first baroclinic mode, the exact distribution must, however, be determined by continuous profils of horizontal velocities.

literature *).

3.3 The two layer model

In strongly stratified lakes with a well mixed epilimnion, the Brunt-Väisälä frequency is practically zero in the epilimnion and small in the hypolimnion, but large in a narrow metalimnion. As an approximation, one may choose $N = 0$ everywhere except at the thermocline in which N is a Dirac pulse. Such an approximation corresponds to the two layer model with an upper layer of constant depth H_1 and density ρ_1 and a second layer with density $\rho_2 > \rho_1$ and (variable) depth H_2. Its equations were derived in section 5 of H1 and are listed as (H1, 5.7). The unknown fields are $\underline{v}_1$, $\underline{v}_2$, the mean layer velocities, and ζ_1, ζ_2, the free surface and the interface (thermocline) deflections.

For constant depth equations (H1, 5.7) permit separation of the barotropic and baroclinic equations. To show this, add α-times the lower layer continuity and momentum equations to the respective upper layer equations. If V_α and ζ_α are defined as

$$\underline{V}_\alpha = H_1 \underline{v}_1 + \alpha H_2 \underline{v}_2, \qquad \zeta_\alpha = \zeta_1 + (\alpha-1)\zeta_2. \tag{3.16}$$

Equations (H1, 5.7) are easily shown to reduce to

$$\begin{aligned} &\frac{\partial \zeta_\alpha}{\partial t} + \nabla\cdot\underline{V}_\alpha = 0, \\ &\frac{\partial \underline{V}_\alpha}{\partial t} + f\,\underline{k}\wedge\underline{V}_\alpha + [\lambda\nabla\zeta_1 + \alpha\varepsilon H_2\nabla\zeta_2] = 0. \end{aligned} \tag{3.17}$$

in which $\varepsilon = (\rho_2 - \rho_1)/\rho_2$ replaces (3.11) and $\lambda = H_1 + \alpha(1-\varepsilon)H_2$. Equations (3.17) are equivalent to a linear one-layer model [see (H_1, 5.26) and (H1, 5.28)] provided that the term in braces is proportional to $\nabla\zeta_\alpha$, say $h_\alpha \nabla\zeta_\alpha$. Thus, we require

$$\begin{aligned} &(H_1 + \alpha(1-\varepsilon)H_2)\nabla\zeta_1 + \alpha\varepsilon H_2\nabla\zeta_2 \\ &\quad = -h_\alpha[\nabla\zeta_1 + (\alpha-1)\nabla\zeta_2], \end{aligned} \tag{3.18}$$

from which both, h_α and α may be determined. Two solutions exist and to lowest order in ε they are

*) *Phillips (1966), Lighthill (1978), Turner (1973), LeBlond & Mysak (1980) all present detailed discussions on these.*

$$
\begin{aligned}
h_e &= H(1+O(\varepsilon)), & h_i &= \varepsilon H_1 H_2/(H_1+H_2),\\
\alpha_e &= 1+O(\varepsilon), & \alpha_i &= -H_1/H_2+O(\varepsilon),\\
\underline{V}_{\alpha_e} &= H_1\underline{V}_1 + H_2\underline{v}_2, & \underline{V}_{\alpha_i} &= H_1(\underline{v}_1-\underline{v}_2),\\
\zeta_{\alpha_e} &= \zeta_1+O(\varepsilon), & \zeta_{\alpha_i} &= -\frac{H_1+H_2}{H_2}\zeta_2+O(\varepsilon),
\end{aligned}
\tag{3.19}
$$

where the indices "e" and "i" refer to the external and internal modes, respectively. It is seen that for barotropic modes the equivalent depth h_e equals H, and that the equivalent baroclinic depth h_i corresponds to the first baroclinic equivalent depth (3.13). Solving $(3.19)_{3,7}$ for the layer velocities, one obtains

$$
\underline{v}_1 = \frac{\underline{V}_{\alpha_e} + \frac{H_2}{H_1}\underline{V}_{\alpha_i}}{H_1+H_2}, \quad \underline{v}_2 = \frac{\underline{V}_{\alpha_e} - \underline{V}_{\alpha_i}}{H_1+H_2}. \tag{3.20}
$$

Suppose now that $\underline{V}_{\alpha_i} = \underline{0}$ (no baroclinic motion); then

$$
\underline{v}_1 = \underline{v}_2 = \frac{1}{H_1+H_2}\underline{V}_{\alpha_e}, \; [\zeta_1 = \zeta_{\alpha_e}, \; \zeta_2 = \frac{H_2}{H_1+H_2}\zeta_{\alpha_e}]. \tag{3.21}
$$

On the other hand, when $\underline{V}_{\alpha_e} = \underline{0}$ (no barotropic motion), then

$$
\begin{aligned}
H_1\cdot\underline{v}_1 &= -H_2\underline{v}_2 = \frac{H_2}{H_1+H_2}\underline{V}_{\alpha_i},\\
[\zeta_1 &= \frac{\varepsilon H_1}{H_1+H_2}\zeta_{\alpha_i}, \quad \zeta_2 = -\frac{H_2}{H_1+H_2}\zeta_{\alpha_i}].
\end{aligned}
\tag{3.22}
$$

For pure barotropic processes the horizontal velocity is uniform with depth and for baroclinic processes the upper layer volume flux equals the lower layer volume flux, but the two have opposite direction, reminiscent of the first baroclinic mode in Figure 4.

The two layer model is an attractive one, in particular in its equivalent depth variant

$$
\frac{\partial\zeta_\alpha}{\partial t} + \nabla\cdot\underline{V}_\alpha = 0, \quad \frac{\partial\underline{V}_\alpha}{\partial t} + f\,\underline{k}\wedge\underline{V}_\alpha + g\,h_\alpha\nabla\zeta_\alpha = 0; \tag{3.23}
$$

however it eliminates all higher order baroclinic modes and in the form (3.23) it requires to substitute a constant equivalent depth. Both assumptions may be illegitimate*). Longuen-Higgins in Mortimer (1952) and Heaps (1961) therefore introduced a three layer model and Lighthill (1969) showed, how a N-layer model can be reduced to the equivalent depth form (3.23).

3.4 Kelvin waves

For lakes rotational effects in gravity waves are usually manifest as modifications of the solutions in which the rotation of the Earth is ignored. To demonstrate this, we start with the equations

$$\frac{\partial u}{\partial t} - fv = -g\frac{\partial \zeta}{\partial x}, \qquad \frac{\partial v}{\partial t} + fu = -g\frac{\partial \zeta}{\partial y},$$
$$\frac{\partial \zeta}{\partial t} + \frac{\partial hu}{\partial x} + \frac{\partial hv}{\partial y} = 0, \tag{3.24}$$

appropriate for a homogeneous fluid, or a stratified fluid with constant depth**). In a long and narrow channel the dominant motion is longitudinal, so that $|v| \ll |u|$. One may, thus, first solve the equations on the left of (3.24) with v ignored in comparison to u. This yields $\zeta = \zeta_0(x,t)$ and $u = u_0(x,t)$; note that ζ_0, u_0 have no y-dependence. In a second step the second equation of (3.24) is solved, thereby again ignoring the term involving v. In other words, one balances the Coriolis with the pressure term (geostrophic balance). With $u \simeq u_0$, this yields

$$\frac{\partial \zeta}{\partial y} = -\frac{f}{g}u_0(x,t) \Rightarrow \zeta = \zeta_0(x,t) - \frac{f}{g}\,u_0(x,t)\,y, \tag{3.25}$$

and demonstrates that for longitudinal motions the surface must be transversely inclined because of Coriolis effects. This approximate procedure i known as *Kelvin wave dynamics*. It was used first by F. Defant (1953) and later by Platzmann & Rao (1964) and others.

Consider now a half infinite basin bounded along $y = 0$; we seek wave solutions of (3.24) of the form

$$u = u_0\,\phi(y)\,e^{i(kx-\omega t)}, \qquad v = 0, \qquad \zeta = \zeta_0\,\phi(y)\,e^{i(kx-\omega t)} \tag{3.25}$$

---

*) *In this case $h = h_\alpha$, and $(u,v) = (U,V)/h_\alpha$ as may immediately be deduced from a comparison of (3.23) and (3.24).*

**) *Bäuerle (1981) and Hutter & Schwab (1982) have made attempts to delimit the conditions for which the equivalent depth model can be used when h_α varies with position, but results are not conclusive. So, until more is known, the model should be used with constant equivalent depth.*

and find straightforwardly that $\phi(y) = \exp(-(f/c)\cdot y)$ and $\zeta_0 = c/g \cdot u_0$, with $c = \sqrt{gh}$. The amplitudes of u and ζ thus decay exponentially as one moves away from the boundary. The rate of decay is given by the *Rossby radius* $R = c/f$; its value depends on the Coriolis parameter and on the phase speed $c = \sqrt{gh}$ whose value depends on the size of the equivalent depth (see (3.19)). Typically $R \simeq 500$ km for barotropic waves and $R \simeq 5$ km for baroclinic waves. A lake needs to have the width of several Rossby radii in order that rotational effects come to bear. Figure 5 depicts this Kelvin wave solution. It has the general form of a *coastal trapped* wave, a class of waves which is treated in greater detail by LeBlond & Mysak (1980) and by Mysak (1980). Notice that the boundary layer structure becomes lesser and lesser when f decreases or c increases. In particular when $f = 0$, then $\phi(y) = 1$ and the Kelvin wave becomes an ordinary long gravity wave. On the other hand, the dispersion relation $\omega = k\sqrt{gh}$ indicates that frequencies of Kelvin waves are not restricted by rotation (f).

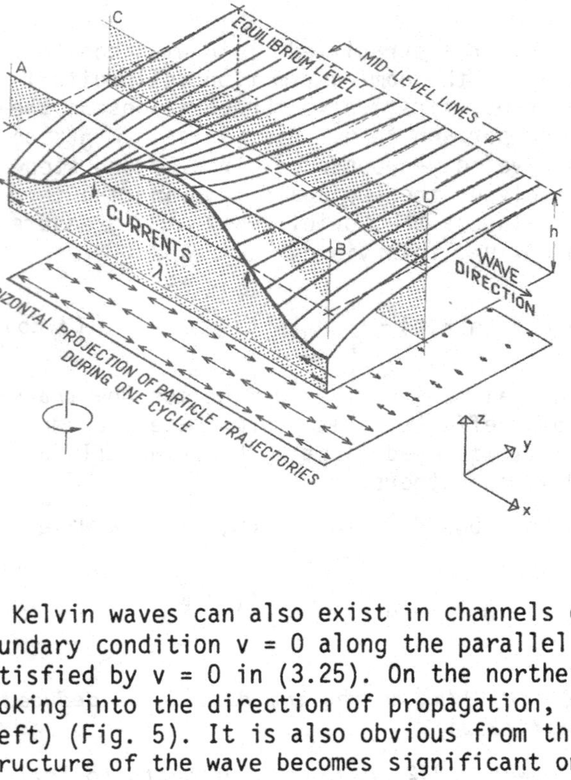

Figure 5

Long Kelvin wave progressing in the positive x-direction in a uniform depth model. Velocities are strictly longitudinal. The vertical planes AB and CD may be the vertical shores of a channel (from Mortimer (1974)).

Kelvin waves can also exist in channels of finite width, because the boundary condition $v = 0$ along the parallel walls $y \equiv 0$ and $y \equiv B$ are satisfied by $v = 0$ in (3.25). On the northern (southern) hemisphere, when looking into the direction of propagation, the wave crests to the right (left) (Fig. 5). It is also obvious from this figure that the transverse structure of the wave becomes significant only when $B > R$.

Any linear combination of wave solutions is again a solution of equations (3.24). We thus combine the *forward* Kelvin wave

$$u_f = u_0 \exp[-\frac{f}{c} y + i(kx-\omega t)],$$

$$\zeta_f = \zeta_0 \exp[-\frac{f}{c} y + i(kx-\omega t)]$$

with the *backward* travelling wave

$$u_b = u_0 \exp[\frac{f}{c}(B+y) - i(kx+\omega t)],$$

$$\zeta_b = \zeta_0 \exp[\frac{f}{c}(B+y) - i(kx+\omega t)]$$

and obtain

$$\begin{aligned} u &= u_0 \{ \exp(ikx-\frac{f}{c}y) + \exp(-ikx+\frac{f}{c}(B+y) \} e^{-i\omega t}, \\ \zeta &= \zeta_0 \{ \exp(ikx-\frac{f}{c}y) + \exp(-ikx+\frac{f}{c}(B+y) \} e^{-i\omega t}. \end{aligned} \tag{3.26}$$

This solution for ζ is graphically displayed on the cover page 39. Alternatively, one may write $\zeta = A(x,y) \exp[i(\phi(x,y)-\omega t)]$, where A is a real amplitude and ϕ the phase angle. Because of the separation of the spatial from the temporal dependence in (3.26), the interfering waves have "standing wave" character. Curves of equal amplitude A and equal phase are given by

$$\begin{aligned} A^2 &= 2\zeta_0^2 \, e^{\frac{f}{c}B} \left[\cos 2kx + \cosh\left(\frac{2f}{c}\left(\frac{B}{2}+y\right)\right)\right] = \text{const}, \\ \tan\phi &= \operatorname{cotan} kx \, \coth\left(\frac{f}{c}\left(\frac{B}{2}+y\right)\right) = \text{const}. \end{aligned} \tag{3.27}$$

The former are called *co-range* or *co-amplitude* lines. Zero amplitudes occur only where the bracket in $(3.27)_1$ vanishes. This is on the channel axis at the points $(x,y) = ((2n+1)\,\pi/2k, -B/2)$ which are spaced a wavelength apart. When $f = 0$ these points become nodal *lines* and extend across the channel. The amplitudes, of which the distribution is shown dashed in Figure 6 grow as one moves away from the nodal points, which are also called *amphidromic points* or *amphidromes*. Lines of constant phase are called *co-tidal* or *co-phase* lines, shown as solid lines in Figure 6. They all pass through the amphidromic point and, as ϕ increases, these lines rotate around the amphidromic point in the anticlockwise direction*) (see footnote on next page). As evident from $(3.27)_2$ when $f = 0$, $\tan\phi = 0$, $\phi = 0$, or $\pm 180^o$. The motion of two points is then either in phase or out-of-phase, depending on whether an even or odd number of nodal lines lie between the two points.

An important property of the solution $(3.26)_1$ is that there is no line across the channel at which the velocity u would identically vanish. It follows that waves in an enclosed basin cannot simply be obtained by superposing Kelvin waves. More on this will be said in Section 3.6.

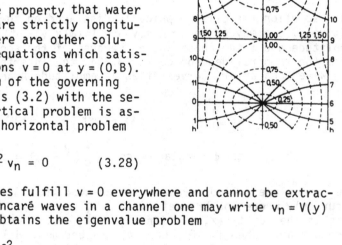

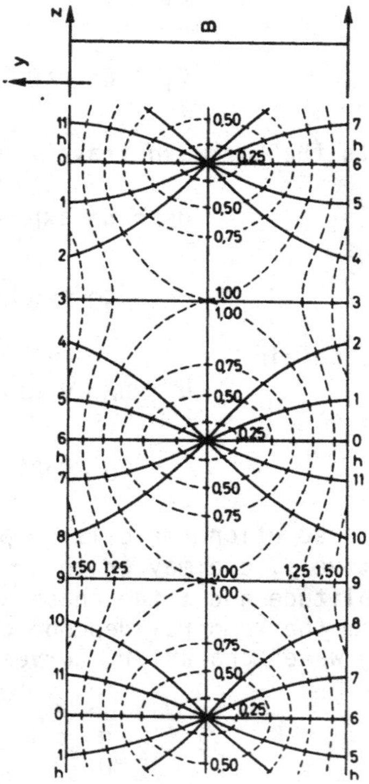

Figure 6

Amphidromic system in an infinitely long canal resulting from superposition of two Kelvin waves of period 12 hours travelling in opposite directions (A. Defant, 1961).

Full lines: co-tidal lines in hours;

broken lines: co-range lines. ($Z_0 = 0.5$, $k = 1$, $f/c = 0.7$ corresponding to a canal width of 400 km, $H = 40$ m). From Krauss (1973).

3.5 Poincaré waves in channels

Kelvin waves have the property that water particle displacements are strictly longitudinal, see Figure 5. There are other solutions of the governing equations which satisfy the boundary conditions $v = 0$ at $y = (0,B)$. The most convenient form of the governing equation to find these is (3.2) with the separation $(3.5)_1$. The vertical problem is assumed to be solved; the horizontal problem is then given by

$$L\, v_n - g\, h_n\, \nabla^2 v_n = 0 \qquad (3.28)$$

(see (3.6b). Kelvin waves fulfill $v = 0$ everywhere and cannot be extracted from (3.28). For Poincaré waves in a channel one may write $v_n = V(y) \exp(i(kx-\omega t))$ and then obtains the eigenvalue problem

$$\begin{aligned} &V'' + \left(\frac{\omega^2 - f^2}{g h_n} - k^2\right) V = 0, \quad 0 \le y \le B, \\ &V = 0, \quad \text{at} \quad y = 0, \quad y = B, \end{aligned} \qquad (3.29)$$

---

*) *This is seen if one writes $(3.27)_2$ for the neighborhood of the amphidromic point, in which case one has*

$$\left(\frac{B}{2} + y\right) = \tan\phi\, \frac{c}{f}\, k\left(\frac{2n+1}{2}\,\pi + x\right),$$

representing a straight line through the point $(-(2n+1)\pi/2, -B/2)$. As ϕ increases this line rotates counterclockwise about the amphidrome. (The direction of rotation depends on the definition of ϕ whose sign is arbitrary and which is only determined up to arbitrary phase shift).

with the solution $V = \overline{V} \sin(m\pi y/B)$ and the dispersion relation

$$\frac{\omega^2 - f^2}{g h_n} - k^2 = \frac{m^2 \pi^2}{B^2}, \quad \text{or} \quad k^2 = \frac{\omega^2 - f^2}{g h_n} - \frac{m^2 \pi^2}{B^2}. \tag{3.30}$$

The frequency is quantized according to the vertical (n) and horizontally transverse (m) mode structure. Clearly, for given ω, f, h_n and B, k^2 is larger than zero only, provided that m is smaller than a maximum value; say m_0. Modes for which $m > m_0$ are exponentially decaying in x when $x \to \infty$ or $x \to -\infty$. These will be important in the reflection problem of Kelvin waves analysed in the next subsection. The propagation condition may also be written in terms of ω,

$$\omega^2 = f^2 + g h_n \left(k^2 + \frac{m^2 \pi^2}{B^2}\right) > f^2 + g h_n \frac{m^2 \pi^2}{B^2} = \omega_c^2. \tag{3.31}$$

The second term on the right is the frequency of the m-th transverse seiche oscillation. For Poincaré waves to propagate, the frequency must not only be larger than the Coriolis parameter and the frequency of the m-th transverse seiche, but ω^2 must be larger than the sum of the square of both. This is called the cut-off frequency. Phase and group velocities are deduced from (3.31) as follows

$$c^2 = \frac{\omega^2}{k^2} = g h_n \left(1 + \frac{m^2 \pi^2}{B^2 k^2}\right) + \frac{f^2}{k^2}, \qquad c_g = \frac{g h_n}{c}. \tag{3.32}$$

They show that c is always larger than the shallow water velocity $\sqrt{g h_n}$; alternatively, c_g is always smaller than $\sqrt{g h_n}$.

Adding a forward and a backward moving Poincaré wave gives a standing wave,

$$v = V(y) \exp(-i \omega t), \tag{3.33}$$

in which V(y) is again given by (3.29)-(3.32). For these, as well as for the progressing waves the u_n-velocities and the ζ_n-displacements can also be determined, see LeBlond & Mysak (1980), Krauss (1973). Here we only discuss this solution qualitatively, see the beautiful Figure 7, due to Mortimer (1974). The motion has cell structure. For $t = 0$ each cell with horizontal wavelengths λ_x and λ_y is subdivided into four subcells in which the elevation has a high and a low, respectively and velocities are longitudinal "to and fro" as shown in the figure. A quarter period later, $t = T/4$, locations of elevation highs and lows are shifted as shown, and the current is transverse and again "to and fro", as indicated in Figure 7. At the time $t = T/2$ the elevation distribution is again as shown on the top of Figure 7 but with highs and lows interchanged, and the same

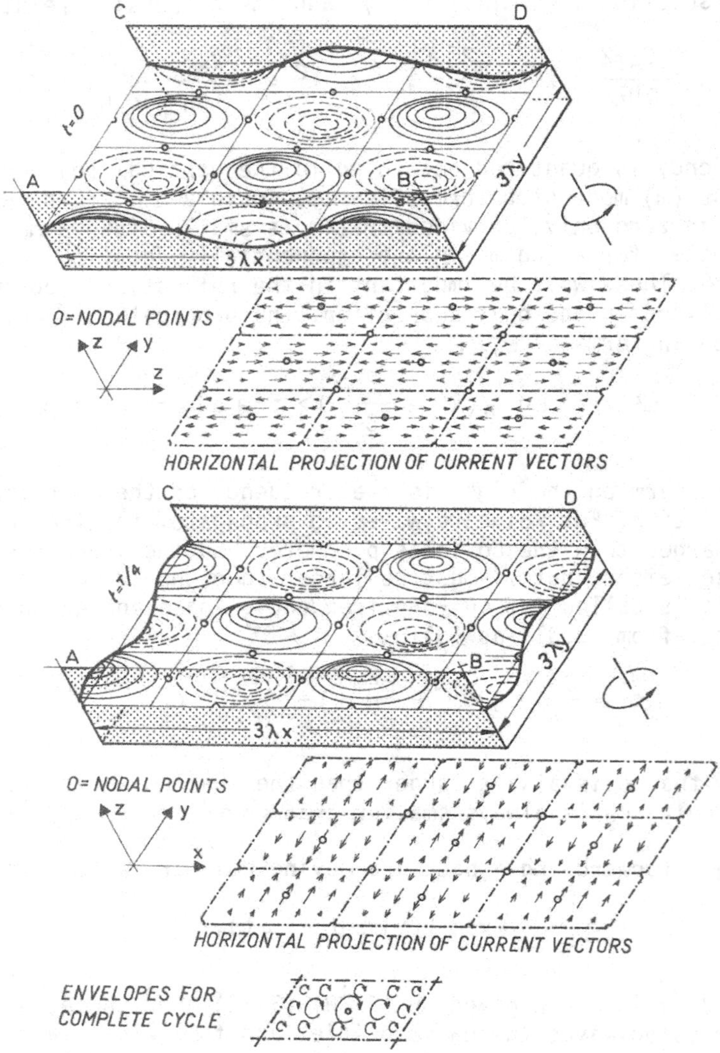

Figure 7 A standing Poincaré wave in a wide, rotating channel of uniform depth. Two phases (separated by 1/4 cycle) of the oscillation are shown for the cross-channel trinodal case, with a ratio of long-channel to cross-channel wavelengths of 2/1). The clockwise rotation and distribution of the current vectors in the Poincaré wave "cells" are explained in the text (from Mortimer, 1974).

also holds for the longitudinal velocities, whose direction is reversed. As time progresses the current vectors rotate in the clockwise direction and particle paths for a harmonic oscillation are elliptical and are traced once within a period. Finally, it should be mentioned that, while the side boundary conditions are met, there is no line across the channel at which the longitudinal velocity would vanish at all points. The above solutions can, therefore, not hold for a closed basin.

3.6 Reflection from the end of a wall

Neither the standing Kelvin waves, nor the standing Poincaré waves discussed in Sections 3.4 and 3.5 satisfy the no-through flux condition at a transverse wall. But a standing Kelvin wave solution plus a whole spectrum of standing Poincaré waves allows the no-through flux condition at a transverse wall to be met. The problem was first attacked by Taylor (1922) and further investigated by van Dantzig & Lauwerier (1969), Rao (1960) and Brown (1973).

The reflection of a Kelvin wave incident onto the end of a channel is solved by adding to this wave a reflected Kelvin wave of the same frequency and same wavenumber and an infinite number of Poincaré waves of the same frequency. The free amplitudes of these Poincaré waves are determined by the condition that there is no flux through the channel wall. If the incoming Kelvin wave has a frequency below the cut-off frequency, ω_c, of Poincaré waves, the Poincaré modes are all evanescent as one moves away from the end wall. Although they are important for the satisfaction of the boundary condition, their presence is felt only near the boundary. For this reason, this type of reflection is called *complete reflection*. Figure 8 shows the amphidromic system and the current ellipses in a rectangular basin closed at one end as determined by G.I. Taylor (1922). The influence of the Poincaré waves is only seen near the channel end. When the frequency of the incoming wave is above the cut-off frequency then this incoming Kelvin wave is reflected by a reflected Kelvin and a series of Poincaré waves the presence of some of which is recognized in the *entire* basin. Reflections are then called *incomplete*.

When the basin is closed by a second wall, an additional reflection condition must be fulfilled at the second wall. This condition will quantize both, the longitudinal wavenumber k and the frequency. The determination of the mode structure and the exploitation of the characteristic equation must be performed numerically. Explicit results are due to Rao (1966).

As is intuitively clear from Figures 5 and 7 Kelvin waves exhibit large currents along the shore, but Poincaré waves do not. On the other hand off-shore currents may be large for Poincaré waves while they are weak for Kelvin waves. Similarly, free surface and thermocline deflections

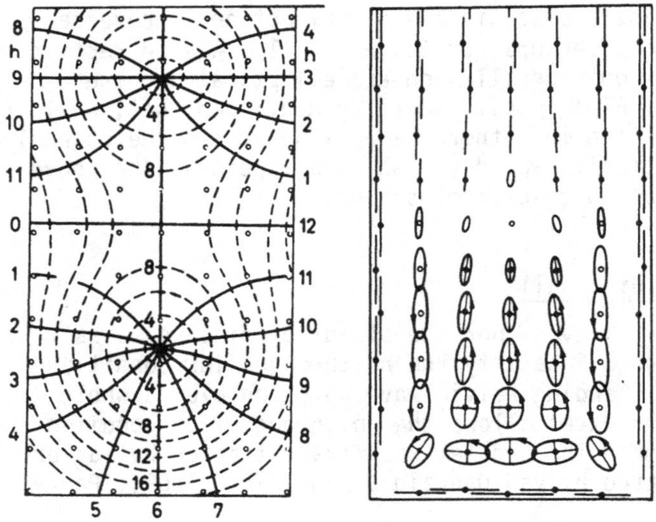

Figure 8

Amphidromic system in a rectangular basin closed at one end.

Left: co-tidal and co-range lines.

Right: current ellipses (G.I. Taylor, 1922)

are shore bound for Kelvin waves, while they are not for Poincaré waves. Provided that the lake is large enough, near shore measurements will thus favor detection of Kelvin waves, while Poincaré wave dynamics may be experimentally detectable by mid-lake measurements. For internal Poincaré waves, this has been beautifilly demonstrated by Mortimer (1963).

4. THE EIGENVALUE PROBLEM FOR AN ENCLOSED BASIN

In natural basins with variable bathymetry, a separation of the vertical mode structure from the horizontal structure is not possible and the general 3-D perturbation eigenvalue problem (H1, 3.18) should be solved. But this problem is too complex to have a chance to be solved numerically for a real lake. The usual procedure is, therefore, to restrict considerations to a 2-layer model and to start from equations (H1, 5.7)*).

A direct application of a harmonic ansatz to (H1, 5.7), $\psi \to \psi \exp(i\omega t)$ yields the system (the indices "1" and "2" indicate the layers)

*) *In a layered model, one approximates not only the stratification, but also the domain of solution, which is restricted to that region of a lake which is common to all layers, see Figure 3 in H1. This is why increasing the number of layers may not improve the accuracy of the model, unless subdomains with different numbers of layers are accordingly connected. But this has never been tried.*

$$
\begin{aligned}
& i\omega \underline{v}_1 + f\hat{\underline{k}}\wedge\underline{v}_1 + g\nabla\zeta_1 = 0\,, \\
& i\omega \underline{v}_2 + f\hat{\underline{k}}\wedge\underline{v}_2 + g\nabla\zeta_2 + g'\nabla(\zeta_2-\zeta_1) = 0\,, \qquad (x,y)\in\mathcal{D} \qquad (4.1)\\
& i\omega\cdot(\zeta_1-\zeta_2) + \nabla\cdot(H_1\underline{v}_1) = 0\,, \\
& i\omega\,\zeta_2 + \nabla\cdot(H_2\underline{v}_2) = 0\,,
\end{aligned}
$$

for the amplitude functions $\underline{v}_1, \underline{v}_2, \zeta_1, \zeta_2$, which in this section will not be differentiated from the original functions. Equations (4.1) must be solved in the lake domain $\mathcal{D}$ subject to the boundary conditions

$$\underline{v}_1\cdot\underline{N} = 0, \quad \underline{v}_2\cdot\underline{N} = 0 \quad \text{at } (x,y)\in\partial\mathcal{D} \qquad (4.2)$$

along the shore line; in (4.2), $\underline{N}$ is the 2-D unit normal vector.

In a finite difference approximation with n mesh points the eigenvalue problem (4.1), (4.2) yields $6n\times 6n$ equations. For usual computers, this is not economical and may cause difficulties because of storage limitations. One may, therefore, solve the momentum equations in (4.1) for the velocities,

$$
\begin{aligned}
\underline{v}_1 &= (\omega^2-f^2)^{-1}\, g\,[\,i\omega\nabla\zeta_1 - f\hat{\underline{k}}\wedge\nabla\zeta_1]\,, \\
\underline{v}_2 &= (\omega^2-f^2)^{-1}\{\, g\,[\,i\omega\nabla\zeta_1 - f\hat{\underline{k}}\wedge\nabla\zeta_1] \qquad (4.3)\\
&\qquad + g'[\,i\omega\nabla(\zeta_2-\zeta_1) - f\hat{\underline{k}}\wedge\nabla(\zeta_2-\zeta_1)]\,\}
\end{aligned}
$$

and eliminate these by substituting (4.3) into the remaining equations (4.1). This yields

$$
\begin{aligned}
&(\omega^2-f^2)(\zeta_2-\zeta_1) - \nabla\cdot(g H_1\nabla\zeta_1) = 0\,, \\
&(\omega^2-f^2)\zeta_2 + \nabla\cdot\{\, g H_2\nabla\zeta_1 + g' H_2\nabla(\zeta_2-\zeta_1)\,\} = 0\,,
\end{aligned} \qquad (4.4)
$$

and the boundary conditions (4.2) may be written as

$$\nabla\zeta_1\cdot\underline{N} = 0, \quad \nabla\zeta_2\cdot\underline{N} = 0, \quad \text{at } (x,y)\in\partial\mathcal{D}. \qquad (4.5)$$

In a finite difference approximation with n grid points, (4.4) and (4.5) will yield $2n\times 2n$ equations and the layer velocities may be obtained a posteriori from (4.3).

Equations (4.4) and (4.5) contain the barotropic and baroclinic gravity modes and also the topographic wave modes, whose time scales are usually far apart. Further simplifications are, therefore, called for.

For instance, for seiches in basins whose width is smaller than the internal Rossby radius, the rotation of the Earth may be ignored; (4.4) and (4.5) are then simply solved with $f = 0$. This automatically eliminates all second class waves. Or one works with a two-layered equivalent depth model, inspite of the nonconstant topography. In this case, one starts from (3.23) and may deduce the eigenvalue problem.

$$(\omega^2 - f^2)\,\zeta_\alpha + \nabla\cdot(g\,h_\alpha \nabla \zeta_\alpha), \quad \text{in } \mathcal{D}$$
$$\nabla\zeta_\alpha \cdot \underline{N} = 0, \quad \text{at } \partial\mathcal{D} \tag{4.6}$$

for the spatial part of the function ζ_α. The volume flux $\underline{V}_\alpha$ follows in a second step from

$$\underline{V}_\alpha = \frac{g\,h_\alpha}{\omega^2 - f^2}\,[\,i\,\omega\nabla\zeta_\alpha - f\,\hat{\underline{k}}\wedge\nabla\zeta_\alpha\,]. \tag{4.7}$$

For a homogeneous fluid (4.6) and (4.7) are exact, even if $h_\alpha = H$ varies with position. For the baroclinic mode h_i must, strictly, be a constant, the equivalent depth, averaged over the lake domain. Bäuerle (1981) and Hutter & Schwab (1982) have shown that one may in (4.6) treat h_i as variable, and will then obtain a better eigenvalue, but in the evolution of the flux (4.7) and the interface deflection (3.22), h_i should still be taken as a mean equivalent depth. Of course, the finite difference approximation of (4.6) and (4.7) is smallest and, for n grid points, leads only to a $n \times n$ matrix eigenvalue equation.

The two-layer equations (4.1) and (4.2) may also be transformed so as to filter out the external gravity modes. This is done by ignoring in $(4.1)_3$ ζ_1 as compared to ζ_2, corresponding to the rigid lid assumption. Adding $(4.1)_3$ and $(4.1)_4$ yields

$$\nabla\cdot(H_1\,\underline{v}_1 + H_2\,\underline{v}_2) = 0. \tag{4.8}$$

This suggests to introduce the barotropic transport stream function ψ according to

$$H_1\,\underline{v}_1 + H_2\,\underline{v}_2 = \hat{\underline{k}}\wedge\nabla\psi$$

and to deduce a coupled system of eigenvalue equations for ψ and ζ_2. This is, in detail explained by Prof. Mysak in his lecture.

5. CHANNEL MODELS

Many lakes are narrow and long. Geometry then favors longitudinal motions. One may, in this situation ignore the transverse dependence alltogether. This yields

5.1 A Chrystal-type two-layer model

Restriction to longitudinal motion requires that the Coriolis effects be ignored. Applying mass and momentum balance to the element shown in Figure 9 leads to the equations

$$\left.\begin{aligned} &\frac{\partial(\zeta_1-\zeta_2)}{\partial t} + \frac{1}{b}\frac{\partial}{\partial x}(S_1 u_1) = 0 ,\\ &\frac{\partial}{\partial t}(S_1 u_1) + g S_1 \frac{\partial \zeta_1}{\partial x} = 0 ,\end{aligned}\right\} \quad \text{upper layer} \qquad (5.1a)$$

$$\left.\begin{aligned} &\frac{\partial \zeta_2}{\partial t} + \frac{1}{b}\frac{\partial}{\partial x}(S_2 u_2) = 0 ,\\ &\frac{\partial}{\partial t}(S_2 u_2) + g S_2 \frac{\partial \zeta_1}{\partial x} + g S_2 \frac{\partial}{\partial x}(\zeta_2-\zeta_1) = 0 .\end{aligned}\right\} \quad \text{lower layer} \qquad (5.1b)$$

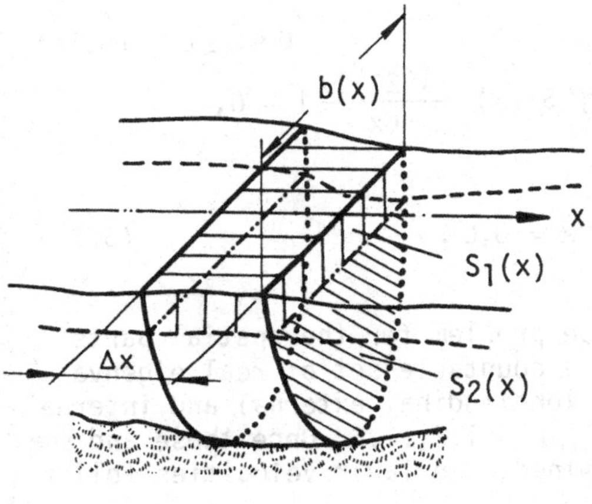

Figure 9

Perspective view of the part of a channel consisting of two layers of fluids, a lighter upper layer and a heavier lower layer. x is the channel axis, $S_1(x)$ and $S_2(x)$ the upper and lower cross sectional areas perpendicular to x and $b(x)$ is the width of the channel at the interface.

In these equations, ζ_1 and ζ_2 are the free and interface surface elevations, u_1 and u_2 are the average longitudinal layer velocities; b is the width of the channel at the thermocline depth, and S_1 and S_2 are the cross sectional areas of the upper and lower layer, respectively. Boundary conditions, which equations (5.1) are subject to, are

$$u_1 = u_2 = 0, \quad \text{at} \quad x = 0, \quad \text{and} \quad x = L \qquad (5.2)$$

and express the no-flux conditions through the channel ends. Equations of the type (5.1), (5.2) for a homogeneous (one layer) fluid were first derived by Chrystal (1904, 1905) and later taken up by F. Defant (1953), Platzmann (1972, 1975) and Platzman & Rao (1964). Eliminating u_1 and u_2 from (5.1) yields the differential equations

$$\frac{\partial^2}{\partial t^2}(\zeta_1-\zeta_2) - \frac{1}{b}\,\frac{\partial}{\partial x}[\,g\,S_1\,\frac{\partial\zeta_1}{\partial x}\,] = 0\,,$$

$$0 \le x \le L \qquad (5.3)$$

$$\frac{\partial^2\zeta_2}{\partial t^2} - \frac{1}{b}\,\frac{\partial}{\partial x}[\,g\,S_2\,\frac{\partial\zeta_1}{\partial x} + g'\,S_2\,\frac{\partial(\zeta_2-\zeta_1)}{\partial x}\,] = 0\,,$$

for ζ_1 and ζ_2, in terms of which (5.2) reads

$$\frac{\partial\zeta_1}{\partial x} = \frac{\partial\zeta_2}{\partial x} = 0, \quad \text{at} \quad x = 0,L. \qquad (5.4)$$

Alternatively, one could also derive an equation for u_1 and u_2. By taking harmonic solutions, $\zeta = \zeta(x)\,\exp(i\,\omega\,t)$, (5.3) and (5.4) become

$$\omega^2(\zeta_1-\zeta_2) - \frac{1}{b(x)}\,\frac{d}{dx}\,[g\,S_1(x)\,\frac{d\zeta_1}{dx}\,] = 0\,,$$

$$0 \le x \le L \qquad (5.5a)$$

$$\omega^2\zeta_2 + \frac{1}{b(x)}\,\frac{d}{dx}[g\,S_2(x)\,\frac{d\zeta_1}{dx} + g'\,S_2(x)\,\frac{d(\zeta_2-\zeta_1)}{dx}\,] = 0,$$

$$\frac{d\zeta_1}{dx} = \frac{d\zeta_2}{dx} = 0, \quad \text{at} \quad x = 0,L\,, \qquad (5.5b)$$

which is a two point-boundary eigenvalue problem for the spatial parts $\zeta_1(x)$ and $\zeta_2(x)$. The problem (5.5) has a countable set of real eigenvalues $\omega_n{}^e$ and $\omega_n{}^i$, corresponding to the longitudinal external and internal seiche periods $T_n{}^e = 2\pi/\omega_n{}^e$, $T_n{}^i = 2\pi/\omega_n{}^i$, $n = 1,2,\ldots$. Once these and the corresponding eigenfunctions are determined, the layer velocities follow from

$$u_1(x) = \frac{ig}{\omega}\,\frac{d\zeta_1}{dx}, \quad u_2(x) = \frac{ig}{\omega}\,\frac{d\zeta_1}{dx} + \frac{ig'}{\omega}\,\frac{d(\zeta_2-\zeta_1)}{dx}\,. \qquad (5.6)$$

These solutions have no transverse structure, but using Kelvin wave dynamics it can approximately be determined. To this end, we write down the transverse geostrophic balance of the two-layer equations (H1, 5.7)

$$f\overset{(0)}{u}_1(x) = -\,g\,\frac{\partial\zeta_1}{\partial y}, \quad f\overset{(0)}{u}_2(x) = -\,g\,[\,\frac{\partial\zeta_1}{\partial y} + \frac{\Delta\rho}{\rho}\,\frac{\partial}{\partial y}(\zeta_2-\zeta_1)\,], \qquad (5.7)$$

in which the upper index (0) indicates that the respective quantity belongs to the channel solution (5.5), (5.6). Straightforward integration of (5.7) yields

$$\zeta_1 = \zeta_1^{(0)}(x) - \frac{f}{g} u_1^{(0)}(x)\, y ,$$

$$\zeta_2 = \zeta_2^{(0)}(x) - \frac{f}{\frac{\Delta\rho}{\rho} g} [\, u_2^{(0)}(x) - (1 - \frac{\Delta\rho}{\rho})\, u_1^{(0)}(x)\,]\, y. \qquad (5.8)$$

These formulas indicate that the transverse structure of the interface is larger than that of the free surface by a factor of $\rho/\Delta\rho = 0(10^3)$. Finally, it should be mentioned that starting from (5.1), one could also deduce an equivalent depth model. Such a model would be exact only for constant S_1, S_2 and b. In times of easy accessibility to computers such a simplification is perhaps a bit outdated. But Mortimer (1979) applies the model to Lake Geneva and Hutter & Schwab (1982) show that thermocline deflections are, in general, well predicted.

5.2 Extended channel models - general concept

Kelvin wave dynamics is not a systematic treatment in the sense of a perturbation expansion, which correctly incorporates rotational effects. Nevertheless, an approach, in which a two-point-boundary value problem is solved and not a full 2-D problem is attractive, because it may be computationally more economical than the full 2-D equations.

The idea is as follows: Considering the fact that the basin is long and narrow, one introduces an orthogonal set of curvi-linear coordinates x,y,z; $y = 0$ defines the lake axis along which x is measured and z is vertical. Any field variable χ (velocity $\underline{v}$, surface elevation ζ) is then expanded in terms of shape functions ϕ_α,

$$\chi(x,y,t) = \sum_{\alpha=0}^{M} \phi_\alpha(y; x)\, \chi_\alpha(x,t). \qquad (5.9)$$

Here ϕ_α are known functions of y (and possibly x) and describe the variation of χ in the transverse direction of the channel. For instance, ϕ_α can be the functions $1, y, y^2, \dots, y^M$. The index M defines the order of the model; it should be small for the model to be computationally economical. The functions $\chi_\alpha(x,t)$ are unknown parameters, which only depend on the longitudinal spatial variable. If the original spatially 2-D boundary value problem can be transformed such that two point boundary value problems for χ_α, $\alpha = 0,1,\dots,M$ emerge, then a systematic extended channel model will be deduceable of which the previous channel model combined with the Kelvin wave dynamics is an unsystematic precursor.

The method to deduce the spatially 1-D equations for χ_α is the *principle of weighted residuals* (Finlayson, 1972). Its application to surface gravity waves has been explained in detail by Raggio & Hutter (1982a,b,c) and Hutter & Raggio (1982) to whom the interested reader is directed. Here we restrict attention to a first order model of surface gravity waves.

5.3 Extended channel model - Surface gravity waves

In a first order model ($M = 1$) the velocity vector and the surface elevations are expanded in the form

$$\begin{aligned} v(x,y,t) &= v^{(0)}(x,t) + v^{(1)}(x,t)y , \\ \zeta(x,y,t) &= \zeta^{(0)}(x,t) + \zeta^{(1)}(x,t)y , \end{aligned} \tag{5.10}$$

and the differential equations which can be deduced to determine $v^{(0)}$, $v^{(1)}$, $\zeta^{(0)}$ and $\zeta^{(1)}$ are listed in Raggio & Hutter (1982a). For a rectangular basin with width B and constant water depth H these equations read

$$\underline{B}\,\underline{x} = \underline{0} \tag{5.11}$$

in which

$$\underline{x} = (\zeta^{(0)}, u^{(0)}, v^{(0)}, \zeta^{(1)}, u^{(1)}, v^{(1)})^T \tag{5.12}$$

is the 6-vector formed with the zeroth and first order surface elevations and velocity components (u is longitudinal, v transversal) and $\underline{B}$ is the matrix operator

$$\underline{B} = \left[\begin{array}{cc|ccc|c} \frac{1}{H} g \frac{\partial}{\partial t} & g \frac{\partial}{\partial x} & 0 & 0 & 0 & 0 \\ g \frac{\partial}{\partial x} & \frac{\partial}{\partial t} & -f & 0 & 0 & 0 \\ \hline 0 & f & \frac{\partial}{\partial t} & g & 0 & 0 \\ 0 & 0 & -g & \frac{1}{12} B^2 \frac{g}{H} \frac{\partial}{\partial t} & \frac{1}{12} B^2 g \frac{\partial}{\partial x} & 0 \\ 0 & 0 & 0 & \frac{1}{12} B^2 g \frac{\partial}{\partial x} & \frac{1}{12} B^2 \frac{\partial}{\partial t} & -\frac{1}{12} B^2 f \\ \hline 0 & 0 & 0 & 0 & \frac{1}{12} B^2 f & \frac{1}{12} B^2 \frac{\partial}{\partial t} \end{array}\right] \tag{5.13}$$

Boundary conditions, Equations (5.12) and (5.13) must be subject to, are the no-through flux conditions, expressible as

$$u^{(0)} = u^{(1)} = 0, \quad \text{at} \quad x = (0,L). \tag{5.14}$$

Physically, the first and fourth component equations in (5.12) are statements of mass balance, the second and fifth equation stem from a longitudinal momentum balance and the remaining two equations correspond to a transverse momentum balance. Coupling effects are easily recognizable from (5.13). For instance, when $f = 0$ the system decouples mainly into two subsystems of which the first 2×2 upper left submatrix in (5.13) is identical with the Chrystal (1904, 1905) model of a rectangular basin. The second 3×3 submatrix in the middle extends the model to a more sophisticated model which includes the transverse pressure gradients and thus adequately accounts for transverse mass flux. For $f = 0$ decoupling of the submatrices is not possible which was to be expected from the Kelvin wave dynamics treated before.

A progressing and standing wave analysis performed with (5.12)-(5.14) indicates that Kelvin- and transversely skew-symmetric Poincaré waves can approximately correctly be predicted including the Taylor reflection problem at a channel wall. In other words, the system (5.13) exhibits all essential features of surface gravity waves on the rotating Earth. More accurate results could be obtained with a higher order model which includes y^2 and even higher order terms, but analytical computations become quickly unwieldy. An electronic application to a model of arbitrary order is given by Raggio & Hutter (1982c) and Raggio (1982).

The above is only a brief presentation of a systematic channel approach which should illustrate that physically two-dimensional problems can, in a systematic fashion be made mathematically one-dimensional without loosing the physically two-dimensional nature. The method is suitable to all problems of mathematical physics which are genuinely multi-dimensional but must be solved in a domain which is long and narrow. The method warrants to be further explored.

6. COMPARISON WITH OBSERVATIONS

Temperature and current records taken at fixed positions within the lake and moored for periods of several weeks to months are the traditional experimental set-up by which the dynamics of a lake is "observed". Direct inspection and/or spectral analysis (auto-spectra, cross correlations and spectra of phase differences between station-pair-time series) are the vehicle for the identification of the conspicuous processes. Below we give evidence of external and internal wave dynamics and demonstrate that under appropriate circumstances Kelvin and Poincaré type waves may form. Cause for the existence of these is meteorological forcing.

6.1 Surface seiches

The literature on gravitational seiches is abundant and proper reference can hardly be made*). The pioneering work is probably that of Forel (1895); however, detailed corroboration of theoretical models by field ob-

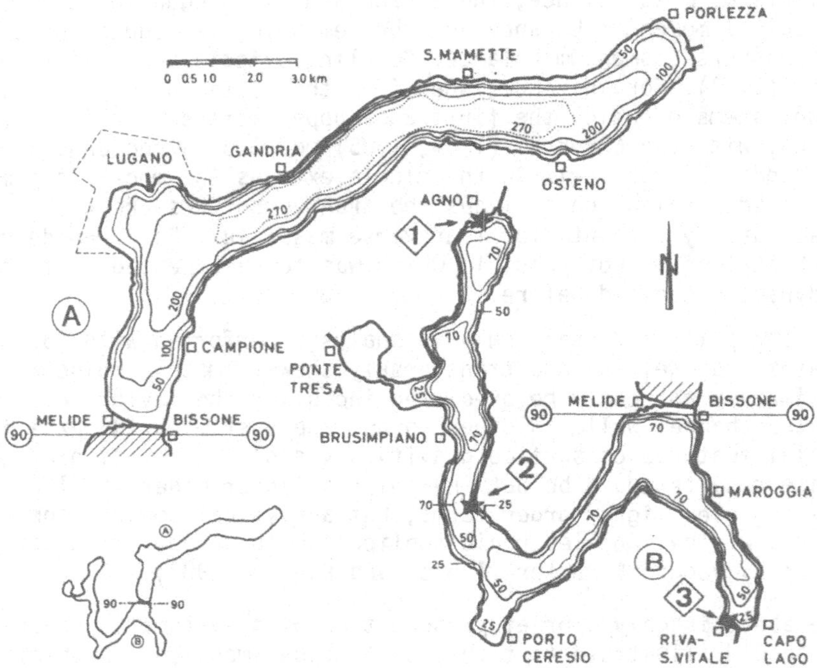

Figure 10 Depth chart of the two basins of the Lake of Lugano. The *north basin* is an extremely deep and L-shaped channel extending to north-east. The *south basin* is substantially curved in the form of a "turned over" S and less deep. Appended to this basin is the small pond at Ponte Tresa, ignored in the dynamical calculations. The symbols 1, 2 and 3 identify stations where seiche limnigraphs were moored in 1982. The location of the north basin relative to the south basin is shown in the insert of the figure.

---

*) *A very detailed bibliography is given by Tison & Tison (1969) with over 200 references. More recent works and work aiming at detailed experimental verification is that of Mortimer (1963, 1974), Mortimer & Fee (1976), Platzman (1972, 1975), Platzman & Rao (1964), Rao & Schwab (1976), Schwab & Rao (1977), Hutter et al. (1982a,b), Hamblin (1974, 1976), Hamblin & Hollan (1978), Mühleisen (1977), Hollan (1979, 1980), Hollan et al. (1980) and others.*

servations has only become possible with the availability of electronic computers. Here we present results for the southern basin of the Lake of Lugano.

Three limnographs were moored at the indicated sites: 1 = Agno, 2 = Morcote, 3 = Riva S. Vitale as indicated in Figure 10, where water fluctuations were measured during February 1982. Figure 11 gives an excerpt of these measurements and indicates fluctuations with approximate periods of 28, 14, 9.6 and 5.8 minutes. The oscillations with the period of 28 minutes are clearly seen at Agno and Riva S. Vitale but not at Morcote, and they are out of phase. Oscillations with a period of approximately 14 minutes are seen at all three stations; here the recordings at Agno and Riva S. Vitale are in phase, but both are out of phase with that at Morcote. The distribution of the energy at certain frequencies and corresponding phase differences between station-pair-time series can best be found from spectral analyses by constructing auto- and cross-spectra of the pair-time series.

Figure 12 shows the coherence, energy density and phase difference spectra of the station pair Agno / Riva S. Vitale for the 12 h episode of

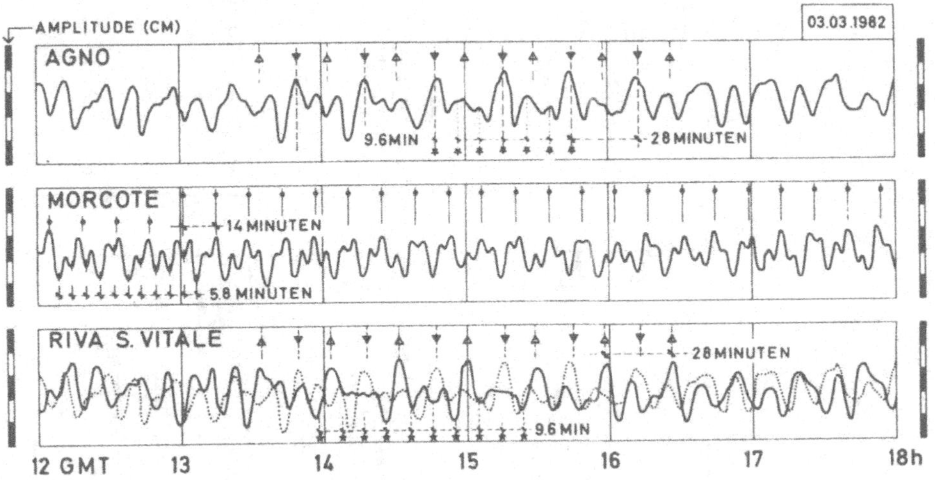

Figure 11 Six-hour episode of the time series for the "Agno", "Morcote" and "Riva S. Vitale" gauges on 3 March 1982 from 12.00 to 18.00 h. The curve of the "Agno" recording is superposed on that of "Riva S. Vitale" (dotted). Full triangles indicate maxima (for "Agno") and minima (for "Riva S. Vitale") with a period of approximately 28 minutes. For the open triangles this role of maxima and minima is reversed. The asteriks indicate a period of approximately 9.6 minutes. Points (· and ~), shown in the recordings for "Morcote" identify periods with 14 and 5.8 minutes.

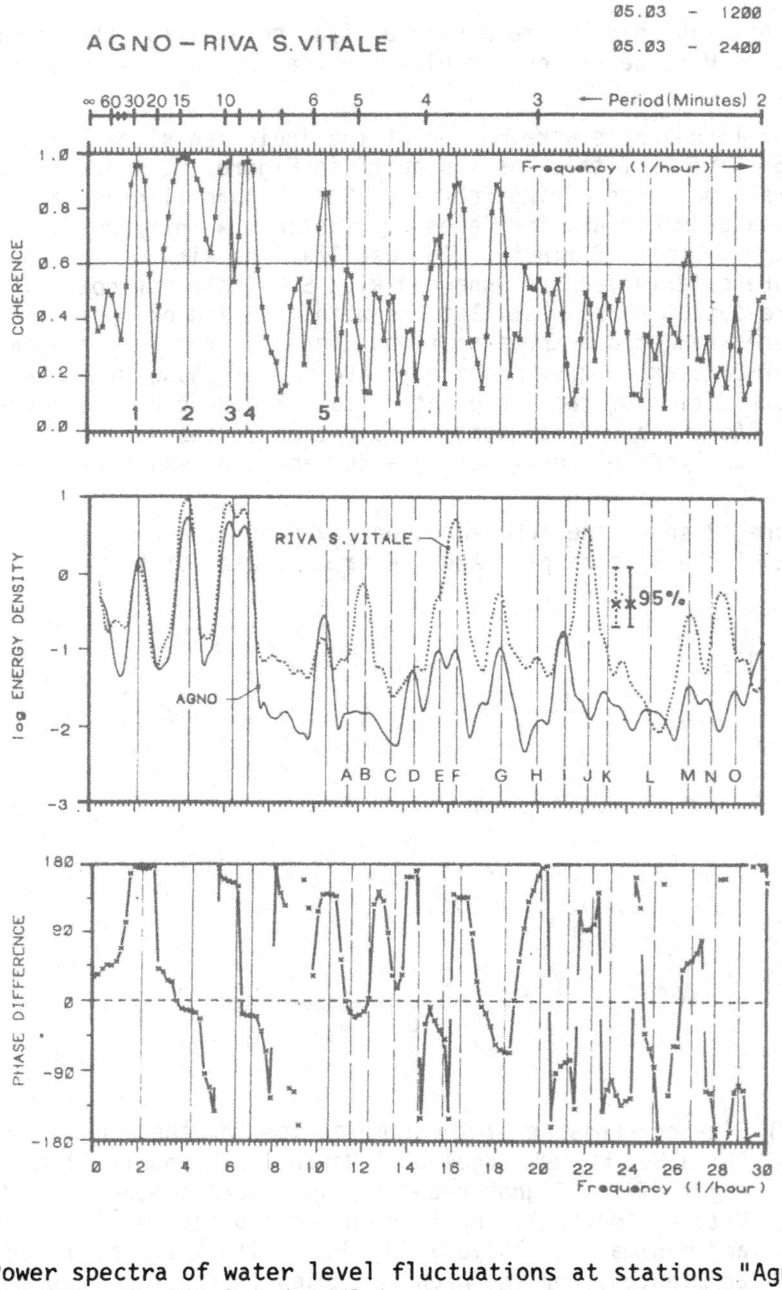

Figure 12 Power spectra of water level fluctuations at stations "Agno" and "Riva S. Vitale" (middle) for a twelve-hour episode on 5 March 1982 and corresponding interstation phase difference (bottom) and coherence (top).

limnigraph recordings on 5 March 1982. There are conspicuous energy maxima at the periods marked by 1,2,...,5,A,...,0. When these maxima arise "simultaneously", coherence is generally large, indicative of good statistical confidence. Phase differences between the recordings at the indicated periods 1 to 5 are nearly 0° or 180°.

The above observations can be explained by a surface seiche model as described by the eigenvalue problem (4.6). This eigenvalue problem was solved for Lake of Lugano using a finite element technique. The first five eigenvalues led to the periods 27.8, 13.6, 9.4, 8.5 and 5.7 minutes. Four of these can also be identified in Figure 11. These periods are also marked by 1, 2,...,5 in Figure 12. Energy maxima in the frequency spectrum thus coincide with the eigenfrequencies of the gravitational oscillations of the system. Phase differences can be understood with the mode structures of the co-range and co-tidal lines. For the first and fourth mode these are shown in Figure 13. Both modes have an amphidrome near

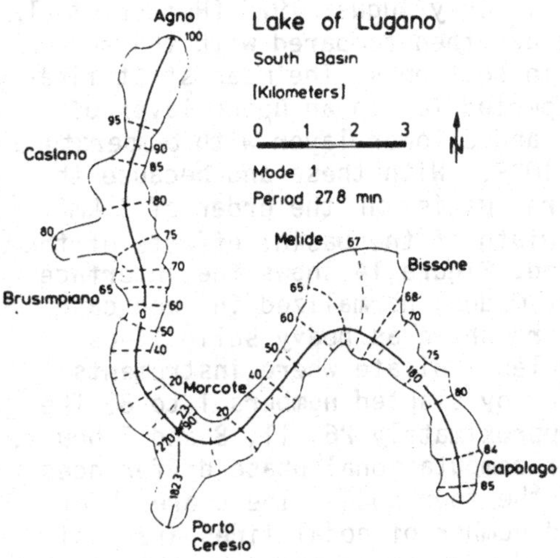

Figure 13 a

Co-range (dashed) and co-tidal lines (solid) for mode one of the south basin of the Lake of Lugano. Mode type is longitudinal with largest amplitudes arising near "Agno" and "Capolago".
Period = 27.8 minutes.

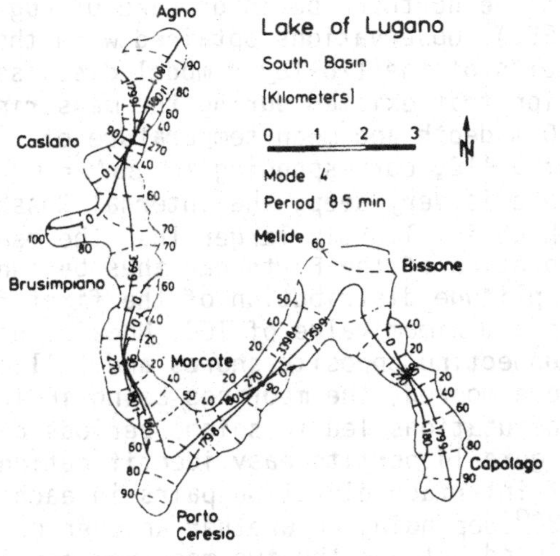

Figure 13 b

Co-range (dashed) and co-tidal lines (solid) for mode four of the south basin of the Lake of Lugano with a period of 8,5 minutes. There are four positive amphidromic points approximately equally distributed along the basin. The largest amplitudes are to be expected at the far ends and in the bays "Porto Ceresio" and north of "Brusimpiano".

Morcote. Since surface elevation amplitudes are small in their neighborhood it is not surprising that oscillating components with periods of 28 and 8.5 minutes are not seen in the Morcote recordings of Figure 11. Co-tidal lines in Figure 13 indicate a 180° phase difference between Agno and Riva S. Vitale for mode 1 and an in-phase relationships for mode 4, in agreement with the phase difference spectra of Figure 12. Hutter et al. (1982b) made many more comparisons between the observations and the theoretical model, and similar comparisons between theoretical modelling and observations have also been made for other lakes. Space limitations do not allow us to go into more detail.

6.2 Internal wave dynamics in small lakes (gravity waves)

Internal wave dynamics can experimentally be detected by thermistor chain temperature recordings and by current measurements. Here, we restrict considerations to temperature records of thermistor chains, moored in the northern basin of Lake of Lugano in July/August 1979 (Hutter et al. 1983). Observations obtained with these are then compared with the results of the two-layer model discussed in section 4. The mean stratification that existed during the measuring period led to an upper layer of 10 m depth and mean temperature of 21°C and a lower layer with temperature of 5.9°C, corresponding to $\Delta\rho/\rho = 1.93 \cdot 10^{-3}$. With these and because the lake is very deep, the internal Rossby radius is on the order of 3 km which is slightly larger than the mean width of the basin; effects of the rotation of the Earth can thus be ignored. Figure 14 shows the interface amplitude distribution of the first four modes, normalized in each case to a maximum value of 100. Nodal lines are shown as heavy solid lines connecting opposite shores and full circles indicate where instruments were moored, the moorings being indicated by slanted numbers 1 to 8. The computations led to seiche periods of approximately 26, 11, 8 and 7 hours. Figure 14 permits easy identification of computational phase differences of interface elevation pairs in each of the four modes. These are 0° or 180° depending on whether an even or odd number of nodal lines are positioned between the two moorings considered. The thermistor-chain-temperature recordings taken at the stations in Figure 14 were processed. Figure 15 shows smoothed energy spectra of time series of representative isotherm depth fluctuations constructed from the temperature time series at the stations 1 to 4 and 8. The two spectra at the bottom are averages of the former, once constructed with the smoothed and once with the non-smoothed spectra. Vertical lines marked by encircled numbers 1 to 10, indicate the computed eigenfrequencies of the two-layer model. The first four modes clearly agree with the relative maxima in the energy spectra. Coherence and phase difference spectra have also been determined for all station pairs. These indicate very rich coherence between data pairs at the mode periods. Furthermore, the phase differences of the data at the

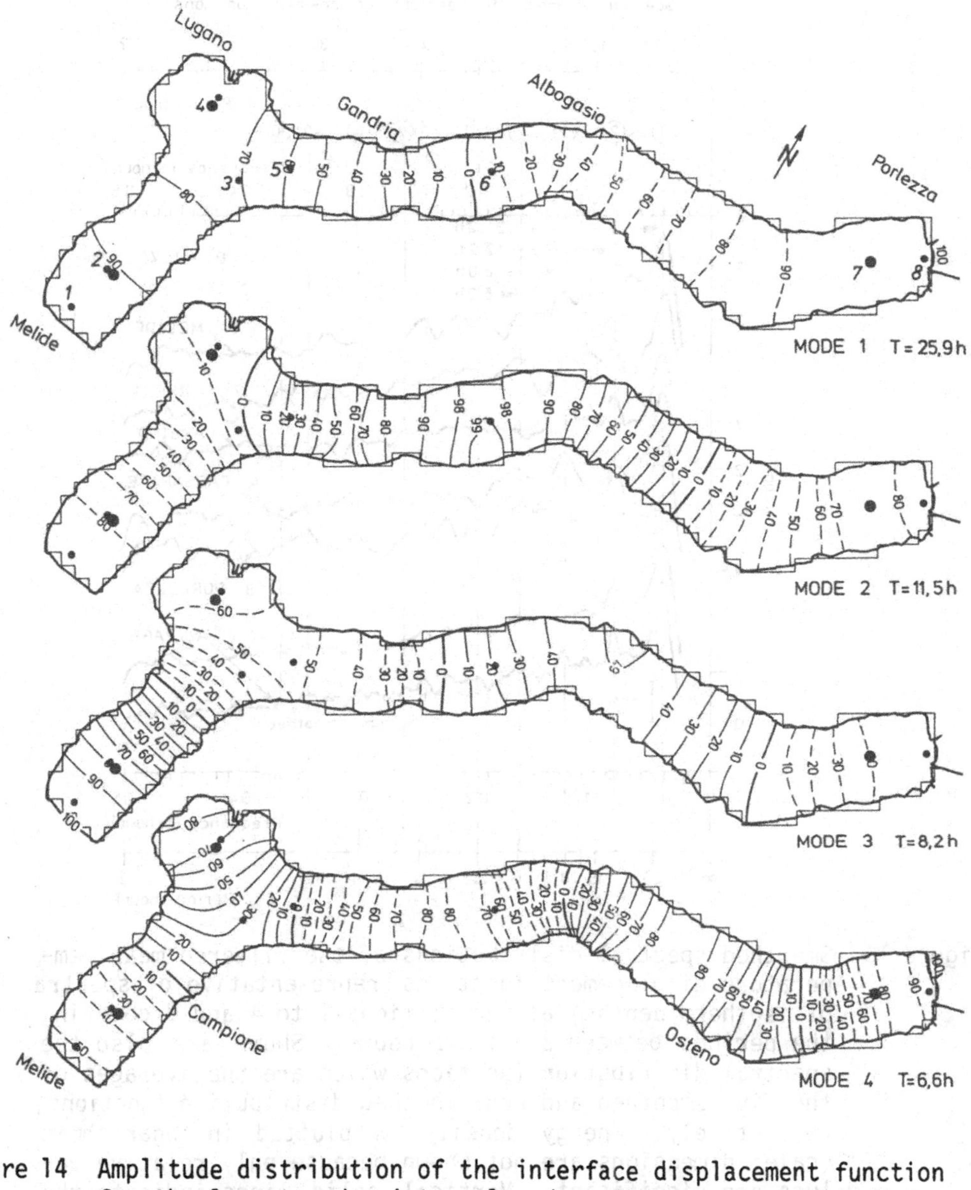

Figure 14 Amplitude distribution of the interface displacement function for the lowest order internal modes. These are normalized to a maximum of value 100. Nodal lines are shown as heavy solid lines, and lines of equal thermocline displacement are shown as light solid and dashed lines, respectively. Positions, where thermistor chains and current meters were moored are indicated by full circles and numbered (slanted) from 1 to 8.

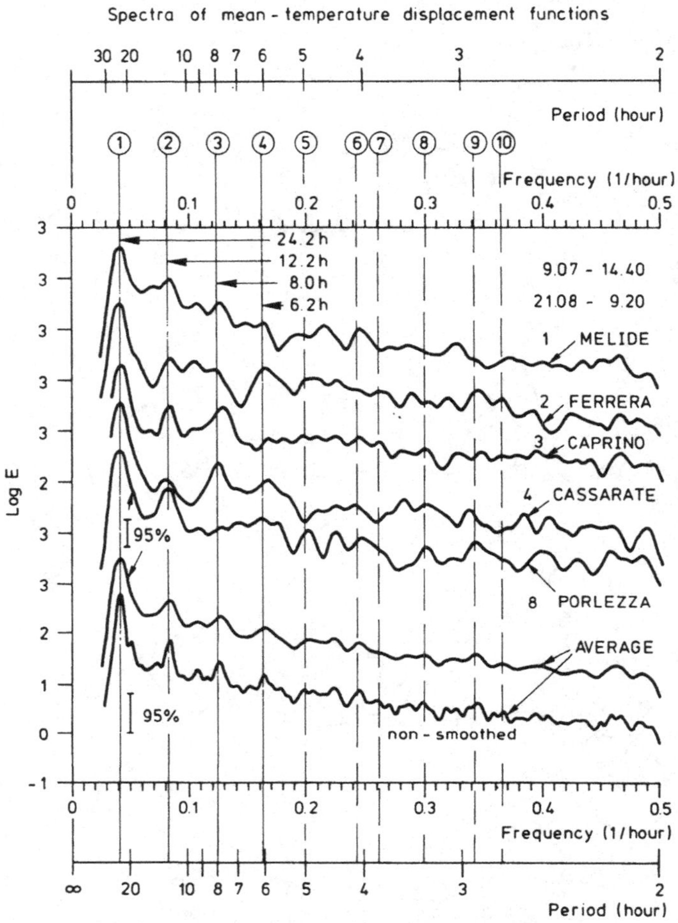

Figure 15 Smoothed spectral distributions of the filtered mean temperature displacement functions (representative of spectra of isotherm depths) at the stations 1 to 4 and 8 covering the periods between 2 and 30 hours. Shown are also the spectral distribution functions which are the averages of the five smoothed and non-smoothed distribution functions, respectively. Energy density is plotted in logarithmic scale; dimensions are not shown because only relative values are significant. Vertical solid lines indicate the periods where the averaged distribution functions have relative maxima, characteristic for the four lowest internal seiche periods. Further, higher order internal seiche periods are shown dashed. The spectral plots are for time series between 9 July and 21 August 1979.

mode periods agree well with those which can be inferred from Figure 14, except where mooring positions are close to a nodal line, in which case phase differences are uncertain anyhow. Hutter et al. (1983) give many more details which show that the two-layer model may explain the key features of the internal dynamics of a stratified lake.

Other works where two layer models or equivalent depth models have been used to explain internal wave dynamics are by Bäuerle (1981), Horn et al. (1983), Kanari (1975), Mortimer (1953, 1963, 1974, 1979), Schwab (1977) and others.

6.3 Evidence of Kelvin waves and Poincaré waves

Rotational effects on gravitational modes can be detected in lakes which are wide as compared, or at least comparable to the Rossby radius $R = c/f$. For barotropic waves R is several hundred km, for internal waves it is just e few km. Evidence for Kelvin wave behavior in barotroipc seiches in Lakes Superior and Michigan is given in a beautiful paper by Mortimer & Fee (1976). The rotational character of the waves can best be seen from the phase progression for a particular mode. In the first mode the phase difference between station pairs is not just 0^{0} or 180^{0}, depending on the location of the nodal line, but according to the position of the limnigraphs and the spacing of the co-tidal lines relative to these. For the Great Lakes the computations (Platzman, 1972, 1975) and the measurements agree remarkably. Figure 16 is a copy of a figure taken from

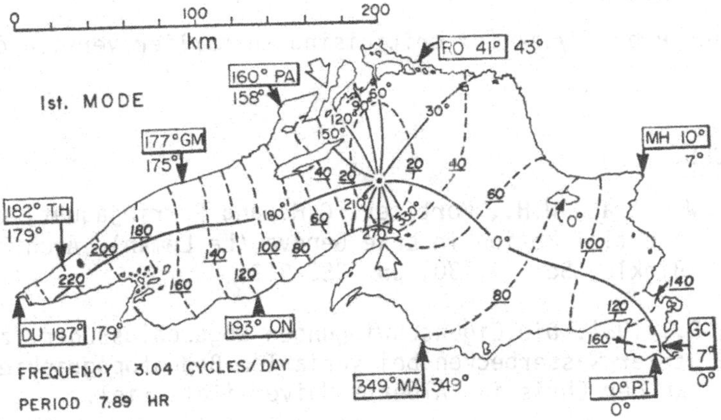

Figure 16 Phase progression of the first longitudinal surface mode in Lake Superior. Observed phases relative to 0^{0} at PI, are shown in boxes and the phases computed by Platzman (1972) are shown outside the boxes. The co-tidal lines and the amplitude distributions are those calculated by Platzman.

Mortimer & Fee (1976). It shows the co-range and co-tidal lines for the lowest mode of the surface seiche for Lake Superior as calculated by Platzman. Measured and calculated phase differences relative to the reference station "PI" are remarkably close to each other and demonstrate the Kelvin wave nature of the oscillatory motion at a period of 7.89 hours.

Lakes of the size of Lake Geneva are too small that rotational effects could be seen in the barotropic mode of surface oscillations. However, by filtering out the barotropic time scales from limnigraph measurements around the shore of Lake Geneva, Mortimer (1963) was able to identify first mode internal Kelvin wave response. Further observational evidence of internal Kelvin waves is given by Csanady and Scott (1974) for Lake Ontario and by Hamblin (1978) in a fjord lake of Canada.

As explained in Section 3.5 Poincaré waves are likely to be seen offshore, if they are excited at all. They have strong transverse structure and, if they are internal, can be detected with thermistor chain arrangements perpendicular to the long direction of the lake. From temperature records taken from a ferry boat crossing Lake Michigan, Mortimer (1974) was able to identify such Poincaré wave behavior, but the measurements are less convincing here than they are for Kelvin waves.

For a thorough treatment of the comparison of field data and models, the reader is referred to Prof. Morimer's lecture.

ACKNOWLEDGEMENTS

I thank Prof. Mysak for criticising an earlier version of this manuscript.

REFERENCES

Bauer, S.W., Graf, W.H,, Mortimer, C.H. and Perrinjaquet, C., 1981. Inertial Motion in Lake Geneva (Le Léman). Arch. Met. Geoph. Biokl., Ser. A, 30, pp. 283-312.

Bäuerle, E., 1981. Die Eigenschwingungen abgeschlossener, zwei-geschichteter Wasserbecken bei variabler Bodentopographie. Dissertation an der Christian Alberts Universität, Kiel.

Brown, P.J., 1973. Kelvin wave reflection in semi-infinite canal. J. Mar. Res., Vol. 31, pp. 1-10.

Courant, R. and Hilbert, D., 1982. Methods of Mathematical Physics. Vol.1, Interscience Publ. John Wiley & Sons, New York.

Csanady, G.T. and Scott, J.T., 1974. Baroclinic coastal jets in Lake Ontario during IFYGL. J. Phys. Oceanogr., Vol. 4, pp. 524-541.

Defant, A., 1961. Physical Oceanography. Oxford University Press, Oxford.

Defant, F., 1953. Theorie der Seiches des Michigansees und ihre Abwandlung durch Wirkung der Corioliskraft. Arch. Met. Geophys. Biokl., Ser. A, Vol. 6, pp. 218-241.

Finlayson, B.A., 1972. The method of weighted residuals and variational principles. Academic Press, New York.

Forel, A.F., 1895. Le Léman. Monographie, Limnologie, tome 1 & 2. Ed. F. Rouge Librairie de l'Université, Lausanne.

Greenspan, H.P., 1968. The Theory of Rotating Fluids. Cambridge University Press.

Hamblin, P.F., 1974. Short period tides in Lake Ontario. Proc. 17th Conf. Great Lakes Res. Internat. Assoc. Great Lakes Res., pp. 789-800.

Hamblin, P.F., 1976. A theory of short period tides in a rotating basin. Phil. Trans. Royal Soc. London, Vol. A 291, pp. 97-111.

Hamblin, P.F., 1978. Internal Kelvin waves in a fjord lake. J. Geophys. Res., Vol. 83, pp. 287-300.

Hamblin, P.F. and Hollan, E., 1978. On the gravitational seiches of Lake Constance and their generation. Schweiz. Z. Hydr., Vol. 40, pp. 119-154.

Hayes, W.H., 1974. Introduction to wave propagation. In: Nonlinear Waves (ed. by S. Leibovich & A.R. Seebass). Cornell University Press, Ithaca, London.

Heaps, N.S., 1961. Seiches in a narrow lake, uniformly stratified in three layers. Geophys. Suppl. J. Roy. Astron. Soc., Vol. 5, pp. 134-156.

Hollan, E., 1979. Hydrodynamische Modellrechnungen über die Eigenschwingungen des Bodensee-Obersees mit einer Deutung des Wasserwunders von Konstanz im Jahre 1949. Schw. Verein für Geschichte des Bodensees und seiner Umgebung, Heft 97, pp. 157-192.

Hollan, E., 1980. Die Eigenschwingungen des Bodensee-Obersees und eine Deutung des "Wasserwunders von Konstanz" im Jahre 1549. Deutsches gewässerkundliches Jahrbuch, ISSN 0344-0788, Abflussjahr 1979.

Hollan, E., Rao, D.B. and Bäuerle, E., 1980. Free surface oscillations in Lake Constance with an interpretation of the "Wonder of the rising water at Konstanz" 1549. Arch. Met. Geophys. Biokl., Ser. A, Vol. 29, pp. 301-325.

Horn, W., Mortimer, C.H. and Schwab, D.J., 1983. Internal wave dynamics of the Lake of Zurich, preprint.

Hutter, K. and Raggio, G., 1982. A Chrystal-model describing gravitational barotropic motion in elongated lakes. Arch. Met. Geophys. Biokl., Ser. A, Vol. 31, pp. 361-378.

Hutter, K. and Schwab, D.J., 1982. Baroclinic channel models. Internal report No. 62, Laboratory of Hydraulics, Hydrology and Glaciology, ETH Zurich.

Hutter, K., Raggio, G., Bucher, C. and Salvadè, G., 1982a. The surface seiches of Lake of Zurich. Schweiz. Z. Hydr., Vol. 44, pp. 423-454.

Hutter, K., Raggio, G., Bucher, C., Salvadè, G. and Zamboni, F., 1982b. The surface seiches of the Lake of Lugano. Schweiz. Z. Hydr., Vol. 44, pp. 455-484.

Hutter, K., Salvadè, G. and Schwab, D.J., 1983. On internal wave dynamics of the Lake of Lugano. J. Geophys. Astrophys. Fluid Dyn., in press.

Kanari, S., 1975. The long period internal waves in Lake Biwa. Limnol.& Oceanogr., Vol. 20, pp. 544-553.

Krauss, W., 1966. Methoden und Ergebnisse der theoretischen Ozeanographie. 2. Intern. Wellen. Borntraeger, Berlin-Stuttgart.

Krauss, W., 1973. Methods and results of theoretical oceanography. 1. Dynamics of the homogeneous and quasihomogeneous Ocean. Borntraeger, Berlin-Stuttgart.

LeBlond, P.H. and Mysak, L.A., 1978. Waves in the Ocean. Elsevier Oceanography Series. Elsevier Scientific Publishing Company, Amsterdam-Oxford-New York.

Lighthill, J., 1969. Dynamic response of the Indian Ocean to outset of the southwest monsoon. Phil. Trans. Royal. Soc. London, Vol. A 265, 1159, pp. 45-92.

Lighthill, J., 1978. Waves in Fluids. Cambridge Uuniversity Press, Cambridge.

Mortimer, C.H., 1952. Water movements in lakes during summer stratification; evidence from the distribution of temperature in Windermere. Phil. Trans. Royal Soc. London, Vol. B 236, pp. 355-404.

Mortimer, C.H., 1953. The resonant response of stratified lakes to wind. Schweiz. Z. Hydr., Vol. 15, pp. 94-151.

Mortimer, C.H., 1963. Frontiers in physical limnology with particular reference to long waves in rotating basins. pp. 9-42. In: Proc. 5th Conf. Freat Lakes Res., Great Lakes Res. Div. Univ. Mich. Publ. 9.

Mortimer, C.H., 1974. Lake hydrodynamics. Mitt. Int. Ver. Theor. Angew. Limnol., 20, pp. 124-197.

Mortimer, C.H., 1975. Substantive corrections to SIL Communications (IVL Mitteilungen) Numbers 6 and 20. Mitt. Int. Ver. Theor. Angew. Limnol., 19, pp. 60-72.

Mortimer, C.H. and Fee, E.J., 1976. Free surface oscillations and tides of Lakes Michigan and Superior. Phil. Trans. Royal Soc. London, Vol. A 281, pp. 1-61.

Mortimer, C.H., 1979. Strategies for coupling data collection and analysis with dynamic modelling of lake motions. In: Lake of Hydrodynamics (eds. W.H. Graf & C.H. Mortimer). Elsevier, Amsterdam, pp. 183-222.

Mühleisen, R. and Kurth, W., 1978. Experimental investigations on the seiches of Lake Constance. Schweiz. Z. Hydr., Vol. 40.

Mysak, L.A., 1980. Recent advances in shelf wave dynamics. Reviews of Geophysics and Space Physics, Vol. 18, pp. 211-241.

Phillips, O.M., 1966. The Dynamics of the Upper Ocean. Cambridge University Press, Cambridge.

Platzman, G.W., 1972. Two-dimensional free oscillation in natural basins. J. Phys. Oceanogr., Vol. 2, pp. 117-138.

Platzman, G.W., 1975. Normal modes of the Atlantic and Indian Ocean. J. Phys. Oceanogr., Vol. 5, pp. 201-221.

Platzman, G.W. and Rao, D.B., 1964. The free oscillations of Lake Erie. Studies on Oceanography (Hidaka Volume, ed. by K. Yoshida) Tokio Univ. Press, pp. 359-382.

Raggio, G. and Hutter, K., 1982a. An extended channel model for the prediction of motion in elongated homogeneous lakes. Part 1: Theoretical Introduction. J. Fluid Mech., Vol. 121, pp. 231-255.

Raggio, G. and Hutter, K., 1982b. An extended channel model for the prediction of motion in elongated homogeneous lakes. Part 2: First order model applied to ideal geometry; rectangular basin with flat bottom. J. Fluid Mech., Vol. 121, pp. 257-281.

Raggio, G. and Hutter, K., 1982c. An extended channel model for the prediction of motion in elongated homogeneous lakes. Part 3: Free oscillations in natural basins. J. Fluid Mech., Vol. 121, pp. 283-299.

Rao, D.B., 1966. Free gravitational oscillations in rotating rectangular basins. J. Fluid Mech. 24, pp. 523-555.

Schwab, D.J., 1977. Internal free oscillations in Lake Ontario. Limnol. & Oceanogr., Vol. 22, pp. 700-708.

Schwab, D.J., 1980. The free oscillations of Lake S. Clair - An application of Lanczos' procedure. Great Lakes Environmental Research Laboratory. No. AA, Tech. Memorandum ERL GLERL-32.

Schwab, D.J. and Rao, D.B., 1977. Gravitational oscillations of Lake Huron, Saginow Bay, Georgian Bay, and the North Channel. J. Geophys. Research, Vol. 82, pp. 2105-2116.

Simons, T.J., 1980. Circulation Models of Lakes and Inland Seas. Canadian Bulletin of Fisheries and Aquatic Sciences, No. 203, Ottawa.

Taylor, G.I., 1922. Tidal oscillations in gulfs and rectangular basins. Proc. London Math. Soc. Ser. 2 20, pp. 148-181.

Tison, L.J. and Tison G. Jr., 1969. Seiches et dénivellations causées par le vent dans les lacs, baies, estuaries. Note Technique No. 102, Organisation Météorologique Mondiale, Genève, Suisse.

Turner, J.S., 1973. Buoyancy Effects in Fluids. Cambridge University Press.

Van Dantzig, D. and Lauwerier, H.A., 1960. The North Sea problem: free oscillations of a rotating rectangular sea. Proc. K. Ned. Akad. Wet., A 63, pp. 339-354.

Witham, G.B., 1974. Linear and Non-linear Waves. Wiley, New York.

Yih, C.S., 1965. Dynamics of Nonhomogeneous Fluids. MacMillan, New York.

Hydrodynamics of Lakes: CISM Lectures
edited by K. Hutter, 1984
Springer Verlag Wien-New York

TOPOGRAPHIC WAVES IN LAKES

Lawrence A. Mysak
Departments of Mathematics and Oceanography
The University of British Columbia
Vancouver, B.C., Canada V6T IW5

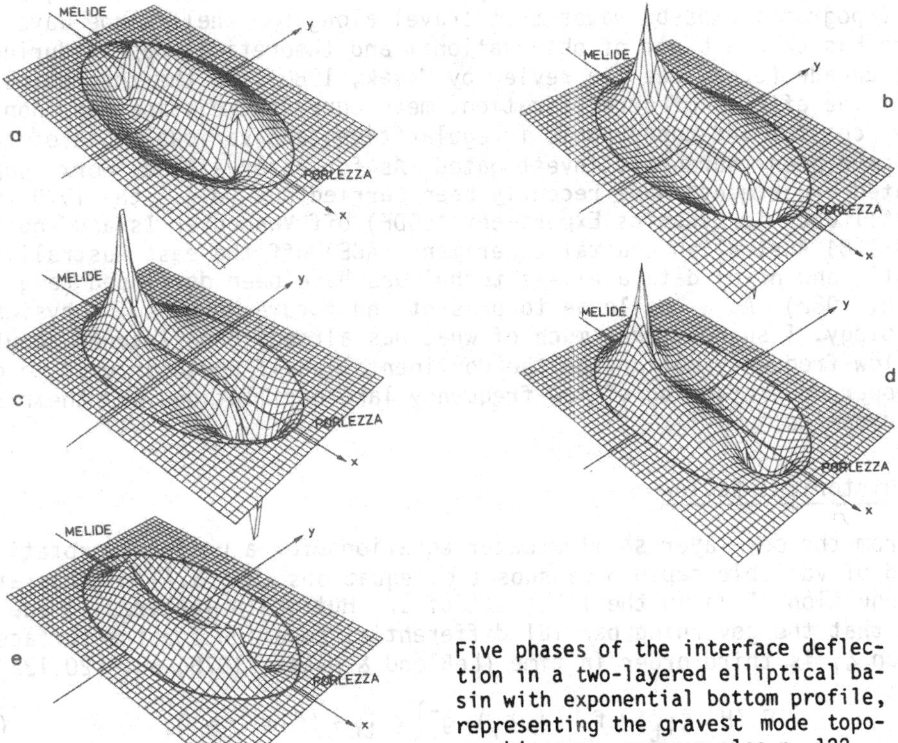

Five phases of the interface deflection in a two-layered elliptical basin with exponential bottom profile, representing the gravest mode topographic wave, compare also p. 122.

1. INTRODUCTION

1.1 Preamble

During the past decade there has been an increasing number of temperature and current observations in various lakes which show pronounced oscillations with a characteristic period of a few days. The existence of such long-period oscillations in a rotating circular basin with a parabolic depth profile was pointed out by Poincaré (1910), and the first explicit solution for these topographic wave modes (also called second class waves, vortex modes and quasi-geostrophic waves) was given by Lamb (1932). However, it is only very recently (e.g., see Saylor et al., 1980) that observations *and* an appropriate theory have been combined to provide a unified account of low-frequency motions in lakes.

On reviewing the literature on topographic waves (hereafter referred to as TW's) in lakes, it soon became apparent to me that there is a tremendous opportunity for more observational and theoretical work to be done on this topic. In the related subject of continental shelf waves, which are topographic Rossby waves that travel along the shelf/slope wave guide, there has been a bloom of observational and theoretical papers during the past decade (e.g., see the review by Mysak, 1980). On the theoretical side, the effects of stratification, mean currents, dissipation, nonlinearity, coastal and topographic irregularities, and various types of forcing have been thoroughly investigated. As for observational work, sophisticated experiments have recently been carried out (e.g., the 1979/1980 Coastal Oceanic Dynamics Experiment (CODE) off Vancouver Island and the 1983/1984 Australian Coastal Experiment (ACE) off the east Australian coast), and novel data analysis techniques have been developed (e.g., Hsieh, 1982). As a challenge to present and future workers in physical limnology, I suggest that much of what has already been learned about TW's and low-frequency motions on the continental shelf could be used to gain a deeper understanding of low-frequency lake oscillations and their related dynamics.

1.2 Historical Review

From the one-layer shallow water equations for a uniformly rotating fluid of variable depth H (a subset of equations (H1, 5.7) – this stands for equation (5.7) in the first set of Dr. Hutters's lectures -) one can show that the governing partial differential equation for the surface elevation ζ_1 is third order in time (LeBlond & Mysak, 1978, eq. (20.13a)):

$$\nabla\cdot(H\nabla\zeta_{1_t}) + fJ(H,\zeta_1) - g^{-1}(\partial_{tt} + f^2)\zeta_{1_t} = 0, \qquad (1.1)$$

where J is the Jacobian operator (see (H1, 4.20)). When H = constant, (1.1) is only second order in time and the solutions represent (first class) surface gravity waves modified by f (see Hutter 2 lectures). However, when H is variable *and* $f \neq 0$, a third time derivative arises and this allows for the possibility of (second class) topographic Rossby waves. As an illustration of these two classes we quote Lamb's classical solution for a parabolic circular basin (Lamb, 1932, Art. 212). For the depth profile

$$H = H_0(1 - r^2/a^2), \quad r \leq a \tag{1.2}$$

and circularly travelling waves of the form

$$\zeta_1 = F_n(r)\, e^{i(m\Theta - \sigma t)}, \quad m = 1,2,\ldots, \tag{1.3}$$

where m is the azimuthal wave number, (1.1) implies that the eigenfrequency equation is a cubic in σ, which can be written as

$$\frac{(\sigma^2 - f^2)a^2}{2g H_0} + \frac{fm}{\sigma} = 2(n+2)(m+n+1) - m, \tag{1.4}$$

where $n = 0,1,2,\ldots$ is the radial mode number giving the number of nodal circles of the amplitude function $F_n(r)$. F_n itself satisfies a hypergeometric equation (see Lamb (1932) for details).

The two roots of (1.4) for which $\sigma^2 > f^2$ represent surface gravity wave seiches which are modified by f. The third root of (1.4) for which $\sigma < f$ represents a TW which (in the northern hemisphere) travels counterclockwise (cyclonically) around the basin with a phase speed of $c_p = a\sigma/m$.

Under the rigid-lid approximation $f^2 a^2/g H_0 \ll 1$, or equivalently,

$$a^2 \ll r_e^2, \tag{1.5}$$

where $r_e = (g H_0)^{1/2}/f$ is the external Rossby radius of deformation, eq. (1.4), for $\sigma^2 < f^2$, simplifies to the TW frequency equation

$$\frac{f}{\sigma} = \frac{2\,(n+2)(m+n+1)}{m} - 1\,. \tag{1.6}$$

It is interesting to note that (1.6) is independent of a, suggesting a type of scale invariance for TW's in circular lakes! For a mid-latitude lake ($f = 10^{-4}\ s^{-1}$) of depth 160 m (e.g., southern Lake Michigan), r_e = 400 km. Therefore, even for the large Laurentian Great Lakes, characterized by $a \lesssim 100$ km, (1.5) is a reasonable approximation, introducing an error of only a few percent. If one invokes (1.5), the governing vorticity equation for TW's is simply (H1, 4.20), which is first order in time.

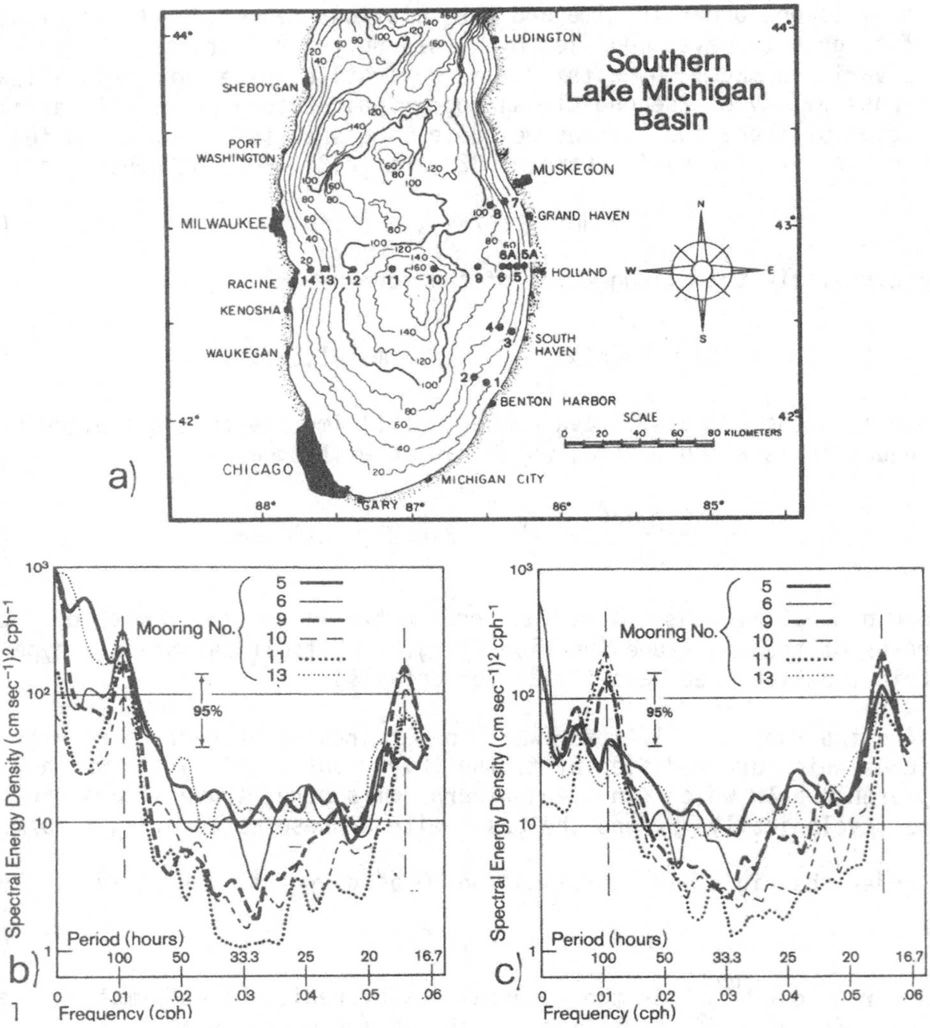

Figure 1

a) Location map showing the bathymetry of southern Lake Michigan and the positions of the 16 current meter moorings deployed during late spring, summer and fall 1976.

b) Kinetic energy spectra of the eastward velocity component (u) recorded at the 25 m level at six stations in southern Lake Michigan. Vertical lines are drawn to accentuate the two conspicuous energy peaks at near-inertial and near 4-day periods. The latter signal represents a TW.

c) As in b) except for the northward velocity component (v). From Saylor, Huang and Reid (1980).

Since lake observations of low-frequency current observations do not show "nodal circles", it is reasonable to put $n=0$ in (1.6). Also, because of the large scale of the forcing wind fields, it is generally argued, that $m=1$ applies, corresponding to only one nodal diameter in the streamline pattern. Thus for this gravest mode ($n=0, m=1$), (1.6) yields $f/\sigma=7$, corresponding to a period of 4-5 days for mid-latitude lakes. Since this is a typical time scale of many subinertial lake oscillations (see Figure 1), it is often argued that this TW mode, with its double gyre streamline pattern slowly propagating counterclockwise, is an important factor in the description of low-frequency lake motions (see Saylor et al., 1980).

Following Lamb's seminal solution for *free* oscillations in a variable depth circular basin, very little work was done on this topic for several decades. However, in 1977 Birchfield and Hickie considered the transient wind generation of the Lamb modes in an attempt to understand the role of first and second class oscillations in wind-driven lake circulation. For the first time they were able to show how the relative contributions of gravity seiches and TW's modulate the seiche response by a slow rotation in the cyclonic sense (direction of TW propagation) of the pattern of coastal jets and return flow across the lake center. Recent new *free* circular basin solutions for second class waves have been found by Wenzel (1978) for H of the form

$$H = a_0 + a_1 r + a_2 r^2, \quad \text{with} \quad a_0 + a_1 + a_2 = 0, \tag{1.7}$$

and by Saylor et al. (1980) for H of the form

$$H = H_0 [1-(r/a)^q], \quad q > 0. \tag{1.8}$$

In Saylor et al. (1980) it was assumed *a priori* that the barotropic stream function ψ (see (H1, 4.19)) has no nodal circles. As a consequence, they were able to find the simple frequency equation

$$\frac{f}{\sigma} = \frac{2q+3m}{m}, \tag{1.9}$$

which for $q=2$ (parabolic depth profile) reduces to the $n=0$ case of (1.6). For $q=1$, a conical basin, it is interesting to note that the period is substantially decreased.

In 1965 Ball prescribed an analytical method for determining the second class modes in an elliptic paraboloid, which is obviously a better model for the many elongated lakes found in nature. However, because of the extensive algebra involved, Ball was able to find only the first two azimuthal modes explicitly. Numerical solutions for a few of the higher modes were found by Bennett and Schwab (1981), but significant truncation errors made the calculation of the modes for especially elongated basins not feasible. In partial response to this difficulty, a new class of solutions for elliptic TW's has recently been found by Mysak (1983), who assumed that the lake shoreline and depth contours form a family of confocal ellipses. As a consequence, for the classical exponential depth profile used in shelf wave theory, closed-form solutions in terms of elementary functions were found for all the azimuthal and radial modes. These solutions will be presented in section 4.

It has recently been shown by Gratton (1982, 1983) that in a two-layer channel or circular basin, barotropic TW's as described above are only slightly affected by the interfacial motion (characterized by ζ_2 in H1, Fig. 3) provided the upper layer is thin ($D_1 \ll D_2$) and the topography is relatively steep. Thus in order to describe TW's and their related motion in a stratified deep lake, one first solves eq. (H1, 4.20) for ψ, the barotropic stream function. This solution is then substituted into the right hand side of the governing equation for ζ_2. One then solves this forced equation for the interfacial oscillations. Once both ζ_2 and ψ are known, the upper and lower layer velocity fields ($\underline{u}_1$, $\underline{u}_2$) can be readily computed. The details of this procedure will be described in section 3.

The first observation of TW's in lakes appears to be due to Csanady (1976) who used a form of the shelf wave dispersion relation of Mysak (1968) to characterize the westward propagation of a continuous spectrum of low-frequency signals along the north shore of Lake Ontario. A more detailed analysis of these waves was carried out by Marmorino (1979) who examined current meter records from nine stations around the whole lake. Evidence for a cyclonically travelling TW wave with a (discrete) 13-day period was found in this data. More recently, a 4-day TW was observed in current meter records from southern Lake Michigan (see Figure 1 and also Huang and Saylor, 1982), and a 3-day temperature oscillation in the Lake of Lugano, Switzerland also appears to be due to a gravest mode elliptical TW (Hutter et al., 1983; Mysak et al., 1983). In addition, it has been suggested (Mysak, 1983) that the 4-day temperature oscillation observed in the Lake of Zurich (Mortimer & Horn, 1982), is a manifestation of a fundamental mode TW. In the Bornholm Sea (in the southern Baltic), which can be modelled by a circular basin, 5-day oscillations have been inferred from satellite and hydrographic data, and these have also been attributet to the presence of TW's (Wenzel, 1978; Kielmann, 1983).

1.3 Critique of Earlier Theoretical Work

As seen in the above review nearly all the theoretical papers on TW's in lakes deal with a barotropic model. Thus the work of Gratton (1982, 1983), who investigated free TW's in a two-layer circular lake, represents a significant step forward in our understanding of second class motion that might occur during summer, when lakes are strongly stratified.

Further, a limitation of the earlier work on forced TW's is that the problems treated are all barotropic (e.g., see Birchfeld and Hickie, 1977; Huang and Saylor, 1982). The wind generation of TW's in a stratified lake has not yet been studied.

Finally, the problem of finding TW modes in elongated lakes has only briefly been tackled. Moreover, the studies involved almost exclusively deal with the barotropic equation (H1, 4.20) in an elliptical basin (e.g., Ball, 1965; Bennett and Schwab, 1981; Mysak, 1983).

Therefore, there is clearly a need for a unified TW theory which takes into account all of the above features: stratification, wind-stress forcing and elongated basin shape. The main purpose of these lectures is to develop such a theory. A secondary purpose is to apply this theory to the Lake of Lugano, Switzerland in a attempt to explain the existence of the 3-day temperature oscillation found there during summer 1979 (Hutter et al., 1983).

2. GOVERNING EQUATIONS

We consider wind-driven small-amplitude motions in a uniformly rotating, two-layer lake as illustrated in Figure 2. We assume that the upper layer is thin ($D_1 << D_2$) and has an equilibrium depth $D_1 < D_S$, the depth at the lake edge. The assumption of vertical side walls is reasonable for many intermontane lakes; it also simplifies the mathematical analysis because depth variations are confined to the lower layer. In such a model with a deep lower layer, bottom-friction effects are negligible (Huang and Saylor, 1982). The governing equations are (H1, 5.7) together with the addition of the surface wind stress, which is modelled as a body force distributed over the upper layer. In scalar form these equations can be written as

$$\left.\begin{aligned} u_{1t} - f v_1 &= -g \zeta_{1x} + \tau^x/\rho_1 D_1 \\ v_{1t} + f u_1 &= -g \zeta_{1y} + \tau^y/\rho_1 D_1 \end{aligned}\right\}, \tag{2.1}$$

$$D_1(u_{1x} + v_{1y}) = \zeta_{2t} - \underline{\zeta_{1t}}, \tag{2.2}$$

$$\left.\begin{aligned} u_{2t} - f v_2 &= -g \zeta_{1x} - \varepsilon g(\zeta_{2x} - \underline{\underline{\zeta_{1x}}}) \\ u_{2t} + f u_2 &= -g \zeta_{1y} - \varepsilon g(\zeta_{2y} - \underline{\underline{\zeta_{1y}}}) \end{aligned}\right\}, \tag{2.3}$$

$$(H_2 u_2)_x + (H_2 v_2)_y = -\zeta_{2t}, \tag{2.4}$$

where $\underline{\tau} = (\tau^x, \tau^y)$ is the surface wind stress vector and $\varepsilon = (\rho_2 - \rho_1)/\rho_2 << 1$,

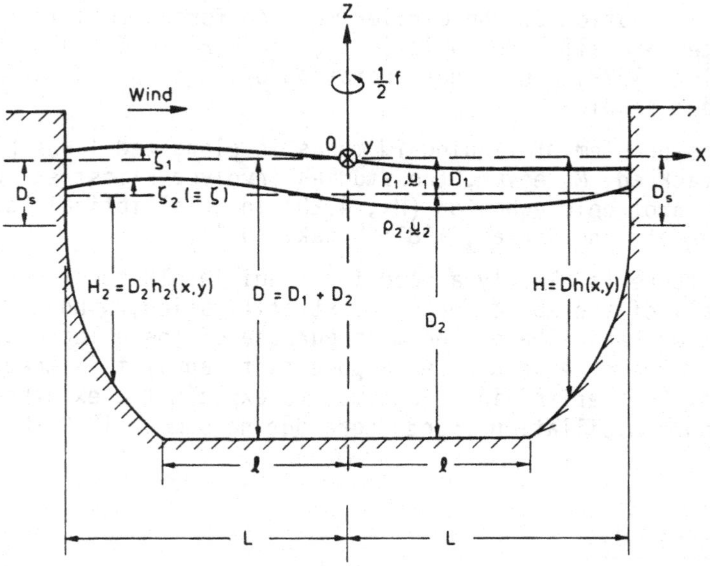

Figure 2 Sketch of side view of rotating two-layer lake model used in the present analysis. The right-handed coordinates (x, y, z) are centered on the surface of the lake. The vector $\underline{u}_j = (u_j, v_j)$ denotes the depth-averaged horizontal velocity in layer j, with x,y components u_j, v_j respectively (from Mysak et al., 1983).

To filter out all surface gravity waves and the barotropic Kelvin wave from the analysis, we invoke the rigid-lid approximation (1.5), which here corresponds to neglecting the underlined term $-\zeta_{1t}$ in (2.2). Since typically $\varepsilon = O(2 \cdot 10^{-3})$ for most lakes during summer, $\varepsilon g \zeta_{1x} \ll g \zeta_{1x}$ and $\varepsilon g \zeta_{1x} \ll g \zeta_{1y}$. Thus we shall also neglect the double underlined terms in (2.3).

Under the rigid-lid approximation, the sum of (2.2) and (2.4) yields

$$(D_1 u_1 + H_2 u_2)_x + (D_1 v_1 + H_2 v_2)_y = 0 \; ; \tag{2.5}$$

hence we can introduce a stream function $\psi = \psi(x, y, t)$ for the depth-averaged flow:

$$-\psi_y = D_1 u_1 + H_2 u_2, \quad \psi_x = D_1 v_1 + H_2 v_2. \tag{2.6}$$

ψ is called the *barotropic* or *mass transport* stream function. To study topographic waves (TW's) in a stratified fluid it is most convenient to work with a pair of coupled equations for ψ and ζ_2, the latter quantity representing the *baroclinic* part of the motion. The derivation of these equations, without $\underline{\tau}$, is given in Allen (1975) and in Wright and Mysak (1977). Here we quote only the final results, with $\underline{\tau}$ now included:

$$\nabla\cdot(H^{-1}\nabla\psi_t) + f(\nabla\Psi \times \nabla H^{-1})\cdot\hat{\underline{z}} = -g' D_1(\nabla\zeta \times \nabla H^{-1})\cdot\hat{\underline{z}} + \frac{1}{\rho_1}[\nabla\times(\underline{\tau} H^{-1}) + \frac{H}{D_1}\underline{\tau}\times\nabla H^{-1}]\cdot\hat{\underline{z}}\,, \tag{2.7}$$

$$H\nabla^2\zeta_t - \frac{H^2}{g' D_1 H_2} L\zeta_t + \frac{D_1}{H_2}\nabla\zeta_t\cdot\nabla H - \frac{f D_1}{H_2}(\nabla\zeta\times\nabla H)\cdot\hat{\underline{z}} = \frac{1}{g' H_2}[\nabla(L\psi)\times\nabla H]\cdot\hat{\underline{z}} - \frac{H}{\rho_1 g' D_1} f(\nabla\times L\underline{\tau})\cdot\hat{\underline{z}}\,, \tag{2.8}$$

where $\hat{\underline{z}}$ = unit vector in z-direction, $L = \partial_{tt} + f^2$, and $g' = \varepsilon g$, the reduced gravity. Also, for convenience we have dropped the subscript on ζ_2.

In the absence of stratification and wind forcing, the right side of (2.7) is zero and the equation reduces to (H1, 4.20), the governing vorticity equation for barotropic TW's. In a stratified fluid with topography, the first term on the right side of (2.7) produces the so-called joint baroclinic-topographic effect. That is, the term acts as a vorticity input for barotropic currents. Holland (1973) has shown that this coupling term combines additively with the windstress terms to produce an enhanced western boundary transport in a wind-driven ocean circulation model of the Gulf stream. For deep lakes, however, this coupling term is negligible, as will be seen from a scale analyis in section 3. For a constant depth fluid and no wind forcing, only the first two terms on the left side of (2.8) survive, yielding the long wave equation for internal gravity waves and Kelvin waves which propagate with speed $c_i = (g' D_1 H_2 / H)^{1/2}$. When H is variable such waves are slightly modified by the relatively small third and fourth terms on the left side of (2.8), and are driven by the barotropic component of the motion (ψ) and the wind stress curl.

Once ψ and ζ are known from the solution of (2.7) and (2.8) together with appropriate boundary and initial conditions, the velocity fields can be found from the relations

$$L\underline{u}_1 = \frac{1}{H}[\hat{\underline{z}}\times\nabla(L\psi) + H_2 g'(\nabla\zeta_t - f\hat{\underline{z}}\times\nabla\zeta) + \frac{H_2}{\rho_1 D_1}(\underline{\tau}_t - f\hat{\underline{z}}\times\underline{\tau})]\,, \tag{2.9}$$

$$L\underline{u}_2 = \frac{1}{H}[\hat{\underline{z}}\times\nabla(L\psi) - D_1 g'(\nabla\zeta_t - f\hat{\underline{z}}\times\nabla\zeta) - \frac{1}{\rho_1}(\underline{\tau}_t - f\hat{\underline{z}}\times\underline{\tau})]\,. \tag{2.10}$$

The first and second terms in (2.9) and (2.10) represent respectively the barotropic and baroclinic parts of the current; the third term is a directly forced wind-driven component.

As a final simplification in this section we invoke the low frequency approximation

$$L = \partial_{tt} + f^2 \simeq f^2 , \qquad (2.11)$$

which filters out the internal gravity waves (e.g., the internal seiches) and the internal Kelvin wave from the analysis. For oscillations of periods 3 days and longer, $\partial_t \sim \omega \leq 2.42 \cdot 10^{-5}\,s^{-1}$ and therefore $\omega^2/f^2 \leq 0.05$ for $f = 1.05 \cdot 10^{-4}\,s^{-1}$ (e.g., the Lake of Lugano). Therefore, the approximation (2.11) introduces an error of only a few percent. For large lakes (e.g., the Laurentian Great Lakes) the omission of the internal Kelvin wave mode may not be permissible (Simons, 1980). However, for smaller lakes we argue that its wavelength is too long for the wave to wrap itself around the inside of the basin. For the Lakes of Lugano and Zurich, a typical internal wave phase speed is c_i = 45 cm s^{-1} (Mortimer and Horn, 1982; Hutter et al., 1983). For a period of 3 days, the corresponding wavelength of a Kelvin wave is 117 km, which is roughly three times the circumference of these Swiss lakes.

3. SCALE ANALYSIS OF THE GOVERNING EQUATIONS

The first scale analysis of the coupled equations (2.7) and (2.8) with $\underline{\tau} = \underline{0}$ was done by Allen (1975) in the context of free coastal trapped waves which propagate along an exponential shelf/slope bottom topography. In his nondimensional version of these equations, there appeared the dimensionless parameter $\lambda = r_i/L_T$, where r_i is the internal Rossby radius of deformation (i.e., $r_i = c_i/f$) and L_T is a characteristic horizontal scale of the bottom topography. Off the coast of Oregon and in other eastern boundary current regimes during summer r_i = O(15 km) and L_T = O(100 km), so that $\lambda \ll 1$. Thus Allen was able to study the strength of the coupling between ψ (shelf waves) and ζ (internal Kelvin waves) through a perturbation expansion in λ. For many lakes, and especially those in Switzerland, r_i = O(4 km) in summer and L_T = O(5 km), so that λ = O(1). Hence Allen's perturbation analysis cannot be used to solve (2.7) and (2.8) for low-frequency lake motions.

Another type of low-frequency wave which can arise out of the solution of (2.7) and (2.8) with $\underline{\tau} = \underline{0}$ is the bottom-trapped topographic Rossby wave studied by Rhines (1970) (see also LeBlond and Mysak, 1978, § 20). In this case the scaling of these equations yields the nondimensional parameter $\delta_s = |\nabla H|\, L_T/D$, where D is the total depth and L_T is the same as defined above. For typical continental slope regions, $|\nabla H|$ = O(10^{-2}), L_T = O(50 km) and D = O(4 km). Thus δ_s = O(10^{-1}) and a perturbation expansion in δ_S can be used to solve the governing system of equations. For the Swiss lakes, we typically have $|\nabla H| = 3\cdot 10^{-2}$, $L_T = 5$ km and $D = 200$ m, yielding $\delta_S = 0.75 = O(1)$. Thus this approach also does not apply.

A suitable scale analysis of (2.7) and (2.8) which can be applied to many lake situations has, however, been recently developed by Gratton (1982, 1983). But before describing this work, it is helpful to first estimate the relative importance of the different wind stress terms since these were not included in Gratton's analysis.

3.1 Wind Stress Forcing Mechanisms

The wind stress terms in (2.7) can be written as (upon neglecting the common factor ρ_1)

$$\begin{aligned} &[\nabla \times (\underline{\tau} H^{-1}) + \frac{H}{D_1} \underline{\tau} \times \nabla H^{-1}]\cdot \hat{\underline{z}} \\ &= [\underbrace{H^{-1} \nabla \times \underline{\tau}}_{(1)} + \underbrace{(\nabla H^{-1}) \times \underline{\tau}}_{(2)} + \underbrace{\frac{H}{D_1} \underline{\tau} \times \nabla H^{-1}}_{(3)}]\cdot \hat{\underline{z}}. \end{aligned} \tag{3.1}$$

The first term on the left represents the driving term that arises in a strictly barotropic lake model with topography (Simons, 1980, p. 99), whereas the second term on the left appears because of the stratification. For lakes, the wind stress curl term (1) on the right side of (3.1) is generally small compared to the topographic driving terms (2) and (3). This is because L_W, the horizontal length scale for $\underline{\tau}$, is generally much larger than the topographic length scale L_T (Simons, 1980, p. 73). Thus the ratio of terms (1)/(2) = $O(L_T/L_W) \ll 1$. In practice, one therefore usually takes $\underline{\tau} = \underline{\tau}(t)$ in lake wave models which include wind forcing (e.g., Birchfield and Hickie, 1977; Huang and Saylor, 1982). For a spatially constant $\underline{\tau}$, it also follows that the wind stress curl term in (2.8) drops out.

Finally, comparing the two topographic driving terms on the right side of (3.1), we find (2)/(3) = $O(D_1/D)$. Thus for a deep lake with a thin upper layer (as in Figure 2), we can neglect term (2) as well, leaving term (3) as the sole wind forcing term in our system. For convenience, we now rewrite here the simpler forms of (2.7) and (2.8) which take into account these approximations and the low-frequency approximation (2.11):

$$\begin{aligned} &\nabla\cdot(H^{-1}\nabla\psi) + f(\nabla\psi \times \nabla H^{-1})\cdot\hat{z} \\ &= -g' D_1(\nabla\zeta \times \nabla H^{-1})\cdot\hat{z} + \frac{H}{\rho_1 D_1}(\underline{\tau} \times \nabla H^{-1})\cdot\hat{z}, \end{aligned} \tag{3.2}$$

$$\frac{1}{H}\nabla^2\zeta_t - \frac{f^2}{g' D_1 H_2}\zeta_t - \frac{D_1}{H_2}\nabla\zeta_t\cdot\nabla H^{-1} + \frac{f D_1}{H_2}(\nabla\zeta \times \nabla H^{-1})\cdot\hat{z}$$
$$= -\frac{f^2}{g' H_2}(\nabla\psi \times \nabla H^{-1})\cdot\hat{z}\,. \tag{3.3}$$

Invoking (2.11) in (2.9) and (2.10) we also find

$$\underline{u}_1 = \frac{1}{H}\left[\hat{z}\times\nabla\psi - \frac{H_2 g'}{f^2}(\nabla\zeta_t - f\,\hat{z}\times\nabla\zeta) + \frac{H_2}{\rho_1 D_1 f^2}(\underline{\tau}_t - f\,\hat{z}\times\underline{\tau})\right], \tag{3.4a}$$

$$\underline{u}_2 = \frac{1}{H}\left[\hat{z}\times\nabla\psi - \frac{D_1 g'}{f^2}(\nabla\zeta_t - f\,\hat{z}\times\nabla\zeta) - \frac{1}{\rho_1 f^2}(\underline{\tau}_t - f\,\hat{z}\times\underline{\tau})\right]. \tag{3.4b}$$

3.2 Gratton's Scaling

Because of the variable depth H, the coupled equations (3.2) and (3.3) have variable coefficients and are not readily solved by analytical methods. However, it was shown by Gratton (1982, 1983) that for a deep fluid with a thin upper layer (i.e., $D_1 \ll D_2$), the first coupling term on the right side of (3.2) is of $O(D_1/D_2)$ compared to the ψ terms on the left side. Therefore to a first approximation, this term can be dropped and the equations become uncoupled. Then one can solve (3.2) for ψ, and substitute this solution into the right side of (3.3). The latter equation can then be integrated to find ζ. Finally, $\underline{u}_1$ and $\underline{u}_2$ can be found from (3.4a) and (3.4b).

We now introduce the following set of nondimensional variables, denoted with primes:

$$\psi = \psi_0\,\psi', \qquad \zeta = \zeta_0\,\zeta', \qquad \underline{\tau} = \tau_0\,\underline{\tau}',$$
$$(x,y) = (L\,x', L\,y'), \qquad t = f^{-1}\,t', \tag{3.5}$$

where τ_0 is a characteristic wind stress scale (determined from observations), L is the lake half-length (see Fig. 2), and ψ_0 and ζ_0 are amplitude scales to be determined later. Substituting (3.5) into (3.2) and (3.3) and using the relations (see Fig. 2)

$$H = D\,h, \qquad H_2 = D_2\,h_2, \tag{3.6}$$

where h, h_2, $|\nabla_{\underline{x}'} h|$ $|\nabla_{\underline{x}'} h_2|$ are all of order unity, we obtain the following scaled equations:

$$\nabla\cdot(h^{-1}\nabla\psi_t)+(\nabla\psi\times\nabla h^{-1})\cdot\hat{z}$$
$$= -C_1(\nabla\zeta\times\nabla h^{-1})\cdot\hat{z}+\left(\frac{L\,D\,\tau_0}{f\,\rho_1\,D_1\,\psi_0}\right)(h\,\underline{\tau}\times\nabla h^{-1})\cdot\hat{z}\,, \tag{3.7}$$

$$\frac{1}{h}\left(\nabla^2\zeta_t-\frac{L^2 h}{r_i^2 h_2}\zeta_t\right)-\frac{D_1}{D_2 h_2}\nabla\zeta_t\cdot\nabla h^{-1}+\frac{D_1}{D_2 h_2}(\nabla\zeta\times\nabla h^{-1})\cdot\hat{z}$$
$$= -C_2\,h_2^{-1}(\nabla\psi\times\nabla h^{-1})\cdot\hat{z}\,, \tag{3.8}$$

where the *coupling coefficients* C_i are given by

$$C_1=\frac{g'\,D_1\,\zeta_0}{f\,\psi_0},\qquad C_2=\frac{f\,\psi_0}{g'\,D_2\,\zeta_0}, \tag{3.9}$$

and $r_i=(g'\,D_1\,D_2/D\,f^2)^{1/2}$ is the internal Rossby radius. Note that in (3.7) and (3.8) we have dropped the primes on the scaled (nondimensional) variables.

Let us now suppose that (3.7) and (3.8) are strongly coupled, i.e., that C_1 and C_2 are both $O(1)$. Then (3.9) implies that

$$\zeta_0 = O\left(\frac{f}{g'\,D_1}\,\psi_0\right) \tag{3.10}$$

and

$$\zeta_0 = O\left(\frac{f}{g'\,D_2}\,\psi_0\right). \tag{3.11}$$

We observe that independent of how ψ_0 is chosen, (3.10) and (3.11) are consistent *only if* $D_1/D_2=O(1)$. Since we are concerned with the case $D_1 \ll D_2$, it follows that C_1 and C_2 cannot both be of order unity, i.e., that (3.7) and (3.8) are only weakly coupled. Suppose we assume that (3.10) applies and thus choose

$$\zeta_0/\psi_0 = f/g'\,D_1 \tag{3.12}$$

as the scaling for the ratio ζ_0/ψ_0. Then $C_1=1$ and $C_2=D_1/D_2\ll 1$. Therefore to $O(D_1/D_2)$, eq. (3.8) reduces to, upon also using $h/h_2=1+O(D_1/D_2)$,

$$\left(\nabla^2-\frac{L^2}{r_i^2}\right)\zeta_t = 0, \tag{3.13}$$

which is not a wave equation. Since observations clearly show wave-like fluctuations for ζ (e.g., see Fig. 3 below), (3.12) is obviously not the correct scaling choice. Moreover, (3.12) leads to an unrealistically large value for the ζ_0 scale (for Lake of Lugano, $\zeta_0 \sim 50$ m, which is several times the upper layer depth - about 10 m!).

Hence we are compelled to choose the scaling (3.11) (Gratton's choice, which was based on data from the Strait of Georgia, B.C.). Putting

$$\zeta_0/\psi_0 = f/g'\, D_2 \,, \tag{3.14}$$

we find $C_2 = 1$ and $C_1 = D_1/D_2 << 1$. We choose the ψ_0 scale by setting the coefficient of the wind stress term in (3.7) equal to unity; this gives

$$\psi_0 = L\, D\, \tau_0 / f\, \rho_1\, D_1 . \tag{3.15}$$

Substituting (3.15) into (3.14) gives an expression for ζ_0 in terms of the wind stress scale τ_0:

$$\zeta_0 = L\, D\, \tau_0 / \rho_1\, g'\, D_1\, D_2 . \tag{3.16}$$

Also, substituting $\psi_0 = L\,D\,U$ into (3.15), where U is a velocity scale for the barotropic currents (see (2.6)), we find

$$U = \tau_0 / \rho_1\, f\, D_1 . \tag{3.17}$$

Using (3.14) and (3.15) in (3.7) and (3.8), we obtain, correct to $O(D_1/D_2)$,

$$\nabla\cdot(h^{-1}\,\nabla\psi_t) + (\nabla\psi \times \nabla h^{-1})\cdot\hat{z} = (h\,\underline{\tau} \times \nabla h^{-1})\cdot\hat{z} \,, \tag{3.18}$$

$$(\nabla^2 - S^{-1})\,\zeta_t = -(\nabla\psi \times \nabla h^{-1})\cdot\hat{z} \,, \tag{3.19}$$

as the appropriate nondimensional equation for ψ and ζ. In arriving at (3.18) and (3.19), we have also used the relation $h/h_2 = 1 + O(D_1/D_2)$ and have introduced the stratification parameter S, defined as

$$S = r_i^{\,2}/L^2 . \tag{3.20}$$

Also, for the derivation of (3.19) it is important that $h_2 \neq 0$ (as illustrated in Fig. 2). If $h_2 = 0$ the third and fourth terms on the left side of (3.8) are not uniformly $O(D_1/D_2)$ and hence could not be neglected.

Substituting (3.5) into (3.4a,b) and using (3.14), (3.15), $\psi_0 = L\,D\,U$ and U for the velocity scale, we obtain the following scaled velocity formulae:

$$\underline{u}_1 = \frac{1}{h}\left(\hat{z} \times \nabla\psi + h_2\left[(\nabla\zeta_t - \hat{z} \times \nabla\zeta) + \frac{D_2}{D}(\underline{\tau}_t - \hat{z} \times \underline{\tau})\right]\right), \tag{3.21}$$

$$\underline{u}_2 = \frac{1}{h}\left(\hat{z}\times\nabla\psi - \frac{D_1}{D_2}\left[(\nabla\zeta_t - \hat{z}\times\nabla\zeta) + \frac{D_2}{D}(\underline{\tau}_t - \hat{z}\times\underline{\tau})\right]\right) . \tag{3.22}$$

To $O(D_1/D_2)$ these can be approximated by

$$\underline{u}_1 = \frac{1}{h}\left[\hat{z}\times\nabla\psi + h(\nabla\zeta_t - \hat{z}\times\nabla\zeta + \underline{\tau}_t - \hat{z}\times\underline{\tau})\right] , \tag{3.23a}$$

$$\underline{u}_2 = \frac{1}{h}\hat{z}\times\nabla\psi . \tag{3.23b}$$

Thus for deep lakes, the lower layer current associated with TW's is essentially barotropic, whereas the upper layer current consists of a barotropic part, a baroclinic part and a contribution directly forced by the wind. Thus we conclude that the current motions are surface intensified.

3.3 Amplitude Scales for Lake of Lugano

To help convince the more skeptical reader of the usefulness of the above scale analysis, we shall now introduce specific values of the parameters which describe the geometry and physics of the Lake of Lugano during summer 1979 and then estimate the amplitude scales τ_0, U and ζ_0 introduced in Section 3.2. Finally these theoretical scales will be compared with the observed amplitudes.

During summer 1979, the stratification of the Lake of Lugano is well described by a two-layer fluid with

$$\begin{aligned} \varepsilon &= (\rho_2 - \rho_1)/\rho_2 = 1.94\cdot 10^{-3} , \\ D_1 &= 10\,\mathrm{m},\ \ D_2 = 270\,\mathrm{m}\ \ (D = 280\,\mathrm{m}) , \end{aligned} \tag{3.24}$$

(Hutter et al., 1983). Since the lake is located at 46°N, $f = 1.05\cdot 10^{-4}\,\mathrm{s}^{-1}$ and therefore

$$\begin{aligned} c_i &= (\varepsilon g D_1 D_2/D)^{1/2} = 0.428\,\mathrm{m\ s}^{-1} , \\ r_i &= c_i/f = 4.08\ \mathrm{km} . \end{aligned} \tag{3.25}$$

The half length L (see Fig. 2) is 8.7 km, and therefore

$$S^{-1} = L^2/r_i^2 = 4.55 . \tag{3.26}$$

Since $D_1/D_2 = 1/27$, neglecting terms of $O(D_1/D_2)$ is indeed a very good approximation. Also, the small size of r_i relative to the lake scale L shows that rotational effects must be important.

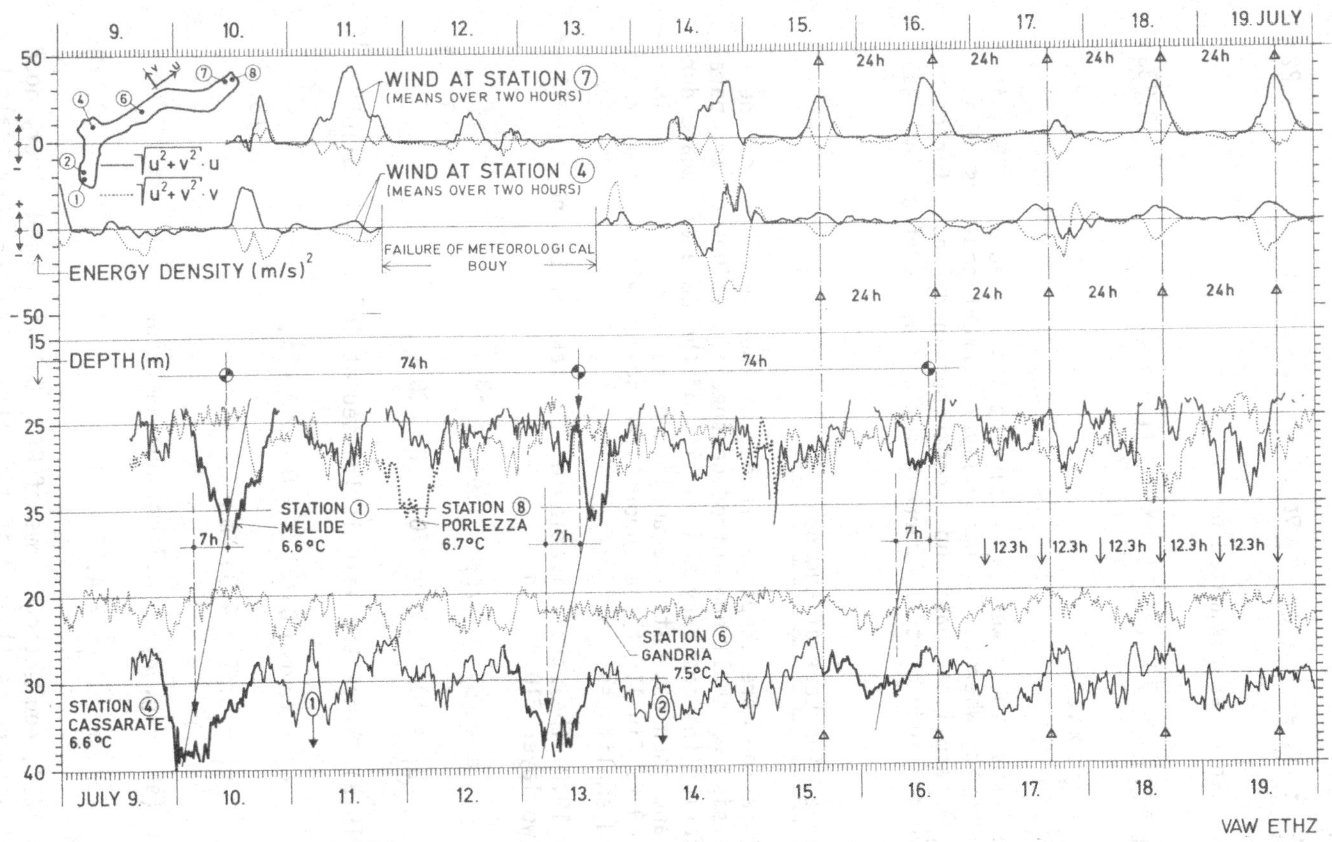

Figure 3 Time series of 2 h means of the longitudinal and transverse components of the wind energy ($\sqrt{u^2+v^2}\,u$, $\sqrt{u^2+v^2}\,v$, direction of u and v components as indicated in the insert in the upper left corner) at the stations ⑦ (Porlezza and ④ (Cassarate) and non-smoothed, unfiltered time series of selected isotherm depths at the station ① (Melide, top solid curve), ⑧ (Porlezza, superposed on the Melide curve, dotted), ⑥ (Gandria, dotted), and ④ (Cassarate, solid). Components of the motion with conspicuous periods are marked by special symbols: Troughs in heavy solid and dashed lines and marked by circles = = 74h; triangles = 24h; arrows = 12h or 8h. Episode is from 9 to 19 July 1979 (from Hutter et al., 1983).

Figure 3 shows selected time series of the wind and isotherm depths collected by Hutter et al. (1983). In addition to the regular 24 h internal seiche oscillation (15 - 19 July), we observe a distinct 74 h (3-day) signal in the first part of the record (9 - 16 July). The signal is first evident at Cassarate ④, and is then seen to propagate to the south of the lake at Melide ①, reflect back northward to Cassarate ④ (amplitude smaller), and finally travel all the way to Porlezza ⑧ in about one day. The signal reflects back southward from Porlezza ⑧, reaching Cassarate ④ again about a day later. Thus the total "circuit" time is about 74 km, and this cycle is repeated again, though with a much smaller amplitude. During the first cycle, the observed isotherm deflection has an unusually large amplitude of 5-6 m; during the second cycle, the amplitude is about 2-3 m. The second estimate is probably a more typical value of the ζ_0 scale introduced in Section 3.2. During this part of July, the low-pass filtered currents from 39 m at Gandria ⑥ also show a 3-day oscillation with an amplitude of 3 cm s^{-1} (see Fig. 15b in Section 4.3). Moreover, the currents are very nearly rectilinear in the longshore (u) direction (see Figures 15b and 16b in Section 4.3). Since the currents in the lower layer are mostly barotropic (see (3.23)), we conclude that the observed amplitude for U is 3 cm s^{-1}.

To estimate ζ_0 and U from the theoretical expressions (3.16) and (3.17) respectively, we also need a value for τ_0. From the wind records at Cassarate ④ (Fig. 3) and Campione ② (not shown), we find that a characteristic value of U_W *at* a 3-day time scale is 4 m s^{-1}. The wind spectrum from Porlezza ⑧ shows no distinct 3-day peak, only the diurnal 24 h peak, which is consistent with Figure 3. However, spectra of the winds from Cassarate ④ and Campione ②, do show a broad 2-3 day peak, with the amplitude being larger at Campione (see Mysak et al., 1983, and Fig. 18). Thus using the formula

$$\tau_0 = \rho_{air}\, c_d\, U_W^2 \qquad (3.27)$$

with $\rho_{air} = 1.29$ kg m^{-3}, $c_d = 1.85 \cdot 10^{-3}$ (an average value for lakes during summer (see Simons, 1980, p. 92)), and $U_W = 4$ m s^{-1}, we find $\tau_0 = 0.038$ N m^{-2} and hence

$$\zeta_0 = 1.79 \text{ m}, \quad U = 3.6 \text{ cm s}^{-1}, \qquad (3.28)$$

which agree very favourably with the typical observed amplitudes ($\zeta_0 = 2$-3 m and $U = 3$ cm s^{-1}). Thus we argue that the low-frequency dynamics of the Lake of Lugano are probably well described by the simple system (3.18), (3.19), (3.23a) and (3.23b).

3.4 Boundary Conditions

In order to solve (3.18) and (3.19) in some domain $\mathcal{D}$ for a given $\underline{\tau}$, we have to prescribe initial values for ψ and ζ and the boundary conditions on $\partial\mathcal{D}$, the boundary of $\mathcal{D}$. The first boundary condition we impose is that the total mass flux normal to $\partial\mathcal{D}$ must vanish: $\hat{n}\cdot(D_1 \underline{u}_1 + D_2 h_2 \underline{u}_2) D^{-1} = 0$ on $\partial\mathcal{D}$, where $\hat{n}$ is a unit vector perpendicular to $\partial\mathcal{D}$. On substituting for $\underline{u}_1$ and $\underline{u}_2$ from (3.21) and (3.22), this reduces to

$$\hat{n}\cdot(\hat{z} \times \nabla\psi) = 0, \quad \text{on} \quad \partial\mathcal{D}. \tag{3.29}$$

Since $\hat{n}\cdot(\hat{z}\times\nabla\psi) = (\hat{n}\times\hat{z})\cdot\nabla\psi = \hat{s}\cdot\nabla\psi$, where $\hat{s}$ is a unit vector tangential to $\partial\mathcal{D}$ (3.29) implies $\partial\psi/\partial s = 0$ on $\partial\mathcal{D}$, or ψ = constant on $\partial\mathcal{D}$. Thus without loss of generality, we take

$$\psi = 0, \quad \text{on} \quad \partial\mathcal{D}. \tag{3.30}$$

Next we require $\hat{n}\cdot\underline{u}_i = 0$ on $\partial\mathcal{D}$ for each layer i. Upon again using (3.21) and (3.22), together with (3.29), we find

$$\frac{\partial}{\partial n}\frac{\partial\zeta}{\partial t} - \frac{\partial\zeta}{\partial s} = -\frac{D_2}{D}(\hat{n}\cdot\underline{\tau}_t - \hat{s}\cdot\underline{\tau}) = -\hat{n}\cdot\underline{\tau}_t + \hat{s}\cdot\underline{\tau} \quad \text{on} \quad \partial\mathcal{D} \tag{3.31}$$

to $O(D_1/D_2)$.

4. BAROTROPIC SOLUTION FOR ELLIPTICAL TOPOGRAPHIC (ET) WAVES

4.1 Analytical Solution

A thorough analysis of the forced equations (3.18) and (3.19) for an elongated lake is beyond the scope of these lectures. As a first step in this direction, we shall present here the recently discovered (Mysak, 1983) free TW modes for the elliptical-shaped lake illustrated in Figure 4. We assume that the shoreline and depth contours of the lake form a *family of confocal ellipses* with foci at $x = \pm a$. In terms of the elliptic cylinder coordinates (ξ,η) defined by (see also Fig. 5)

$$\begin{aligned} x &= a \cosh\xi \cos\eta, \quad 0 < \xi < \infty, \\ y &= a \sinh\xi \sin\eta, \quad 0 < \eta < 2\pi, \end{aligned} \tag{4.1}$$

the shoreline is denoted by the ellipse $\xi = \xi_s$ and the lake bottom along main axis of lake corresponds to the degenerate ellipse $\xi = 0$ $(y=0, |x| \le a)$. Elimination of η from (4.1) yields the family of ellipses

$$x^2/a^2 \cosh^2\xi + y^2/a^2 \sinh^2\xi = 1, \tag{4.2}$$

which are illustrated in Figure 5. The orthogonal curves are the family of hyperbolas

$$x^2/a^2 \cos^2\eta - y^2/a^2 \sin^2\eta = 1, \tag{4.3}$$

which are also illustrated in Figure 5. Since the shoreline $\xi = \xi_s$ has the

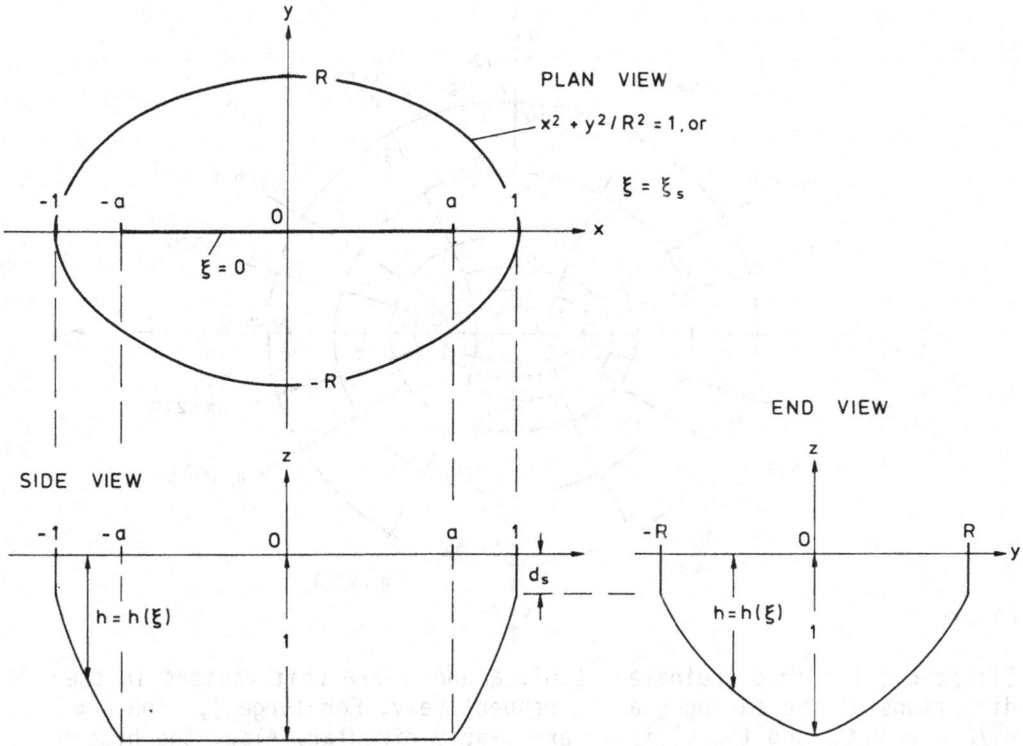

Figure 4 Elliptical lake model in terms of nondimensional coordinates (x,y,z). With reference to Figure 2, the horizontal coordinates have been scaled by L and the vertical coordinate by D. Therefore, $d_s = D_s/D$ and $a = \ell/L$. The loci ξ = constant represent a family of confocal ellipses, with foci at $x = \pm a$. (From Mysak, 1983).

equation $x^2 + y^2/R^2 = 1$ (see Fig. 4), it follows from (4.2) that

$$1 = a \cosh \xi_s, \quad R = a \sinh \xi_s \tag{4.4}$$

and hence that

$$R = \tanh \xi_s \,. \tag{4.5}$$

The relationship between ξ_s and R(= W/L, where W and L are the dimensional semi-minor and semi-major lake axes respectively) is given in Table 1 for various W, L pairs.

In terms of the (ξ,η) coordinates, (3.18) with $\underline{\tau} = \underline{0}$ reduces to

$$\frac{\partial}{\partial \xi}\left(\frac{1}{h}\,\frac{\partial^2 \psi}{\partial \xi\,\partial t}\right) + \frac{1}{h}\,\frac{\partial^3 \psi}{\partial^2 \eta\,\partial t} - \frac{d}{d\xi}\left(\frac{1}{h}\right)\frac{\partial \psi}{\partial \eta} = 0\,, \tag{4.6}$$

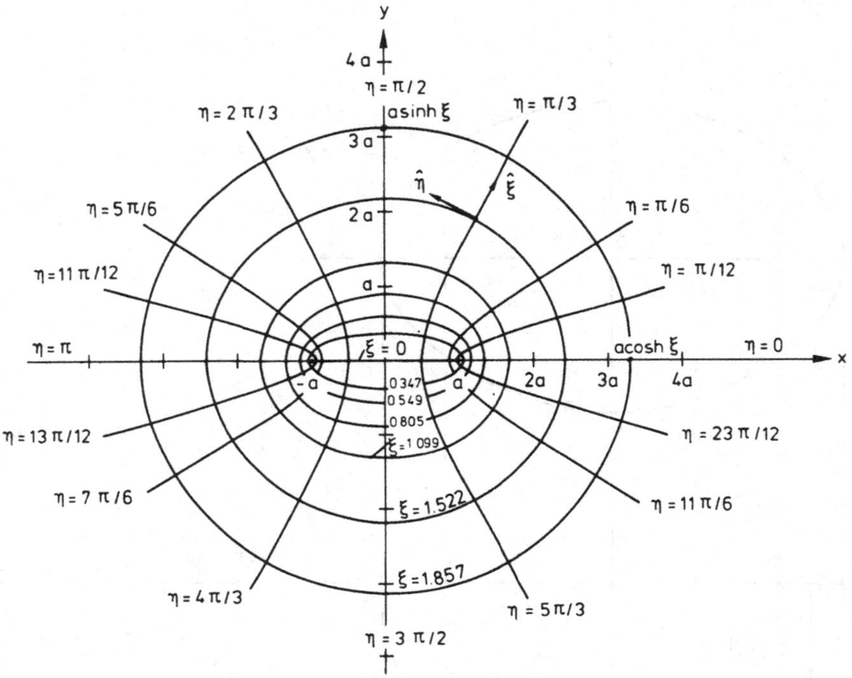

Figure 5

Elliptic cylinder coordinates (ξ,η). $\hat{\xi}$ and $\hat{\eta}$ are unit vectors in the directions of increasing ξ and η respectively. For large ξ, $\sinh\xi \simeq e^{\xi}/2 \simeq \cosh\xi$, and the ellipses are nearly circular. Also, the hyperbolas tend to straight lines. Thus we refer to ξ as a radial coordinate, and to η as an angular coordinate.

W:L	R (W/L)	ξ_s ($\tanh^{-1} R$)	a ($1/\cosh \xi_s$)
1: ∞	0	0	1
1:10	0.100	0.100	0.995
1:5	0.200	0.203	0.980
1:3	0.333	0.347	0.943
1:2	0.500	0.549	0.866
1:1.5	0.667	0.805	0.745
1:1.25	0.800	1.099	0.600
1:1.10	0.909	1.522	0.417
1:1	1	∞	0

Table 1

Numerical relations between W, L, R, ξ_s and a.

where $h = h(\xi)$ only because of our assumption about the distribution of the depth contours. Remarkably, (4.6) is independent of the scale factors h_ξ, h_η for elliptic coordinates, which are given by

$$h_\xi = h_\eta = a\,J(\xi,\eta) \equiv a\,(\sinh^2\xi + \sin^2\eta)^{1/2}. \tag{4.7}$$

Therefore, we can readily find separable solutions of (4.6) of the form

$$\psi = \psi_{mn}(\xi,\eta,t) = F_n(\xi)\,e^{i(m\eta - \sigma t)}, \tag{4.8}$$

where $m = 1,2,\ldots$ is the azimuthal mode number and $n = 1,2,\ldots$ is the radial mode number. Equation (4.8) represents an elliptically travelling wave whose phase moves along the ellipse $\xi = \xi_s$ with speed

$$c_p = h_\eta(\xi_0,\eta)\,\frac{d\eta}{dt} = a(\sinh^2\xi_0 + \sin^2\eta)^{1/2}\,\frac{\sigma}{m}. \tag{4.9}$$

Note that c_p varies with η, the angular coordinate! For $\xi_0 >> 1$ (a "circular" ellipse) $\sinh^2\xi_0 >> \sin^2\eta$, and $c_p \simeq$ constant $= a\sinh\xi_0\,\sigma/m$ for all η, as in the case of a circularly travelling wave (see discussion following (1.4)).

Substitution of (4.8) into (4.6) yields

$$(h^{-1}F')' + (m\sigma^{-1}(h^{-1})' - m^2h^{-1})\,F(\xi) = 0, \quad 0 < \xi < \xi_s, \tag{4.10}$$

where $F \equiv F_n$ for convenience. At the outer boundary $\xi = \xi_s$, $\psi = 0$ (see (3.30)); hence (see (4.8))

$$F(\xi_s) = 0\,. \tag{4.11}$$

At $\xi = 0$, we also take $\psi = 0$, i.e.,

$$F(0) = 0\,. \tag{4.12}$$

Equation (4.12) implies that the cross-isobath barotropic velocity component (u_ξ) is zero along the lake center. This condition approximates the low-frequency current observations from both the Lake of Zurich (W. Horn, personal communication, 1983) and the Lake of Lugano (see Fig. 15b below), which show that the particle motions are nearly rectilinear along the lake center.

For a given depth profile $h(\xi)$ and azimuthal mode number m, (4.10) - (4.12) represents an eigenvalue problem with σ^{-1} as eigenvalue. The solution of this problem can then be used in (4.8) to find the corresponding streamline patterns. Also, from ψ the *barotropic* velocity components along $\hat{\xi}$ and $\hat{\eta}$ (see Fig. 5) can be determined using the relations

$$u_\xi = -\frac{1}{h\,a\,J}\,\frac{\partial\psi}{\partial\eta}, \qquad u_\eta = \frac{1}{h\,a\,J}\,\frac{\partial\psi}{\partial\xi}\,. \tag{4.13}$$

We refer to u_ξ and u_η as the cross-isobath and alongshore components respectively.

From (4.10)-(4.12) we readily obtain the Rayleigh quotient

$$\frac{m}{\sigma} = \frac{\int_0^{\xi_s} h^{-1}(F'^2 + m^2 F^2)\, d\xi}{\int_0^{\xi_s} (h^{-1})' F^2 \, d\xi}, \tag{4.14}$$

which shows that $m/\sigma > 0$ provided $(h^{-1})' > 0$ (which is true for most lakes). Thus (4.9) implies that $c_p > 0$ and hence the waves travel anticlockwise around the lake, with the shoreline (shallow water) on the right. In the southern hemisphere the time scale $f^{-1} < 0$ and the waves travel clockwise around the lake.

For the exponential depth profile

$$h = e^{-b\xi}, \quad b > 0, \tag{4.15}$$

which is familiar to a generation of shelf wave proponents (e.g., see Mysak, 1980), the solution of (4.10)-(4.12) can be written down in terms of elementary functions (Mysak, 1983):

$$F_n = e^{-b\xi/2} \sin(n\pi\xi/\xi_s), \quad n = 1,2,\dots \tag{4.16}$$

The corresponding eigenfunctions are simply given by

$$\frac{1}{\sigma} \equiv \frac{1}{\sigma_{mn}} = \frac{1}{mb}\left[m^2 + \frac{b^2}{4} + \left(\frac{n\pi}{\xi_s}\right)^2\right]. \tag{4.17}$$

Note that $\sigma/m > 0$, in accordance with (4.14). Using (4.16) in (4.8), the real form of ψ is found to be

$$\psi_{mn} = e^{-b\xi/2} \sin(n\pi\xi/\xi_s)\cos(m\eta - \sigma_{mn} t). \tag{4.18}$$

Substituting (4.18) into (4.13) we obtain the corresponding velocity field:

$$u_\xi = (m/aJ)\, e^{b\xi/2} \sin(n\pi\xi/\xi_s)\sin(m\eta - \sigma_{mn} t), \tag{4.19}$$

$$u_\eta = (1/aJ)\, e^{b\xi/2}\Big[-(b/2)\sin(n\pi\xi/\xi_s) + (n\pi\xi/\xi_s)\cos(n\pi\xi/\xi_s)\Big]\cos(m\eta - \sigma_{mn} t), \tag{4.20}$$

which show that u_ξ and u_η are respectively in quadrature and in phase with ψ. Contours of ψ_{mn} and the corresponding velocity profiles are illustrated in Section 4.2.

Although the exponential depth profile (4.15) is not too realistic for some lakes (it has the wrong curvature), the analytical simplicity of the solution *for all the modes of an elliptical lake* enables us to proceed with ease to an analysis of the motions in a stratified lake. A better depth profile for a number of lakes is given by $h = 2 - e^{\beta\xi}$ $(\beta > 0)$, but the corresponding eigenfunctions involve hypergeometric functions (Mysak, 1983). Thus the utility of the solution is limited, especially for an analysis of stratified lakes.

4.2 Numerical Results

As an illustration of the analytical solution for the exponential depth profile (4.15), we compute, for the first few modes, the periods, streamline patterns and associated velocity fields for a model elliptic lake with shoreline $\xi_s = 0.805$, corresponding to $W:L = 1:1.5$ and $a = 0.745$ (see Table 1).

An important free parameter in our model is d_s, the nondimensional depth at the edge of the lake (see Fig. 4). Since $0 < d_s < 1$, it is interesting to consider the limits $d_s \lesssim 1$ (weak topography) and $0 < d_s << 1$ (strong topography). In the first case, the bottom slope is relatively small, corresponding to a weak restoring force for TW's; thus the periods are generally very long (several weeks). In the second case, the bottom slope (restoring force) is relatively large and the periods are much shorter. For most lakes, the second case applies. From (4.15), we find $d_s = e^{-b\xi_s}$, or

$$b = b(\xi_s, d_s) = (1/\xi_s)\ln(1/d_s). \tag{4.21}$$

Thus for fixed ξ_s, $b\downarrow$ as $d_s\uparrow$, so that weak topography corresponds to "small" b, and strong topography, to "large" b. As examples of these limiting cases, Tables 2 and 3 give the periods $T_{mn} = 2\pi/\sigma_{mn} f$ for $b = 0.28$ $(d_s = 0.8)$ and $b = 2.86$ $(d_s = 0.1)$ respectively. For the weak topography case (Table 2), the periods are generally longer than about 20 days and the gravest mode has $T_{11} = 40.2$ days. Since most lake observations indicate the presence of the gravest TW mode with a corresponding period of a few days, a small b value is not applicable. But for $b = 2.86$ (Table 3) the periods range from a few days and up, with $T_{11} = 4.4$ days, which is characteristic of many observed periods (e.g., see Fig. 1).

m \ n	1	2	3	4	5
1	40.2	153.1	341.3	604.8	943.5
2	23.8	80.3	174.4	306.1	475.5
3	20.0	57.6	120.4	208.2	321.1
4	19.3	47.5	94.6	160.5	245.2
5	19.9	42.5	80.1	132.8	200.6

Table 2 $T_{mn} = 2\pi/\sigma_{mn} f$ in days, with $f = 1.05 \cdot 10^{-4}\ s^{-1}$ (46° N) and $b = 0.28$ ($d_s = 0.8$). σ_{mn} is given by (4.17).

m \ n	1	2	3	4	5
1	4.42	15.48	33.90	59.70	92.86
2	2.57	8.10	17.31	30.21	46.79
3	2.12	5.80	11.95	20.54	31.60
4	2.10	4.78	9.38	15.83	24.12
5	2.04	4.26	7.94	13.10	19.73

Table 3 T_{mn} in days with $f = 1.05 \cdot 10^{-4}\ s^{-1}$ (46° N) and $b = 2.86$ ($d_s = 0.1$).

At a fixed latitude, we note that T_{mn} is a function of the two independent parameters ξ_s and d_s (see (4.17) and (4.21)). The behavior of T_{11} with ξ_s and d_s is illustrated in Table 4. In Ball's solution for the gravest mode TW in an elliptic paraboloid, the only free topographic parameter in the formula T_{11} is W/L, or equivalently, ξ_s (see Table 5). Comparison of Tables 4 and 5 shows that for strong topography ($d_s = 0.1 - 0.2$), the period of our model roughly agrees with Ball's T_{11}. More specifically, for strong topography our model has slightly longer periods for narrow lakes ($\xi_s << 1$) and slightly shorter periods for wide lakes ($\xi_s \gtrsim 1$). The same general results also hold for T_{21}.

ξ_s / d_s	0.1 1:10	0.203 1:5	0.347 1:3	0.805 1:1.5	1.522 1:1.1
0.1	33.7	16.6	9.8	4.4	2.7
0.2	45.3	22.4	13.2	6.0	3.6
0.3	58.9	29.1	17.2	7.8	4.7
0.5	99.8	49.3	29.1	13.2	8.1
0.7	192.3	95.0	56.0	25.4	15.6
0.9	649.1	320.8	189.1	85.8	52.6

Table 4 T_{11} (in days) with $f = 1.05 \cdot 10^{-4}\ s^{-1}$ (46° N) as a function of d_s and ξ_s. For a given ξ_s, the equivalent W:L ratio is given underneath.

	W:L				
	1:10	1:5	1:3	1:1.5	1:1.1
T_{11}	22.2	11.6	7.6	5.2	4.9

Table 5 T_{11} (in days) with $f = 1.05 \cdot 10^{-4}\ s^{-1}$ for an elliptic paraboloid (Ball, 1965):

$$T_{11} = \frac{2\pi}{f} \left(\frac{49 - 9\,\Theta^2}{1 - \Theta^2}\right)^{1/2}$$

where $\Theta = [\,1-(W/L)^2] \,/\, [\,1+(W/L)^2\,]$.

Figure 6 shows the time evolution of the contours of ψ_{11} as given by (4.18) through one half of the wave period. The double cell structure of this mode as seen in Figure 6a is reminiscent of the pattern which also arises in Ball's gravest mode for an elliptic paraboloid (see Fig. 7a). However in the Mysak model the nodal line across the major axis of the ellipse does not rotate anticlockwise as the wave progresses; in Ball's model it does (see Fig. 7b,c). Nonetheless, both models show the variable angular speed of propagation of the cells, with the speed being largest at the top and bottom of the ellipse ($\eta = \pi/2$ and $3\pi/2$). For the Mysak model this can be easily seen from (4.9), which implies that c_p is a maximum at $\eta = \pi/2$, $3\pi/2$ and a minimum at $\eta = 0, \pi$.

Figure 6 (a-e) Stream function contours of ψ_{11} with $b = 2.86$ ($d_s = 0.1$) at $\sigma_{11} t$ =

a) $-\pi/2$,

b) $-\pi/4$,

c) 0 ,

d) $\pi/4$,

e) $\pi/2$.

(From Mysak, 1983)

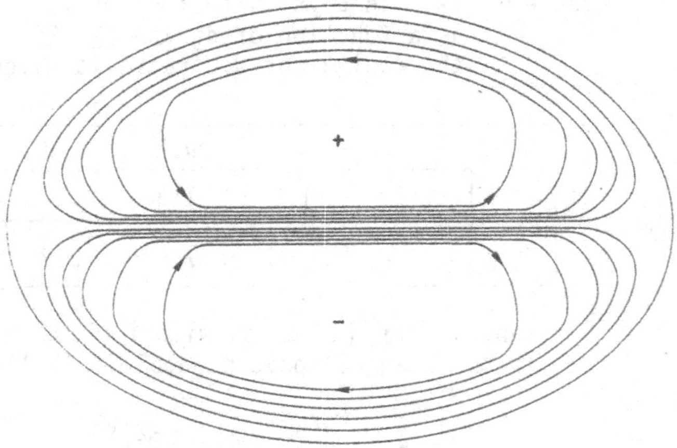

a)

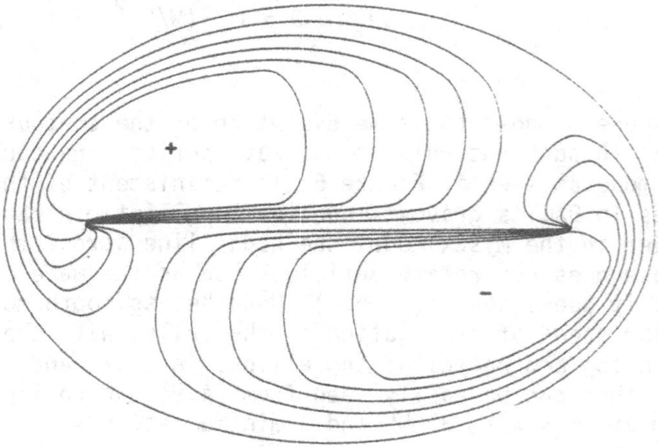

b)

Cont. Figure 6

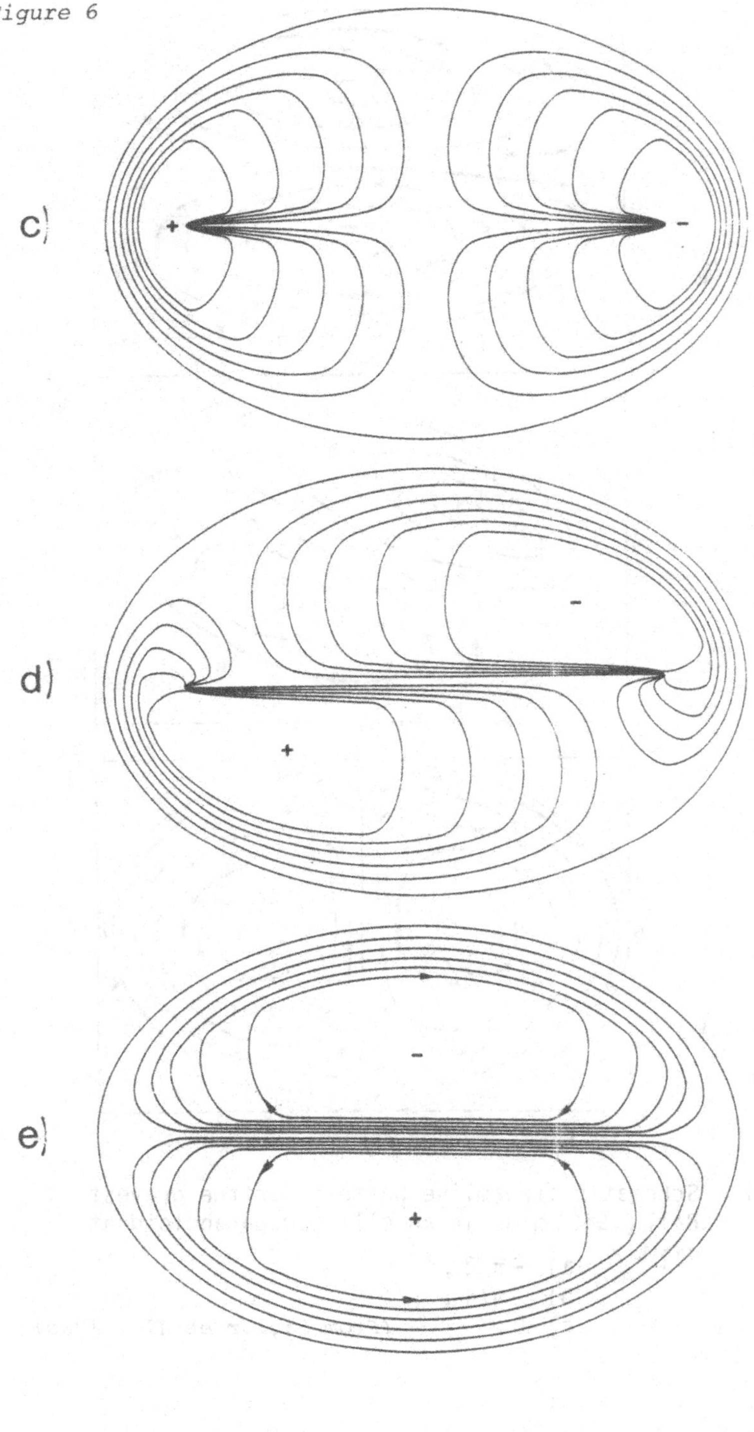

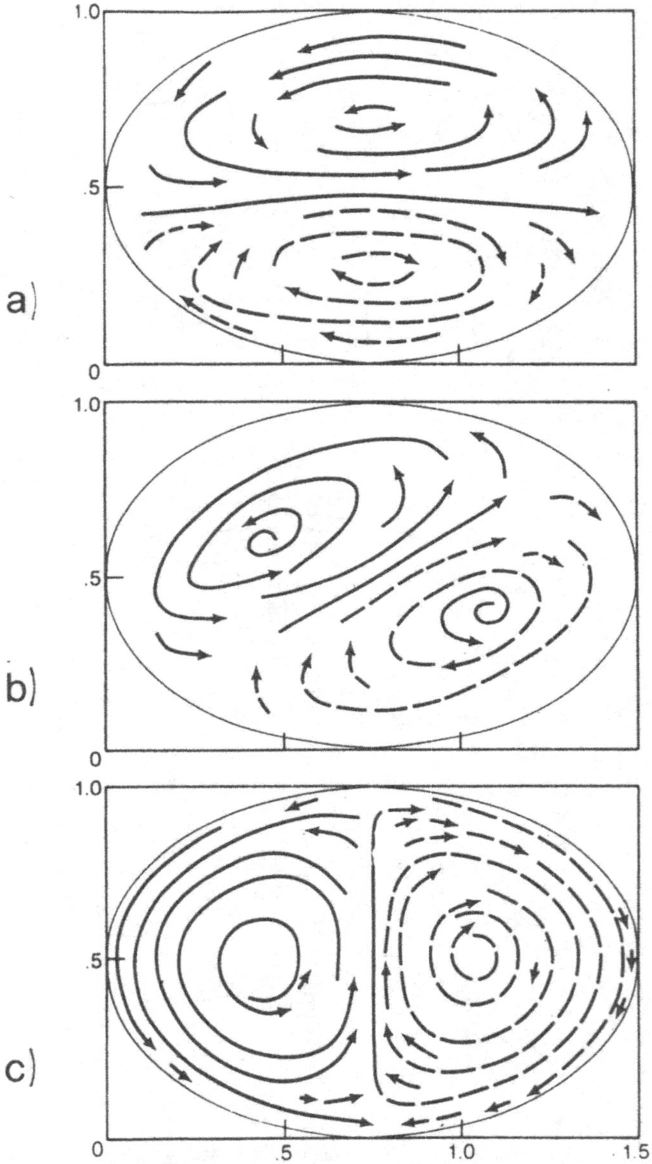

Figure 7 Schematic streamline pattern for the gravest Ball (1965) mode in an elliptic paraboloid at $\sigma_{11}t =$ a) $-\pi/2$, b) $-\pi/4$, c) 0. *(From Saylor et al., 1980)*

Figure 8 shows how the streamline patterns change with b, the topographic parameter. For weak topography (Fig. 8a) the jet across the lake center is weakened considerably, but the coastal flows change little. For strong topography (Fig. 8b), the depth is effectively zero near the coast and accordingly, the coastal currents are weakened, and also the center jet is intensified.

Figure 8 (a,b) Stream function contours of ψ_{11} at $\sigma_{11} t = -\pi/2$:
a) $b = 0.28$ ($d_s = 0.8$ - weak topography)
b) $b = 5.72$ ($d_s = 0.01$ - strong topography).
(From Mysak, 1983)

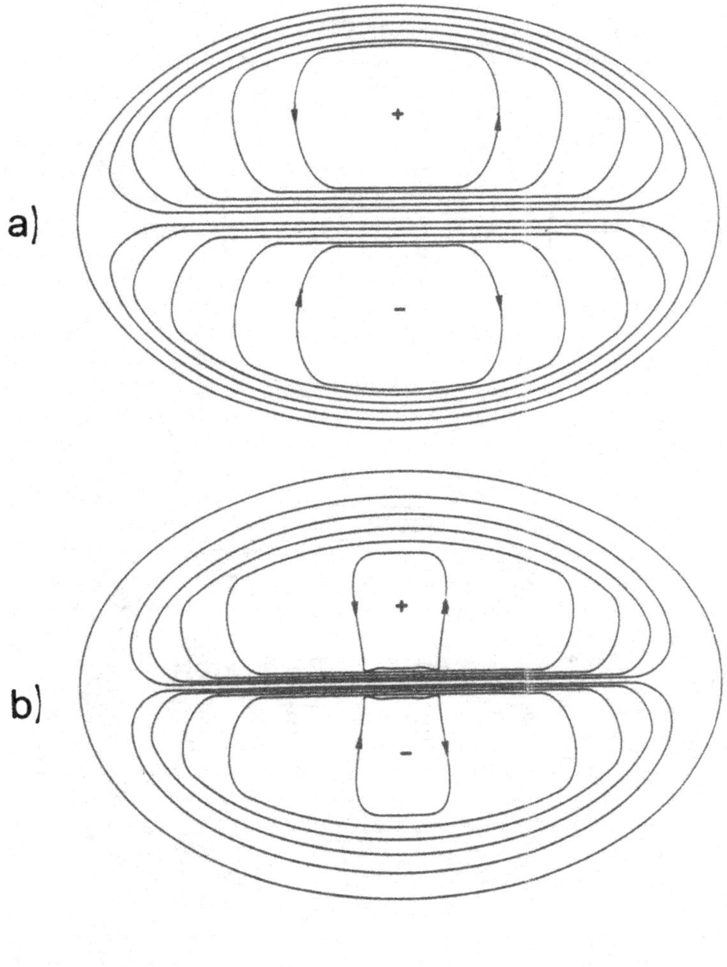

Figure 9 indicates that the currents are largely alongshore ($u_\xi \ll u_\eta$), except near the foci where the cross-isobath component can be moderately large. The time evolution of $\underline{u}_{11}$ through half a wave period is shown in Figure 10. The large arrows seen at the foci in Figures 10 b,c,d are due to the vanishing here of the scale factor a J, which appears in the denominator of u_ξ and u_η (see (4.19) and (4.20)). It is important to note that near the center of the lake the current vector rotates anticlockwise, whereas at the sides of the lake the rotation is clockwise.

Figure 9 (a,b) Profile of a): u_ξ (cross-isobath component) and b): u_η (alongshore component) for the gravest mode velocity at $\sigma_{11} t = -\pi/2$. In both cases $b = 2.86$ ($d_s = 0.1$), as in Figure 6. *(From Mysak, 1983)*

a)

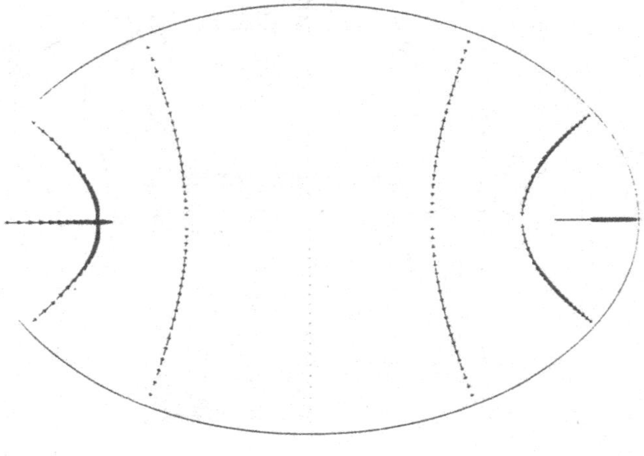

b)

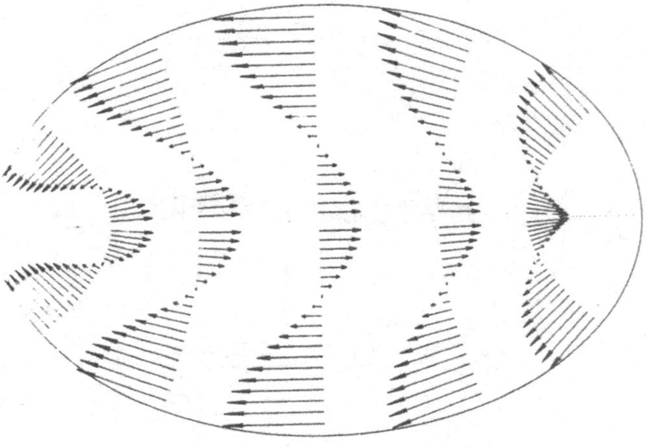

Figure 10 (a-e) Greatest mode velocity field $\underline{u} = (u_\xi, u_\eta)$ with $b = 2.86$ at $\sigma_{11} t =$

a) $-\pi/2$, c) 0,

b) $-\pi/4$, d) $\pi/4$,

e) $\pi/2$.

(From Mysak, 1983)

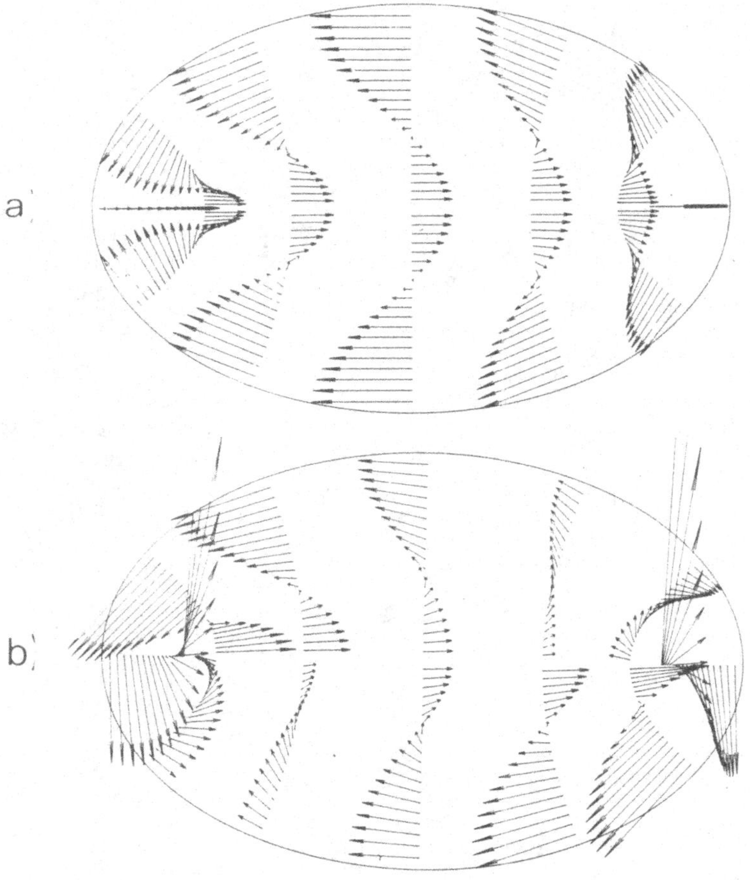

Figure 11 shows the time evolution of ψ_{12}, the first higher mode with a nodal ellipse. The patterns are beautiful but unlikely to be seen in nature, except possibly as a pattern on a sea shell! The corresponding current field $\underline{u}_{12}$ is shown in Figure 12, where a banded structure of alternating flows along the central portion of the lake is evident.

Cont. Figure 10

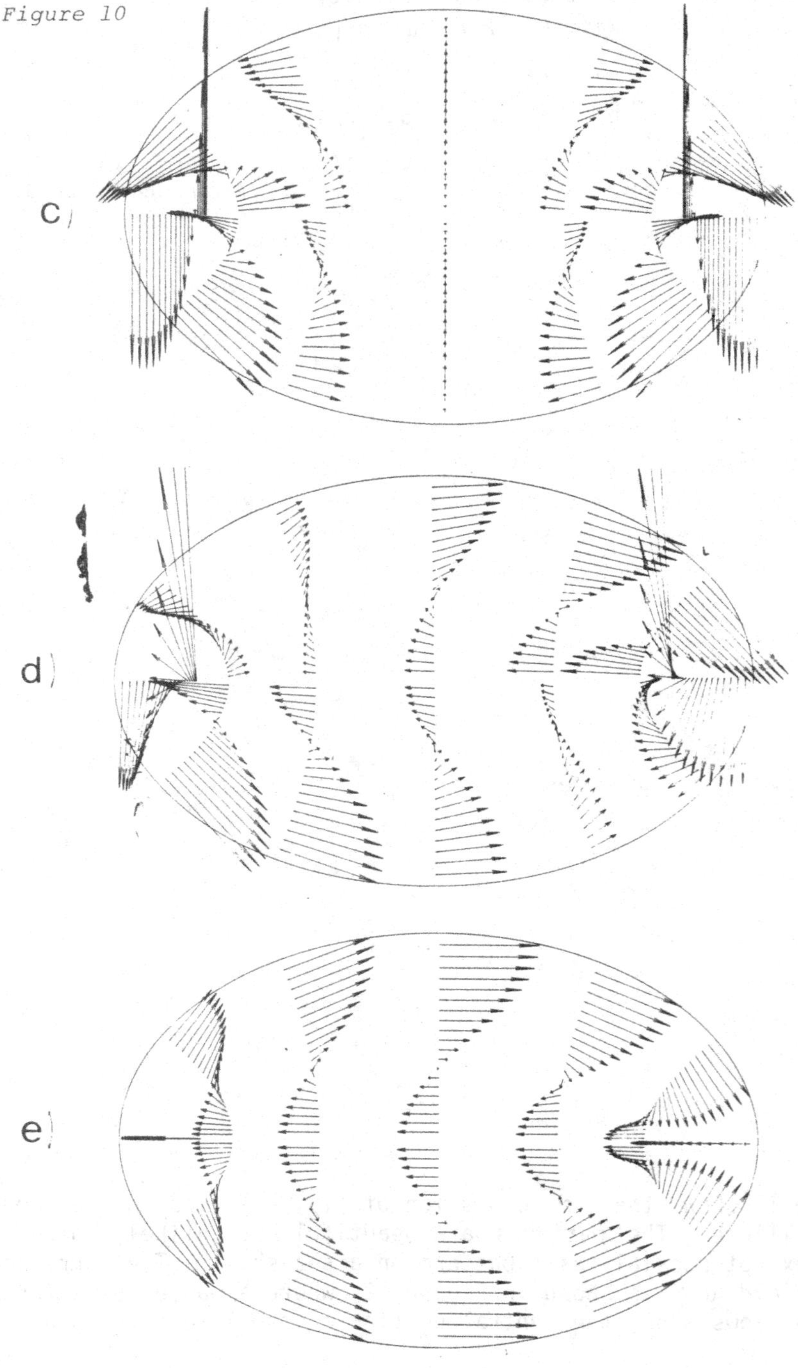

Figure 11 (a-c)

Stream function contours of ψ_{12} with $b = 2.86$ at $\sigma_{12} t =$

a) $-\pi/2$,

b) $-\pi/4$,

c) 0 .

(From Mysak, 1983)

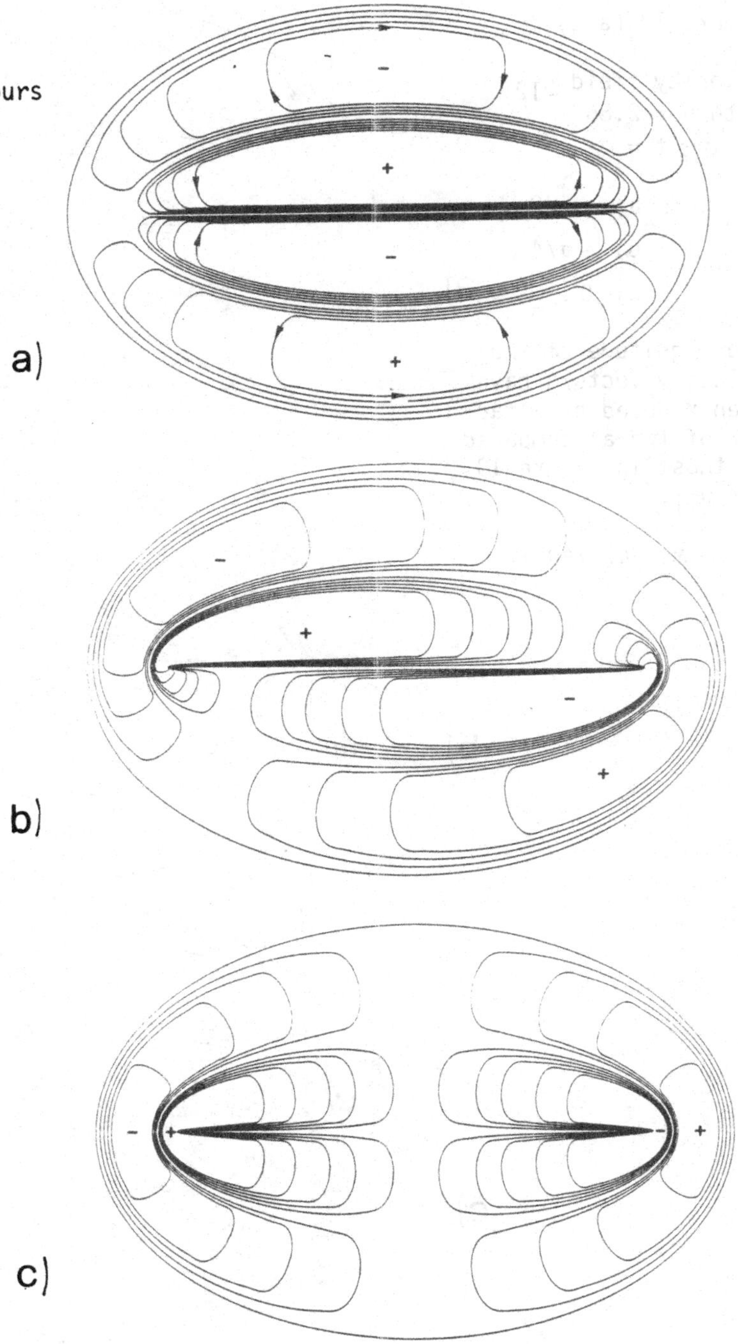

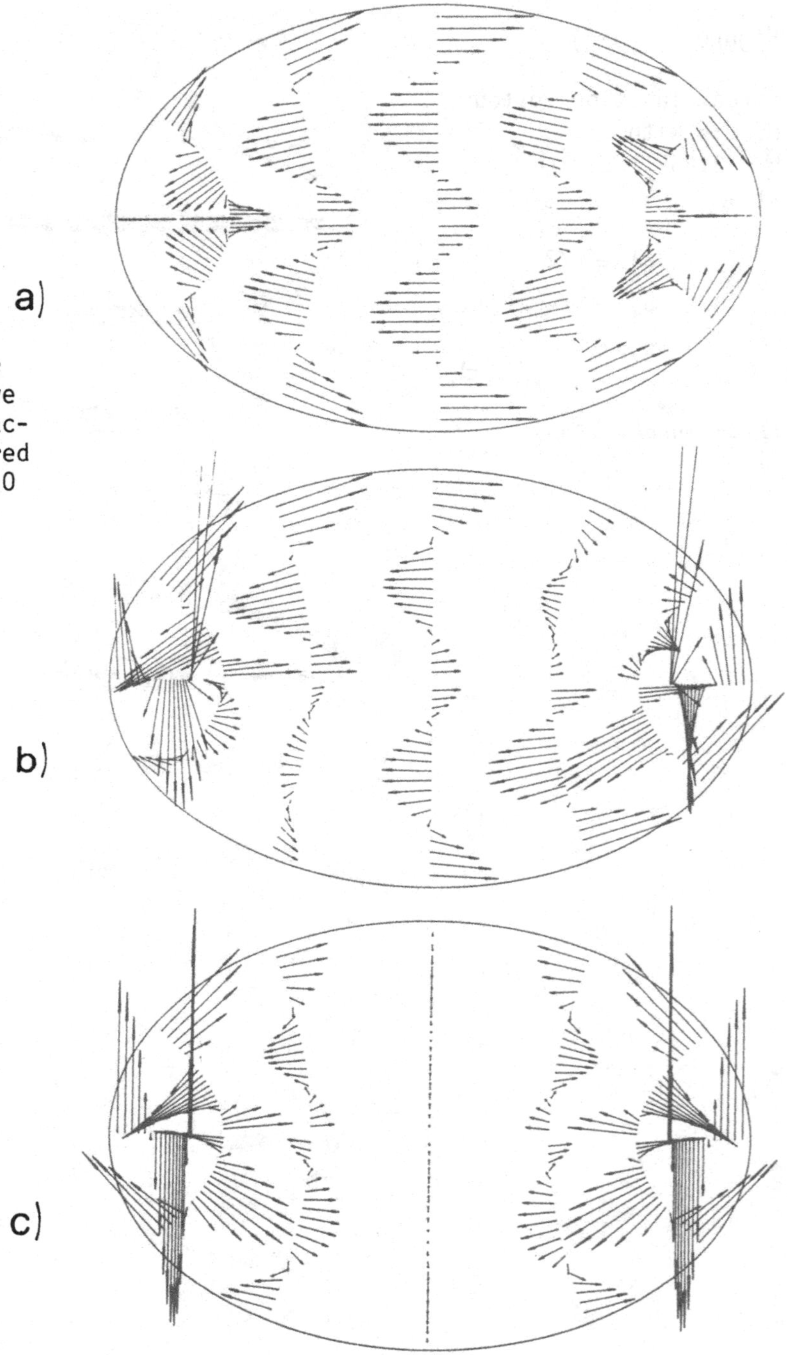

Figure 12 (a-c)

Velocity field $\underline{u}_{12}$ with b = 2.86 at $\sigma_{12} t$ =

a) $-\pi/2$,

b) $-\pi/4$,

c) 0 .

The magnitude of the velocity vectors have been reduced by a factor of two as compared to those in Figure 10 for $\underline{u}_{11}$.

(From Mysak, 1983)

Finally, Figure 13 shows the time evolution of ψ_{21}, the first higher mode with two complete waves around the basis. An interesting feature of this mode is the development of the strong shear along the major axis of the basin (see Figs. 14b,c).

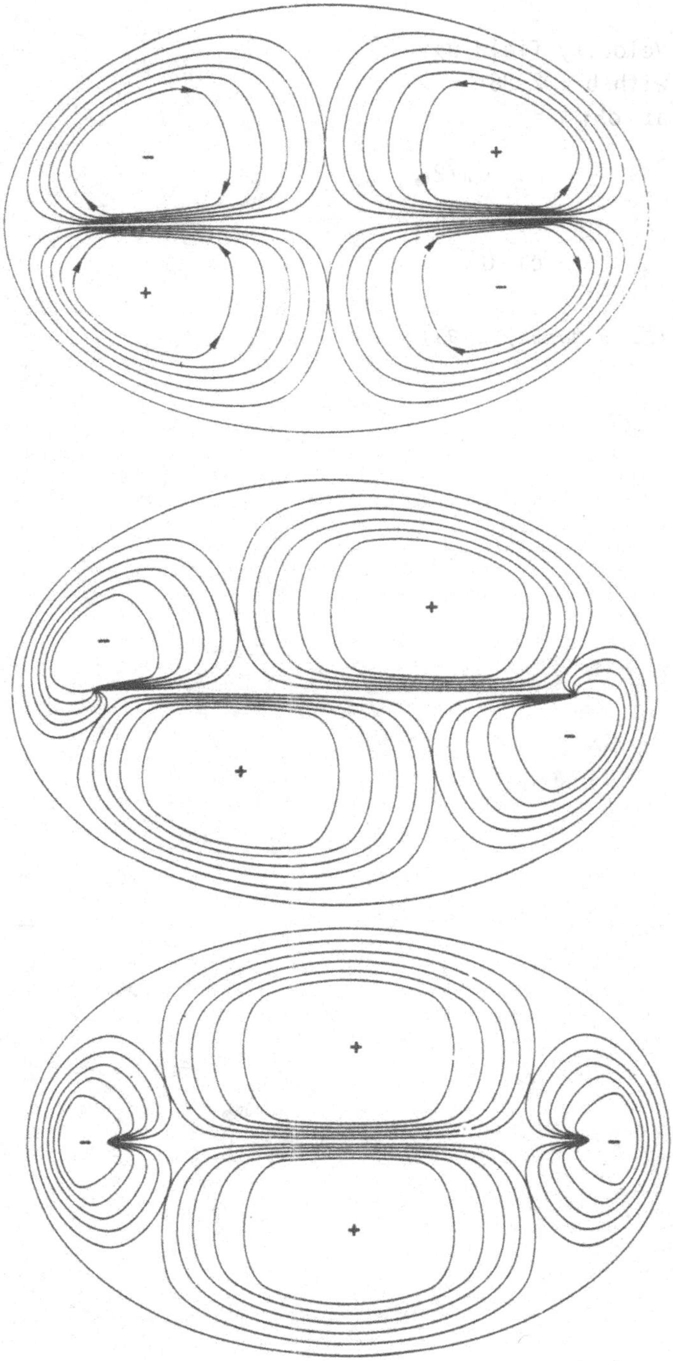

a)

This arises because of our (approximate) boundary condition (4.12) which does not allow any flow across the degenerate ellipse $\xi = 0$.

b)

Figure 13 (a-c)

Stream function contours of ψ_{21} with $b = 2.86$ at $\sigma_{21} t =$

a) $-\pi/2$,

b) $-\pi/4$,

c) 0 .

(From Mysak, 1983)

c)

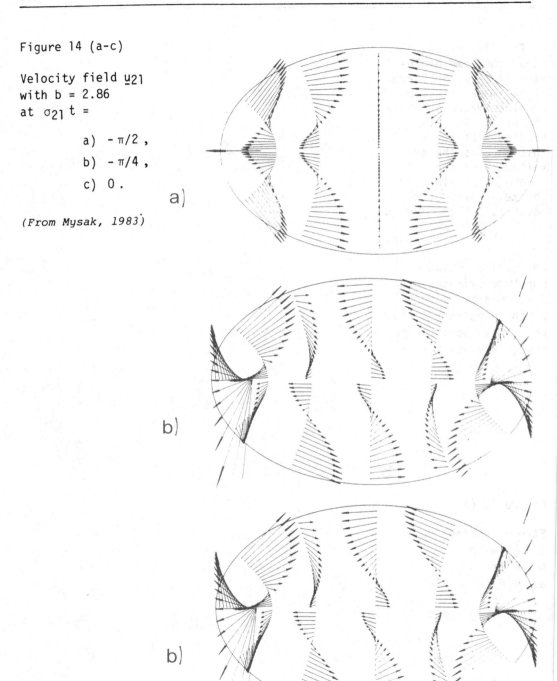

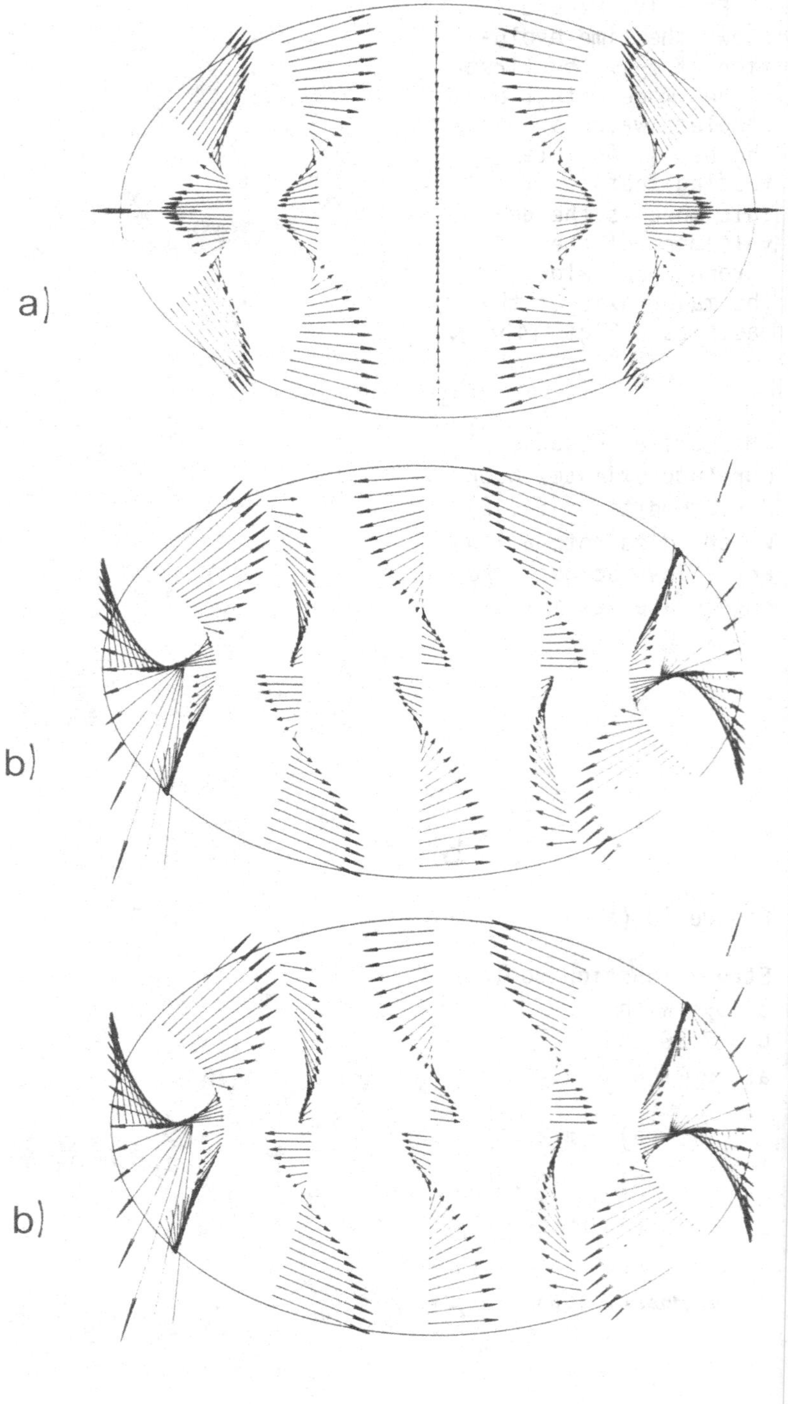

Figure 14 (a-c)

Velocity field $\underline{u}_{21}$
with b = 2.86
at $\sigma_{21}t$ =

a) $-\pi/2$,
b) $-\pi/4$,
c) 0 .

(From Mysak, 1983)

4.3 Application to Lake of Lugano

Some of the physical parameters which characterize the summer 1979 situation in Lake of Lugano were given in Section 3.3. Here we quote the topographic parameter values used by Mysak (1983) to estimate the period T_{11} for the gravest mode TW in this lake. From the depth profile along the Talweg he found

$$L = 8.7 \text{ km}, \quad \ell = 5 \text{ km}, \quad D = 280 \text{ m}, \quad D_s = 25 \text{ m}. \tag{4.22}$$

Thus $a^{-1} = L/\ell = 1.74$ and hence $\xi = 1.15$ upon using $1 = a \cosh \xi_s$ (see (4.4)). This ξ_s value corresponds to $W/L = 1:1.225$, which is a rather wide model lake! However, a large value of ξ_s is needed to accurately represent the topography at the ends of the lake. From (4.22) we also obtain $d_s^{-1} = D/D_s = 11.2$. Hence (4.21) yields $b = 2.10$, which is comparable to the value used in most of Section 4.2 ($b = 2.86$). Thus we find (see (4.17))

$$T_{11} = \frac{2\pi}{fb}\left[1 + \frac{b^2}{4} + \left(\frac{\pi}{\xi_s}\right)^2\right] = 3.15\,\text{d}\,(75.6\,\text{h}) \tag{4.23}$$

for $f = 1.05 \cdot 10^{-4}\,\text{s}^{-1}$. This theoretical value compares very favourably with the observed period of 74 h (see Fig. 3).

The low-passed (30 h cut-off) current vectors from the Gandria mooring at the lake center are illustrated in Figure 15. Unfortunately, only 3 week records were collected due to instrument malfunction; nevertheless both the 15 and 39 m depth records show three cycles of the 3-day oscillation (from 16 July to 25 July). Moreover, the deeper record (Fig. 15b), which is representative of the barotropic currents since 39 m is well into the lower layer [see (3.23) and recall that $D_1 = 10$ m in our model], clearly shows that the current fluctuations are very nearly in the alongshore direction. Thus our hypothesis that $u_\xi = 0$ at $\xi = 0$, or equivalently, that $\psi = 0$ at $\xi = 0$ (see (4.12)) is a reasonable approximation. Also the magnitude of the fluctuations is $3\,\text{cm s}^{-1}$, which agrees with the estimate of the wind-forced current given in Section 3.3, i.e., $3.6\,\text{cm s}^{-1}$.

Figure 15 (a,b) Low passed current record from Gandria mooring 6 (see Fig. 16b for location), Lake of Lugano, July 1979,

(see next page)

a) at depth of 15 m,

b) at depth of 39 m.
The components u and v are alongshore and offshore-onshore respectively.

(From Mysak et al., 1983)

Figure 15

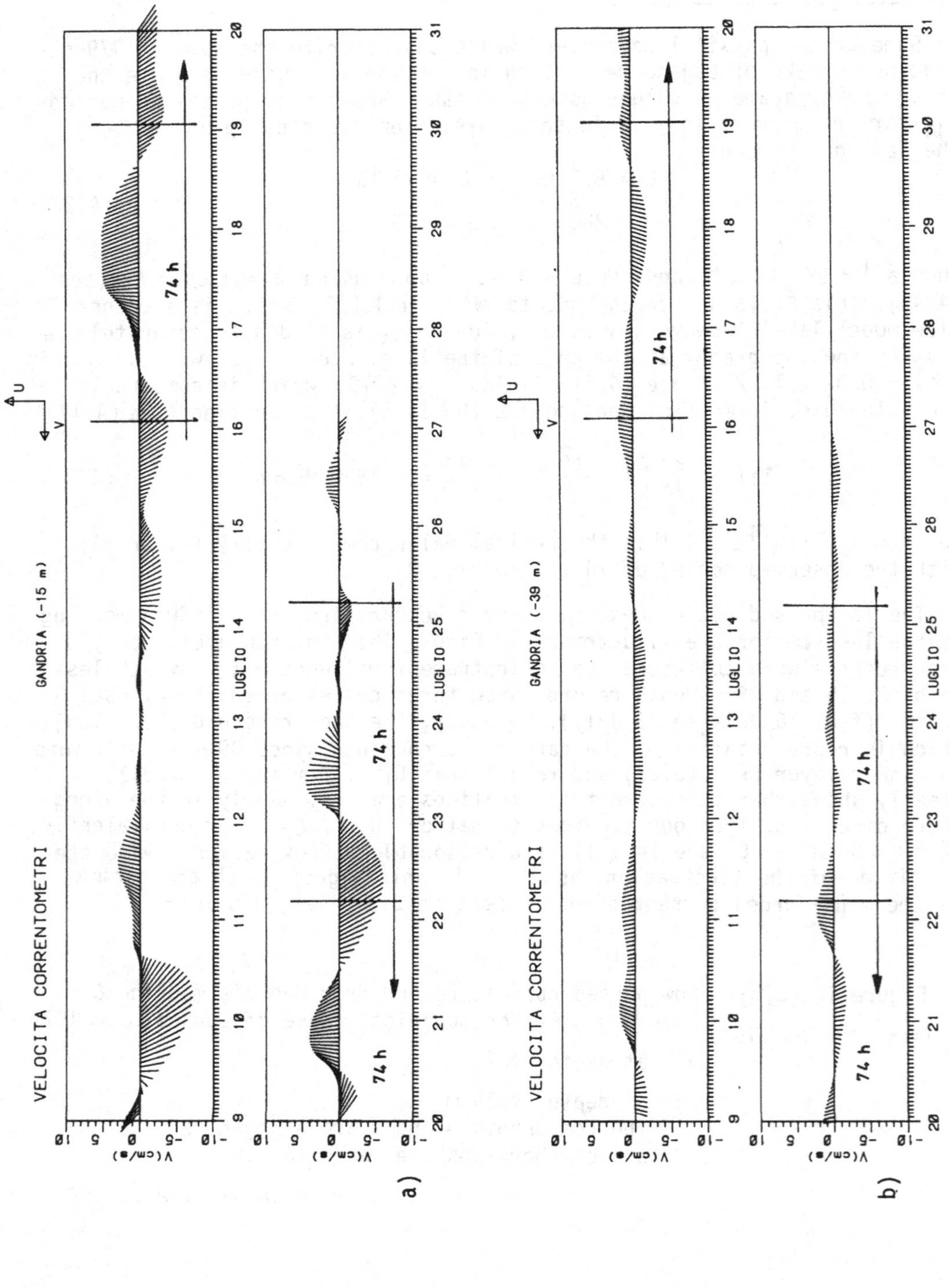
VELOCITA CORRENTOMETRI
GANDRIA (-15 m)
GANDRIA (-39 m)
LUGLIO
74 h
V(cm/s)
U
V
a)
b)

Figure 16 shows the progressive vector diagrams of the currents illustrated in Figure 15. We observe that the motions are generally anticlockwise at both depths, and at 39 m, they are nearly rectilinear. These two properties of the 39 m record precisely describe the time-dependent behavior of the gravest mode currents at the lake center in our model (see Fig. 10).

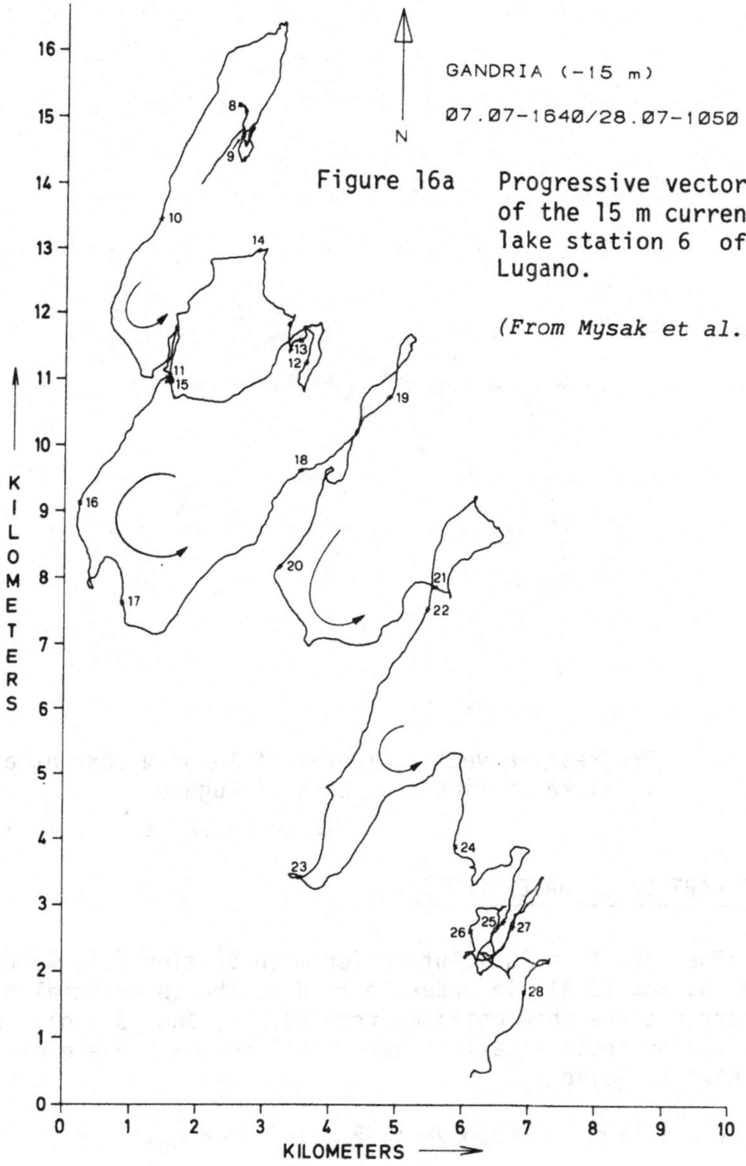

Figure 16a Progressive vector diagram of the 15 m current at mid-lake station 6 of Lake of Lugano.

(From Mysak et al., 1983)

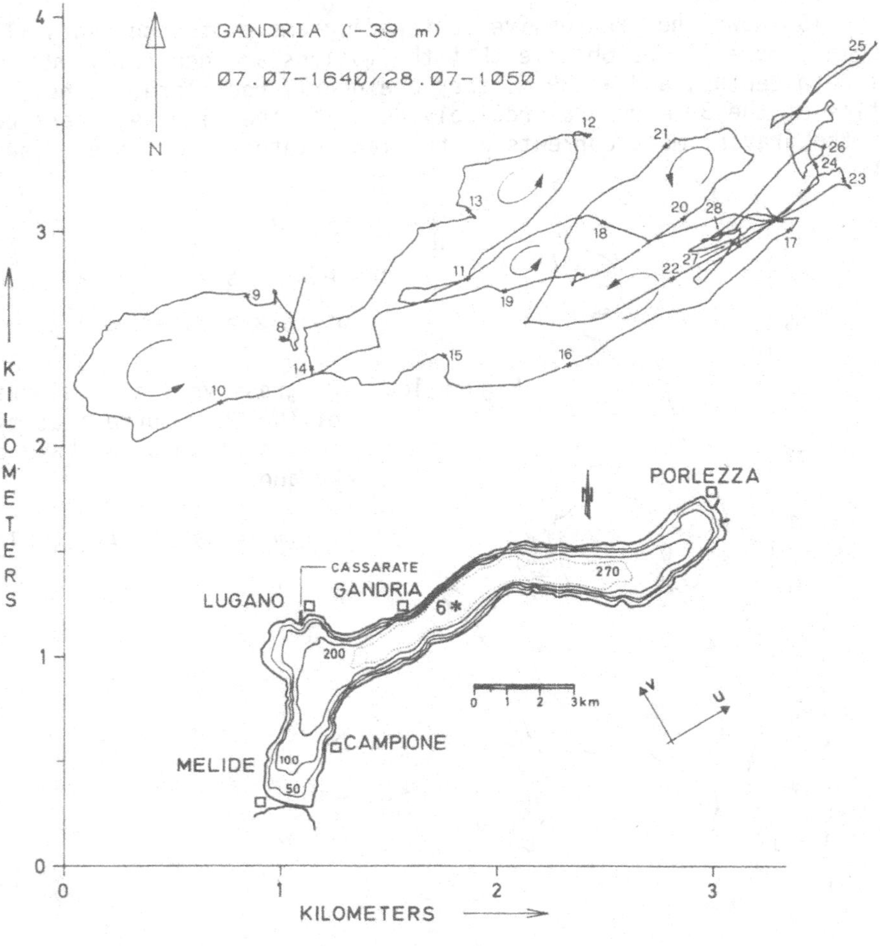

Figure 16b Progressive vector diagram of th 39 m current at mid-lake station 6 of Lake of Lugano

(From Mysak et al., 1983)

5. BAROCLINIC PART OF ET WAVE SOLUTION

Having obtained the free TW solution for ψ in Section 4.1, we can now proceed to (3.19) and (3.31) in order to find ζ, the interfacial displacement. The currents are then obtained from (3.23a) and (3.23b). For convenience, we rewrite these equations here, setting $\underline{\tau} = 0$ where necessary. To find ζ we have to solve

$$(\nabla^2 - S^{-1})\,\zeta_t = -(\nabla\psi \times \nabla h^{-1})\cdot\hat{z} \quad \text{in} \quad \xi < \xi_s, \tag{5.1}$$

subject to the boundary condition

$$\frac{\partial^2 \zeta}{\partial n\, \partial t} - \frac{\partial \zeta}{\partial s} = 0 \quad \text{on} \quad \xi = \xi_s . \tag{5.2}$$

The currents in each layer are then given by

$$\underline{u}_1 = h^{-1}\left[\hat{z} \times \nabla\psi + h(\nabla\zeta_t - \hat{z} \times \nabla\zeta)\right], \tag{5.3}$$

$$\underline{u}_2 = h^{-1}\, \hat{z} \times \nabla\psi . \tag{5.4}$$

For a given TW mode ψ_{mn}, eq. (5.1) takes the form of a forced modified Helmholtz equation after integrating with respect to t. This equation could be solved by using the relevant Green's function, or by an appropriate eigenfunction expansion. In either case, Mathieu and modified Mathieu functions would be involved. Since these special functions are not very well known to most limnologists, it is desirable to find another approach for obtaining ζ. This is outlined in Section 5.1.

5.1 Geometric Optics Approximation

For the Lakes of Lugano and Zurich, the inverse stratification parameter S^{-1} is fairly large, viz.

$$S^{-1} = L^2/r_i^2 = 4.6 \quad \text{(Lugano)} \qquad\qquad = 5.0 \quad \text{(Zurich)}. \tag{5.5}$$

Therefore $S^{-1} >> \nabla^2 = O(1)$, so that the dominant term on the left side of (5.1) is $-S^{-1}\zeta_t$. Because the LHS of (5.1) has the form of a modified Helmholtz equation with S^{-1} as the large "wave number", the simplification

$$S^{-1}\,\zeta_t = (\nabla\psi \times \nabla h^{-1})\cdot\hat{z} \tag{5.6}$$

can be described as a geometric optics approximation for ζ. With $h = h(\xi)$, (5.6) reduces to

$$\zeta_t = -\frac{S}{h_\xi\, h_\eta}\,\frac{\partial\psi}{\partial\eta}\,\frac{d}{d\xi}\left(\frac{1}{h}\right), \tag{5.7}$$

which implies that

$$\zeta(\xi_s, \eta, t) = 0 \tag{5.8}$$

since $\psi(\xi_s, \eta, t) = 0$. Is this consistent with the boundary condition (5.2)? Yes, provided we neglect the "small" $\partial^2\partial/\partial n\,\partial t$ term; this can be done since $\partial t \sim \sigma = O(10^{-1})$ for the TW's under consideration. Therefore, (5.2) can be approximated by $\partial\zeta/\partial s = \partial\zeta/\partial n = 0$, or equivalently, $\zeta(\xi_s, \eta, t) = 0$.

For the exponential depth profile solution (4.18), (5.7) gives

$$\zeta_{mn} = \frac{S\,[\,m + b^2/4 + (n\pi/\xi_s)^2]}{a^2\,J^2}\, e^{b\xi/2} \sin(n\pi\xi/\xi_s) \cos(m\eta - \sigma_{mn} t) \qquad (5.9)$$

Note that ζ_{mn} has the same phase and radial nodes as ψ_{mn}! However, since $J^2 = \sinh^2\xi + \sin^2\eta$, ζ_{mn} has a simple pole at $\xi = 0$ and $\eta = 0$ or π, i.e., at the foci. Therefore we would expect to observe large thermocline displacements near these points in a lake. The patterns of phase propagation seen in the ψ_{mn} contour plots also would occur in the contours for ζ_{mn}. This is borne out in the cover picture showing the amplitude distribution of ζ_{11} for the elliptical basin satisfying the Lake of Lugano parameters.

The lower layer current $\underline{u}_2$ is barotropic (see (5.4)) and hence no further calculation is required - the components of $\underline{u}_2$ are given by (4.19) and (4.20). The computation for $\underline{u}_1$ is straightforward and is therefore left as an exercise for the reader. But to be consistent with the geometric optics solution for ζ, we must first neglect the low-frequency term $\nabla\zeta_t$ in (5.3). Thus the upper layer velocity can be approximated simply by

$$\underline{u}_1 = h^{-1}\,\hat{z} \times \nabla\psi - \hat{z} \times \nabla\zeta\,. \qquad (5.10)$$

The structure of $\underline{u}_1$ will be discussed in the lectures. Here we only remark that according to (5.10), ζ behaves like a stream function for the baroclinic part of the upper layer current.

5.2 Application to Lake of Lugano

A detailed comparison between the theory and the summer 1979 Lake of Lugano observations of temperature, upper layer currents and winds is given in Mysak et al. (1983). Here we shall present only a very brief summary.

Figure 17 shows the auto- and cross-spectra of the low-passed isotherm depths (temperature fluctuations) from five moorings: Porlezza to the north, and Cassarate, Ferrera, Melide and Caprino to the south. From the auto-spectra we observe that Ferrera has the largest amplitude and Melide the smallest. This is consistent with the structure of ζ_{11} as given by (5.9) which is large near $\xi = 0$, $\eta = \pi$ (the approximate location of Ferrera) and very small for ξ near ξ_s and $\eta = \pi$ (the location of Melide). For intermediate ξ-values and other $\eta \neq 0, \pi$, ξ_{11} lies between these extremes, as do the variance spectra from the other three moorings (Porlazza, Cassarate and Caprino). Second, observe that as we progress anticlockwise around the lake, the phase difference at 3 days increases (see Figure 17a, b,c), in agreement with TW propagation for a 3-day signal. Only in Figure 17d does the argument break down - the observed phase difference is only

182°, rather that something like 210° as would be expected from theoretical considerations. The 182° value may be due to interference of the "downward" travelling wave through Cassarate and the returning "upward" travelling wave through Caprino.

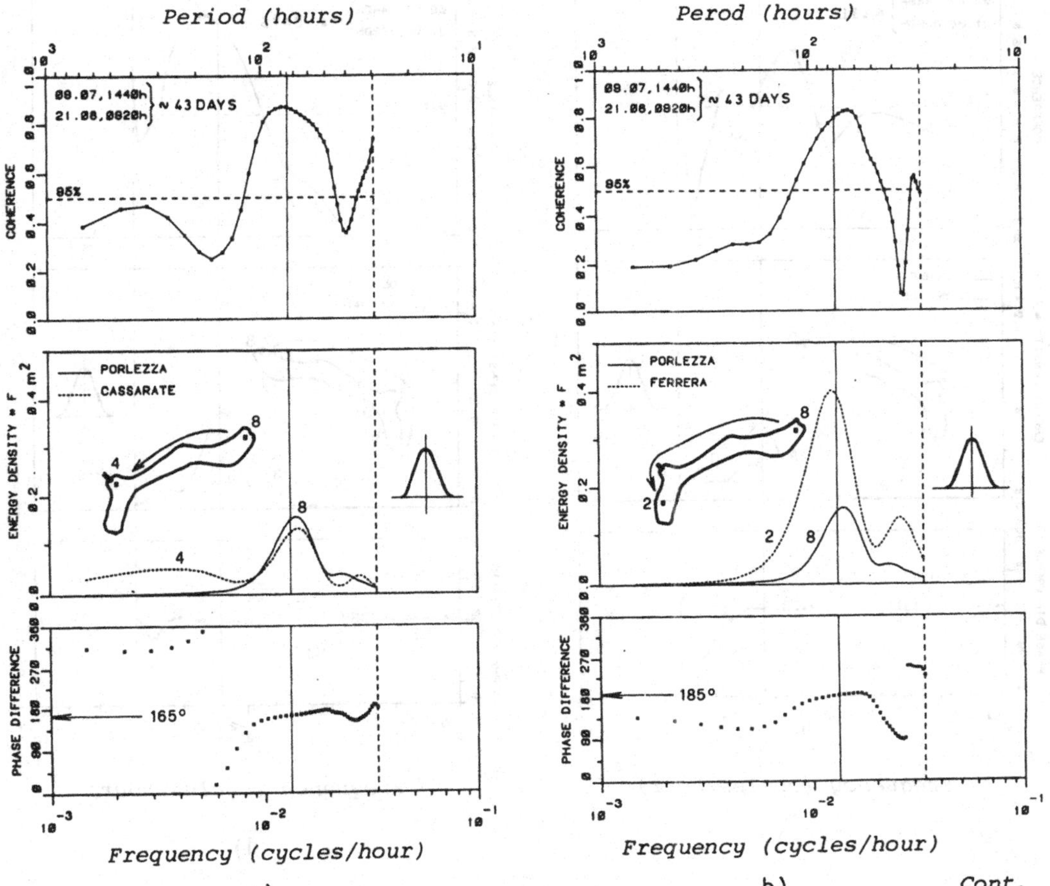

a) b) *Cont.*

Figure 17 (a-d) The variance spectra, coherence and phase between the low passed (30 h cut-off) fluctuations of isotherm depths from four moorings in the Lake of Lugano, summer 1979. Starting with Porlezza at the north and proceeding anticlockwise around the lake, the station names and corresponding mooring numbers (see Fig. 3 and Fig. 16b) are as follows: Porlezza (8), Cassarate (4), Ferrera (2), Melide (1) and Caprino (3) (near Campione). A positive phase indicates that Porlezza leads the paired station. *(From Mysak et al., 1983)*

(see p. 96 and 120)

Cont. Figure 17

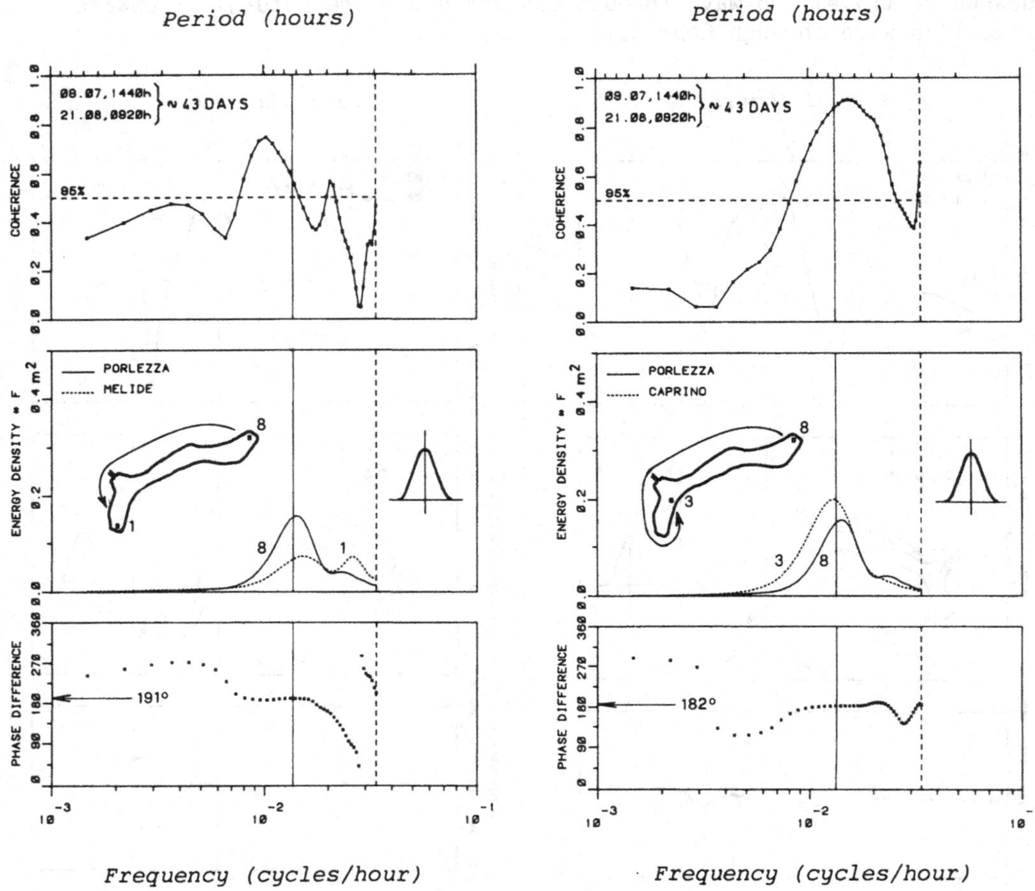

c) d)

The middle of Figure 18 shows that the energy for the 3-day temperature oscillation could come from the wind at Campione, at the southern end of the lake. Although the wind variance is largest at around 2 days, there is still significant energy at 3 days, and at this period, the wind and temperature are coherent at the 95 % level (see Figure 18, top). It is also clear from Figure 18, however, that the 3-day temperature oscillation is not a resonance response of the lake to a 3-day wind oscillation. Rather, bursts of winds every few days probably kick the TW's into motion.

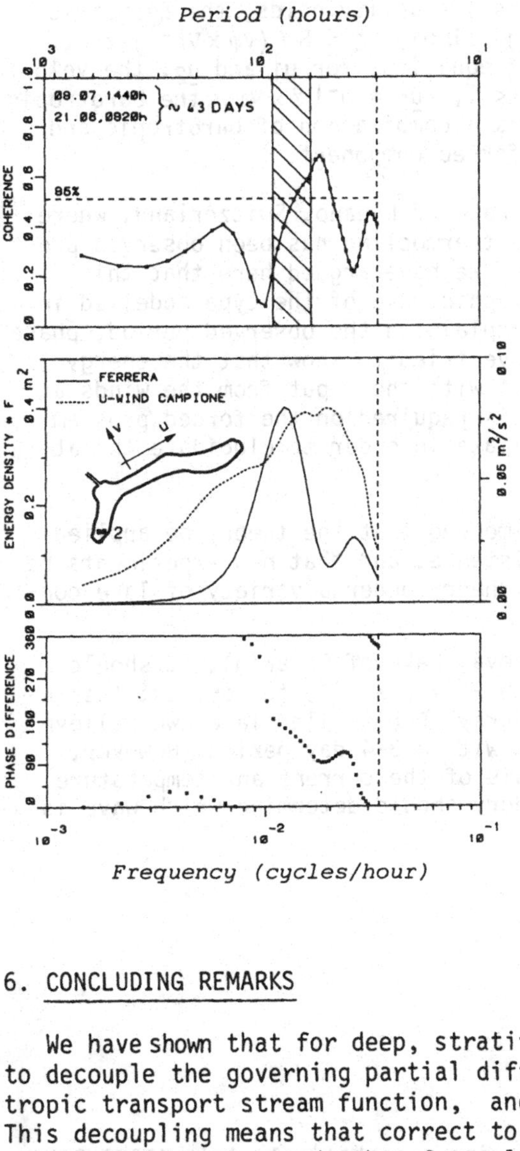

Figure 18

Middle:

Temperature variance spectra at Ferrera (2) and alongshore wind variance spectra at Campione, the neighboring meteorological station.

Top:

Coherence spectrum.

Bottom:

Phase difference. Positive phase means temperature leads the wind.

(From Mysak et al., 1983)

6. CONCLUDING REMARKS

We have shown that for deep, stratified elongated lakes it is possible to decouple the governing partial differential equations for ψ, the barotropic transport stream function, and ζ, the interfacial displacement. This decoupling means that correct to $O(D_1/D_2)$, where D_1 and D_2 are the upper and lower layer depths of our lake model, we can first solve the usual TW equation for ψ. In general, this equation will also have a wind-forcing term proportional to $\underline{\tau} \times \nabla h^{-1}$. The latter term determines the amplitude scale for ψ and hence U, the barotropic current scale. The solution for ψ is then substituted into the right side of the ζ equation; therefore ψ directly drives the interfacial displacement. A further simplification of this equation, which is a modified Helmholtz equation, can be made if $S^{-1} = L^2/r_i^2 >> 1$. If this inequality holds, the geometric optics approximation applies, yielding simply $\zeta = S \int (\nabla\psi \times \nabla h^{-1}) \cdot \hat{z}\, dt$. Finally, ψ and ζ are used in a pair of equations for $\underline{u}_1$ and $\underline{u}_2$, the velocities in each layer. Because $D_1/D_2 << 1$, $\underline{u}_2 \simeq h^{-1}\, \hat{z} \times \nabla\psi$, the barotropic current. The upper layer current $\underline{u}_1$ is a combination of barotropic and baroclinic parts, plus a direct wind-forced component.

The theory has been applied to the Lake of Lugano, Switzerland, where a 3-day temperature oscillation in the thermocline has been observed propagating anticlockwise around the lake. We have argued here that this signal is due to a gravest mode topographic wave of the type modelled in this set of lectures. In addition to explaining the observed period, phase propagation and nodal structure, we have tried to show that the energy level of this oscillation is consistent with the input from the winds at this time scale. However, future work is required on the forced problem for topographic waves in a stratified lake in order to elucidate the atmosphere-lake dynamical mechanisms.

In conclusion, it is strongly recommended that the theory be applied to other lake data sets already in existence, and that new experiments be conducted to test the validity of the theory under a variety of lake conditions.

For larger lakes (e.g., Lake of Geneva, Lake of Ontario), it should also be possible to use the theory with $L = \partial_{tt} + f^2$ to sort out the internal Kelvin/topographic wave controversy. For smaller lakes we believe that no internal Kelvin wave can exist with a 3-4 day period. However, for the larger lakes, a careful analysis of the current and temperature data in a manner suggested by this theory should determine which wave is dominant.

ACKNOWLEDGEMENTS

These lecture notes were prepared during a sabbatical visit (1982-83) to the limnology group of the Laboratory of Hydraulics, Hydrology and Glaciology (VAW), Swiss Federal Institute of Technology, Zurich. The author is grateful to Dr K. Hutter and his group for their kind hopitality and to Mr. Scheiweiller for computing the pictures shown in Figure 6 and Figures 8 - 14. This work was supported by the VAW and the Izaak Walton Killam Memorial Fund for Advanced Studies.

REFERENCES

Allen, J.S., 1975. Coastal trapped waves in a stratified ocean. J. Phys. Oceanogr., 5, pp. 300-325.

Ball, F.K., 1965. Second-class motions of a shallow liquid. J. Fluid Mech., 23, pp. 545-561.

Bennett, J.R. and D.J. Schwab, 1981. Calculation of the rotational normal modes of oceans and lakes with general orthogonal coordinates. J. Comput. Phys., 44, pp. 359-376.

Birchfield, G.E. and B.P. Hickie, 1977. The time dependent response of a circular basin of variable depth to a wind stress. J. Phys. Oceanogr., 7, pp. 691-701.

Csanady, G.T., 1976. Topographic waves in Lake Ontario. J. Phys. Oceanogr., 6, pp. 93-103.

Gratton, Y., 1982. Low-frequency motion on strong topography. Paper presented at the 16th Annual Congress of the Can. Meteorol. Oceanogr. Soc., Atmosphere - Ocean, 16 (Congress Issue), p. 57.

Gratton, Y., 1983. Low-frequency vorticity waves over strong topography. Ph D thesis, Univ. of British Columbia, 143 pp.

Holland, W.R., 1973. Baroclinic and topographic influences on the transport in western boundary currents. Geophys. Fluid Dyn., 5, pp. 187-210.

Hsieh, W.W., 1982. On the detection of continental shelf waves. J. Phys. Oceanogr., 12, pp. 414-427.

Huang, J.C.K. and J.H. Saylor, 1982. Vorticity waves in a shallow basin. Dyn. Atmos. Ocean, 6, pp. 177-196.

Hutter, K., G. Salvadè and D.J. Schwab, 1983. On internal wave dynamics in the northern basin of Lake of Lugano. Geophys. Astrophys. Fluid Dyn., in press.

Kielmann, J., 1983. The generation of eddy-like structures in a model of the Baltic Sea by low frequency wind forcing. Tellus, submitted.

Lamb, H., 1932. Hydrodynamics, 6th edition. Cambridge Univ. Press., 738 p.

LeBlond, P.H. and L.A. Mysak, 1978. Waves in the Ocean. Elsevier, 602 p.

Marmorino, G.O., 1979. Lowfrequency current fluctuation in Lake Ontario. J. Geophys. Res., 84, pp. 1206-1214.

Mortimer, C.H. and W. Horn, 1982. Internal wave dynamics and their implications for plankton biology in the Lake of Zurich. Vierteljahresschrift der Naturforschenden Gesellschaft in Zürich, 127, pp. 299-318.

Mysak, L.A., 1968. Edgewaves on a gently slòping continental shelf of finite width. J. Marine Res., 26, pp. 24-33.

Mysak, L.A., 1980. Recent advances in shelf wave dynamics. Rev. Geophys. Space Phys., 18, pp. 211-241.

Mysak, L.A., 1983. Elliptical topographic waves. Geophys. Astrophys. Fluid Dyn., submitted.

Mysak, L.A., G. Salvadè, K. Hutter and T. Scheiwiller, 1983. Topographic waves in a stratified elliptical basin, with application to the Lake of Lugano. Phil. Trans. Roy. Soc. London, submitted.

Poincaré, H., 1910. Leçons de mécanique céleste, iii-94, Paris.

Rhines, P.B., 1970. Edge-, bottom-, and Rossby waves in a rotating stratified fluid. Geophys. Fluid Dyn., 1, pp. 273-302.

Saylor, J.H., J.S.K Huang and R.O. Reid, 1980. Vortex modes in southern Lake Michigan. J. Phys. Oceanogr., 10, pp. 1814-1823.

Simons, T.J., 1980. Circulation models of lakes and inland seas. Can. Bull. Fish. Aquat. SCI. 203, 146 p.

Wenzel, M., 1978. Interpretation der Wirbel im Bornholmbecken durch topographische Rossby Wellen in einem Kreisbecken. Diplomarbeit, Christian Albrechts Universität Kiel, 36 p. and figures.

Wright, D.G. and L.A. Mysak, 1977. Coastal trapped waves, with application of the northeast Pacific Ocean. Atmosphere, 15. pp. 141-150.

Hydrodynamics of Lakes: CISM Lectures
edited by K. Hutter, 1984
Springer Verlag Wien-New York

NONLINEAR INTERNAL WAVES

Lawrence A. Mysak
Departments of Mathematics and Oceanography
The University of British Columbia
Vancouver, B.C., Canada V6T iW5

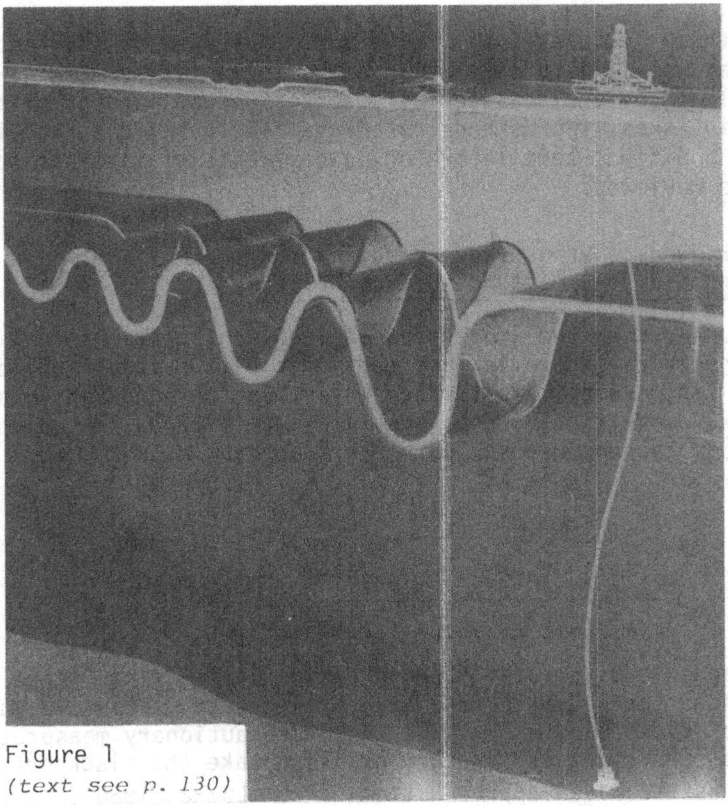

Figure 1
(text see p. 130)

1. INTRODUCTION

1.1 Preamble

For centuries sea-going explorers and navigators have been aware of the dangers associated with large-amplitude surface gravity waves. Particularly frightening are the giant waves commonly found around the southern tips of South Africa and South America. Today's explorer searching for submarine resources such as oil and minerals has to be on the look out for an additional potentially dangerous denizen in the ocean - the large-amplitude internal wave (see Fig. 1). Such slowly propagating underwater waves can have amplitudes of 100 m and more, and associated current speeds of up to a few meters per second. With the approach of a train of these waves, a vessel conducting exploratory drilling must take the slack out of its drilling riser (protective covering of the drill shaft) and cease drilling operations. Clearly, as well as having an intrinsic beauty for all of us to behold, large (nonlinear) internal waves have a tremendous impact on technology and engineering developments in the ocean.

In these lectures I will give a brief review of the theory of nonlinear internal waves, with particular emphasis on long *solitary* waves. This will be followed by a discussion of observations of large-amplitude internal waves in lakes, fjords and continental shelf regions, and in the atmosphere. In addition, some laboratory experiments of internal solitary waves will be reviewed.

1.2 Korteweg-de Vries (KdV) equation for long finite-amplitude surface waves

To understand the properties of long nonlinear internal waves, it is first necessary to review the rudiments of the theory of long (shallow-water) finite-amplitude surface waves. In the nondimensional finite-depth water wave equations (e.g., see (12.41) - (12.43) in LeBlond and Mysak, 1978), two nondimensional parameters appear:

$$\varepsilon = a/H \,, \tag{1.1}$$

$$\mu = (H/\lambda)^2 \,, \tag{1.2}$$

Figure 1 *(Figure is placed on p. 129)*

Signaled on the surface by choppy waves called rips, large internal waves move under the water surface at a slow, steady speed and occur in many parts of the ocean and in lakes. As a precautionary measure, oil company vessels conducting exploratory drilling take the slack out of the drilling risers as the waves approach them. (From Locke, 1980).

where a and λ are respectively the wave amplitude and horizontal length scales, and H is the quiescent water depth. Nonlinearity (wave steepening) and linear phase dispersion are measured by ε and μ respectively, and the Ursell number

$$U = 3\varepsilon/\mu = 3a\,\lambda^2/H^3 \tag{1.3}$$

gives the relative significance of these two effects. For the shallow water regime $\mu \simeq \varepsilon \ll 1$, the governing water wave equations can be solved via a perturbation expansion in ε and μ. To lowest order one obtains, for $\partial_y \equiv 0$, the familiar linear, nondispersive long wave equation (recast here in dimensional form)

$$\eta_{tt} - c_0^2\,\eta_{xx} = 0, \tag{1.4}$$

where $c_0 = (gh)^{1/2}$ and η is the surface displacement. To the next order, one finds, for waves travelling in the positive x-direction, the celebrated Korteweg-de Vries (KdV) equation (see LeBlond and Mysak, 1978, pp. 99-102 for the derivation):

$$\eta_t + c_0\,\eta_x + c_1\,\eta\,\eta_x + c_2\,\eta_{xxx} = 0\,, \tag{1.5}$$

where $c_0 = (gH)^{1/2}$ (as in (1.4)), $c_1 = 3c_0/2H$, and $c_2 = c_0 H^2/6$. Described by Kruskal (1978) as "... the simplest partial differential equation ... not covered by classical methods", its derivation actually goes back to the doctoral thesis of de Vries (1894) (see also Korteweg and de Vries (1895)). An interesting historical account of this equation is given by Miles (1981), and its mathematical properties have been reviewed by Miura (1976). The solitary wave solution of (1.5), however, was observed over a century ago by Russell (1844).

Note that (1.5) is written in dimensional form. If one reverts to suitable nondimensional variables x' and t', it can be shown that the wave amplitude is in fact a function of the slow-time variable $\tau = \varepsilon^{3/2}t'$ and the phase variable $\Theta = \varepsilon^{1/2}(x' - c_0' t')$, which describes slow spatial modulations in a frame which moves with linear (nondimensional) long-wave phase speed c_0'. In terms of τ and Θ, (1.5) can be written in the normal form

$$\eta_\tau' + c_1'\,\eta'\,\eta_\Theta' + c_2'\,\eta_{\Theta\Theta\Theta} = 0,$$

where $\eta'(\tau,\Theta)$, c_1', c_2' are all nondimensional.

For the strongly nonlinear case $c_2 = 0$ (or equivalently $U \gg 1$), the wave solutions of (1.5) steepen ahead of their crests and a hydraulic bore or shock wave is formed. For the infinitesimal amplitude case $c_1 = 0$ ($U \ll 1$), the wave solutions undergo phase dispersion according to the relation

$$\omega = c_0\,k(1 - \beta k^2)\,, \tag{1.6}$$

where $\beta = H^2/6$ and $(kH)^2 \ll 1$. Indeed, (1.6) can be readily derived from the finite-depth linear dispersion relation $\omega = (g\,k \tanh k\,H)^{1/2}$ under the shallow-water approximation $k\,H \ll 1$.

Korteweg and de Vries (1895) obtained periodic progressive cnoidal wave solutions of (1.5) which can be expressed in terms of the Jacobi elliptic function cn (see Miles, 1981, for details). In the limit of long wavelengths, this solution reduces to the familiar solitary wave solution, a nonlinear wave of permanent form and maximum amplitude (as compared to the cn solution):

$$\eta = a\,\text{sech}^2\,[\,(x - ct)/L\,], \qquad (1.7)$$

where a is a prescribed amplitude and

$$c = c_0 + \frac{1}{3}\,a\,c_1 = c_0(1 + \frac{a}{2H}), \qquad (1.8)$$

$$L = (12\,c_2/a\,c_1)^{1/2} = (4\,H^3/3a)^{1/2}\,. \qquad (1.9)$$

The solution (1.7) is illustrated in Figure 2. Inspection of (1.8) reveals that nonlinearity increases the propagation (phase) speed. Also, note that (1.9) implies that $L \propto a^{-1/2}$. From (1.7) and Figure 2 we observe that the effective "wavelength" of the solitary wave is $\lambda = 2L$.

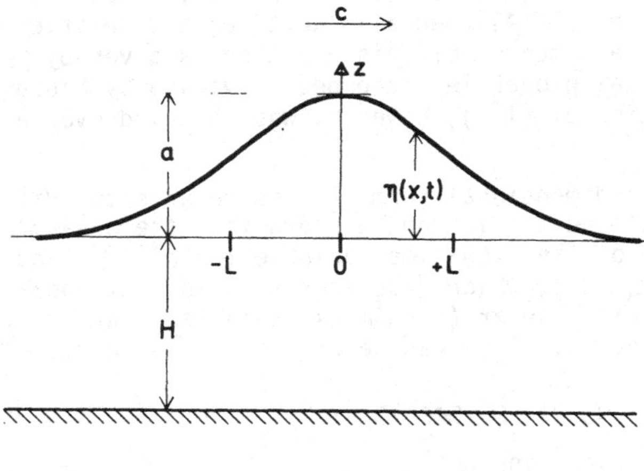

Figure 2

Surface solitary wave with amplitude a moving to the right with phase speed c in water of depth H. For illustrative purposes the amplitude has been exaggerated in this figure.

For a sufficiently smooth and localized initial wave form $\eta(x,0)$, as illustrated in Figure 3a, the asymptotic solution of (1.5) for $t \to \infty$ will consist of a group of *solitons*, trailed by a linear dispersive wave train as shown in Figure 3b (Gardiner et al., 1967). Solitons are nonlinear wave solutions of (1.5) which have the curious property that when two such waves "collide", (i.e., interact nonlinearly) and then emerge, they retain their original shape and amplitude, having undergone only a relative phase shift.

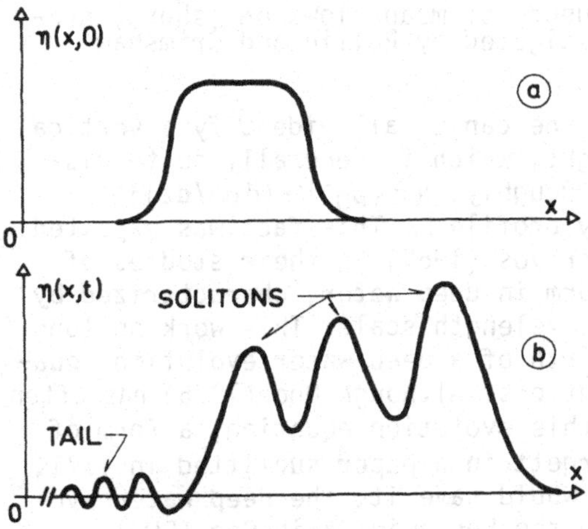

Figure 3

A sufficiently localized initial wave profile η(x,0) shown in ⓐ evolves into ⓑ, a group of solitons and a dispersive linear wave train or "tail".

The leading soliton in Figure 3b always has the largest amplitude, and accordingly, travels the fastest (see (1.8)). The second soliton has the second largest amplitude, and so on. Thus the soliton group tends to spread out, with each wave eventually taking the character of the solitary wave solution (1.7). The number of solitions that emerges from any initial profile η(x,0) is given by the number of zeros of the solution of a Schrödinger equation in which the potential well is given by η(x,0)! An elementary account of soliton wave properties is given in Osborne and Burch (1980). For a discussion of a new powerful technique for solving nonlinear evolution equations of the form (1.5), namely, the inverse scattering transform, see Miles (1981) and the recent book by Ablowitz and Segur (1981).

1.3 Classification of nonlinear internal wave theories

The theory of nonlinear internal waves is extensive, and any categorization of its different aspects is certainly not unique. In view of the above discussion of the KdV equation, it is simplest to begin with a discussion of shallow-water nonlinear internal waves. The theory of long solitary waves in a two-layer fluid was pioneered by Keulegan (1953) and Long (1956). Long solitary and cnoidal waves (undular bores) in a continuously stratified fluid have been studied by Benjamin (1966). Long internal waves which are dominated by nonlinearity are steepened ahead of the crest and are commonly found in lakes (Farmer, 1978) and also produce internal tidal bores (Cairns, 1967).

The Stokes expansion which is well known for finite depth (i.e., short) progressive surface waves (e.g., see LeBlond and Mysak, 1978, pp. 93-96) has been carried out for internal waves by Thorpe (1968), who also compared the theory with experiment. Holyer (1979) carried out the Stokes expansion for interfacial waves to very high order and calculated wave profile shapes for progressive waves moving at the interface between two deep fluids of different density. The influence of mean flows on (short) progressive internal waves has been investigated by Pullin and Grimshaw (1983 a,b).

In a continuously stratified fluid one can usually identify a vertical scale h, the stratification scale height, which is generally quite distinct from the total fluid depth H. [Roughly, $h \sim (\rho_0^{-1} \cdot (d\rho_0/dz))^{-1}$, where $\rho_0(z)$ is the basic state density profile]. This fact was expoited by Benjamin (1967) and by Davis and Acrivos (1967) in their studies of finite-amplitude waves of permanent form in deep water, characterized by $h/\lambda << 1$ and $\lambda/H \to 0$, where λ is the wavelength scale. This work on long nonlinear internal waves led to the birth of a deep-water evolution equation with algebraic solitary wave solutions. Although Ono (1975) has often been credited with the derivation of this evolution equation, a form of it was apparently first derived by Gargett in a paper submitted in 1971, but not published until 1976. As fate would have it, the deep water evolution equation is now commonly called the Benjamin-Davis-Ono (BDO) equation.

Joseph (1977) and Kubota et al. (1978) generalized the BDO equation to obtain the evolution equation for long internal waves of permanent form in a finite depth fluid (i.e., $\lambda \sim H$). Solitary waves for this case where also found by Joseph (1977).

Finally, each of these evolution equations (for shallow-water, deep water and finite-depth-water) have been generalized to include such effects as shear flow, variable topography, dissipation and compressibility. Grimshaw (1983) has provided a review of some of this work, and highlights will be given in Section 3. In Section 2 that follows, the (long-wave) evolution equations for the three basic cases given above are presented and their solitary wave solutions are described.

2. GOVERNING EVOLUTION EQUATIONS FOR LONG NONLINEAR INTERNAL WAVES

2.1 Generalized evolution equation

The various evolution equations describing long nonlinear internal waves of permanent form can be derived, to first order, from a single generalized evolution equation commonly called Whitham's equation (Whitham, 1967):

$$\eta_t + c_1 \eta\eta_x + \frac{\partial}{\partial x} \int_{-\infty}^{\infty} \eta(x',t) \left(\frac{1}{2\pi} \int_{-\infty}^{\infty} c(k) e^{ik(x-x')} dk \right) dx' = 0, \qquad (2.1)$$

where η measures the internal wave displacement field (e.g., $\eta \simeq \int w\,dt$, where w is the vertical velocity, or η is the interfacial displacement at a density discontinuity), c(k) is the linear phase speed (with c_0 being the linear *long-wave* phase speed), and c_1 is a functional of the k = 0 (vertical) eigenfunction.

According to Koop and Butler (1981), we can categorize the existing analytical nonlinear (solitary/soliton) wave studies in the following manner. Let λ be a horizontal scale for the wave (e.g., $\lambda = 2L$ for the solitary wave (1.7)), H be the total fluid depth and h be the scale height of the stratification (e.g., the thermocline thickness). In terms of these quantities, the three limiting cases of interest are:

(i) shallow-water theory

$$\lambda/H \gg 1, \qquad h/H \lesssim O(1) \; ; \tag{2.2}$$

(ii) deep water theory

$$\lambda/H \to 0, \qquad \lambda/h \gg 1 \; ; \tag{2.3}$$

(iii) finite-depth theory

$$\lambda/h \gg 1, \qquad h/H \ll 1 \, , \qquad (\text{i.e.}, \lambda \sim H). \tag{2.4}$$

Note that in addition to the most familiar classification according to the wavelength/depth ratio, the inequalities (2.2) - (2.4) also imply that for each case $\lambda/h \gg 1$: the waves are *long* compared to the intrinsic (vertical) scale of the stratification.

2.2 Korteweg-de Vries equation for shallow-water waves

For shallow-water internal waves that are characterized by (2.2), the linear dispersion law for waves travelling in the positive x-direction can be written as (Benjamin, 1966):

$$c(k) = c_0 - c_0 \, \beta \, k^2, \tag{2.5}$$

which of course is identical in form to the shallow-water surface wave case (1.6). Substitution of (2.5) into (2.1) gives the KdV equation

$$\eta_t + c_0 \, \eta_x + c_1 \, \eta\eta_x + c_2 \, \eta_{xxx} = 0, \tag{2.6}$$

where $c_2 = \beta c_0$. For a continuously stratified fluid there exists a countably infinite number of eigenspeeds $c^{(1)} < c^{(2)} < c^{(3)} < \dots$ corresponding to the different vertical baroclinic modes. Accordingly, (2.6) will then apply for each mode, with c_0, c_1, c_2 taking on the appropriate values for that mode. In the special case of the two-layer fluid (see Fig. 4), there is only one vertical mode (the gravest), and the coefficients c_0, c_1, c_2 take the form (correct to $O(\Delta\rho/\rho_2)$)

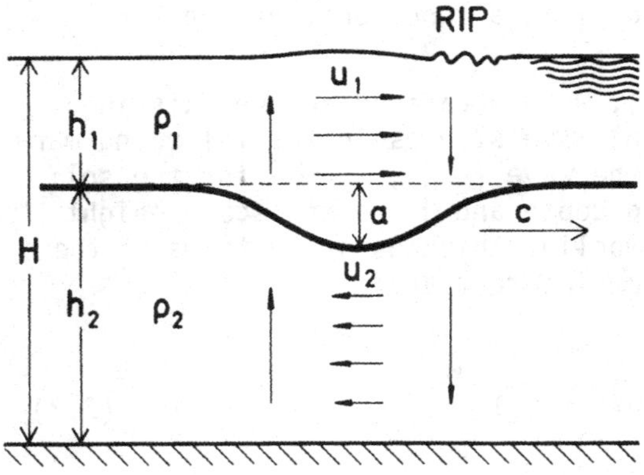

Figure 4

Internal solitary wave in a two-layer fluid with $h_2 > h_1$. Arrows indicate current pattern within the internal wave, resulting in the surface rip which leads the wave. When $h_2 < h_1$, the solitary wave is a wave of elevation, rather than a depression wave.

$$c_0 = (g' h_1 h_2/H)^{1/2}, \tag{2.7}$$

$$c_1 = -3 c_0(h_2 - h_1)/h_1 h_2 , \tag{2.8}$$

$$c_2 = c_0 h_1 h_2/6 , \tag{2.9}$$

where $g' = g(\rho_2 - \rho_1)/\rho_2 = g\Delta\rho/\rho_2$ is the reduced gravity. Comparison of (2.7)-(2.9) with the corresponding c_i's for the surface wave KdV equation (1.5) reveals that $c_1 = 0$ for the internal wave case when $h_1 = h_2$. That is, the nonlinear term in (2.6) vanishes when the layer depths are equal! This special case has been examined only recently (Segur and Hammack, 1983).

Another interesting and important point is that $c_1 < 0$ when $h_2 > h_1$, which is characteristic of most thermocline regions in the ocean and in lakes. As a consequence, the solitary wave solution for this case is a depression wave (Fig. 4). (When $h_2 < h_1$, $c_1 > 0$ and the solitary wave travels like a hump on the interface, as in Fig. 2). Explicitly, for $h_2 > h_1$ we have (Keulegan, 1953)

$$\eta = -a \,\mathrm{sech}^2 [(x-ct)/L], \tag{2.10}$$

where

$$c = c_0 - \frac{1}{3} a c_1 \tag{2.11}$$

$$L = (-12 c_2/a c_1)^{1/2}, \tag{2.12}$$

in which c_0, c_1, c_2 are given by (2.7)-(2.9). Since $c_1 < 0$, the effect of nonlinearity is again to increase the phase speed c. The upper and lower layer horizontal velocities corresponding to (2.10) take the form (Osborne and Burch, 1980)

$$u_1 = (c_0 a/h_1) \,\mathrm{sech}^2 [(x-ct)/L] \tag{2.13}$$

$$u_2 = -(c_0 a/h_2)\,\text{sech}^2\,[\,(x-ct)/L\,], \qquad (2.14)$$

which imply $h_1 u_1 + h_2 u_2 = 0$, as required for baroclinic motion. Associated with internal solitary waves are steep surface gravity waves which preceed the peak of the internal wave. Such "rips" occur because of nonlinear interactions between the internal and surface waves, and provide a means of identifying the internal waves from satellite images (Fig. 5).

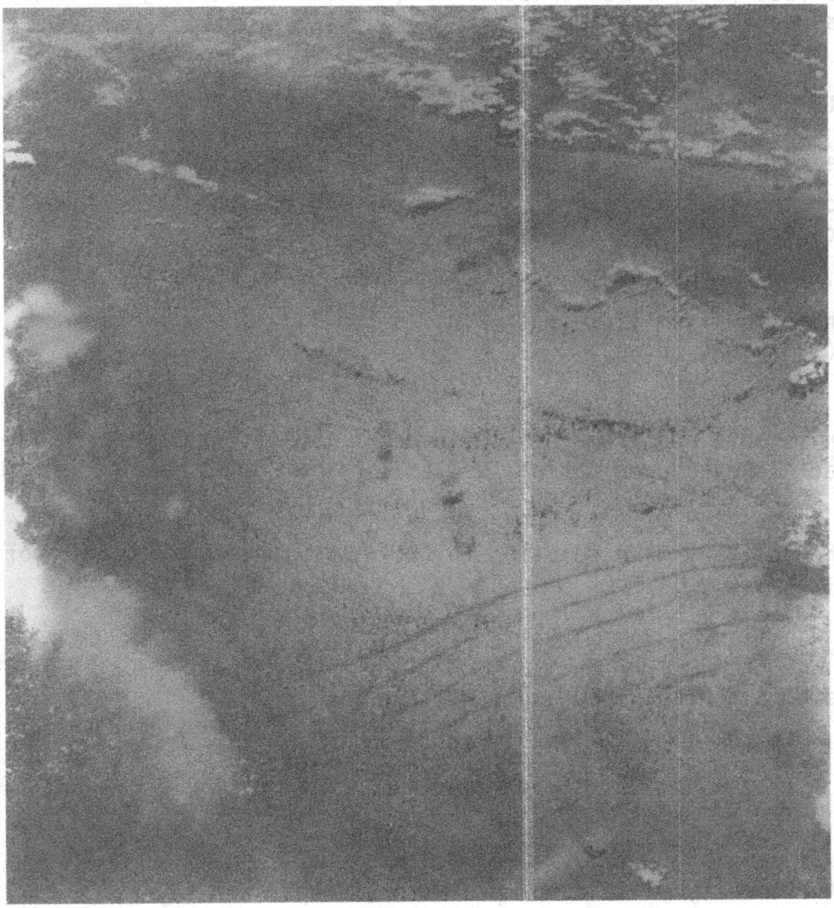

Figure 5

Satellite photographs, like this one taken miles above the earth by the Apollo-Soyuz vehicle, show the pattern on an internal wave group in the Andaman Sea, west of the Malay Peninsula. The waves, which occur regularly in these waters, have crests that are about 100 miles long and are moving northeast, toward the extreme southern shore of Thailand. Typically, the first wave is the largest and the most powerful. (From Locke, 1980).

2.3 Benjamin-Davis-Ono equation for deep-water waves

For deep water long internal waves characterized by (2.3), the linear dispersion law for small kh takes the form (Benjamin, 1967)

$$c = c_0 - c_0 \gamma |k| , \tag{2.15}$$

which is *linear* in k, in contrast to the quadratic law for shallow-water waves (see (2.5)). Substitution of (2.15) into (2.1) gives the Benjamin-Davis-Ono (BDO) equation

$$\eta_t + c_0 \eta_x + c_1 \eta\eta_x + c_2 \frac{\partial}{\partial x}\left(\int_{-\infty}^{\infty} \eta(x',t) \frac{1}{2\pi} \int_{-\infty}^{\infty} |k| e^{ik(x-x')} dk \right) dx' = 0, \tag{2.16}$$

where $c_2 = - c_0 \gamma$.

For the special case of a two-layer fluid with a deep lower layer, c_0 and c_1 are given by

$$c_0 = (g' h_1)^{1/2} , \qquad c_1 = - 3 c_0/h_1 , \tag{2.17}$$

which follow from (2.7) and (2.8) in the limit $h_2 \to \infty$. As shown by Benjamin (1967, eq. (2.5)),

$$c_2 = - c_0 \rho_1 h_1/2 \rho_2, \tag{2.18}$$

which is obtained from the finite-depth two-layer dispersion law for interfacial waves upon taking the limit $h_2 \to \infty$ and then invoking the long-wave approximation $k h_1 \ll 1$ (see (2.3)).

The solitary wave solution to (2.16) is the *Laourentzian* profile given by

$$\eta = \frac{a L^2}{(x-ct)^2 + L^2} , \tag{2.19}$$

where a is a prescribed amplitude and

$$c = c_0 + \frac{1}{4} a c_1, \tag{2.20}$$

$$L = - 4 c_2/c_1 a > 0. \tag{2.21}$$

For the two-layer fluid example with a deep lower layer, $c_2/c_1 > 0$ (see (2.17) and (2.18)) and therefore (2.21) implies that $a < 0$. Therefore (2.19) represents a depression wave, which is consistent with the shallow-water solitary wave solution for $h_2 > h_1$. Also, for this example it follows that $c > c_0$: finite-amplitude effects again increase the phase speed.

From (1.9) or (2.12) one can obtain the solitary-wave amplitude-wavelength scaling ($a L^2$) for shallow-water theory. This scaling implies

$$L/H = O[(a/H)^{-1/2}] = O(\varepsilon^{-1/2}). \tag{2.22}$$

For the deep-water solitary wave, on the other hand, we find (e.g., see (2.21), (2.17) and (2.18)

$$L/h = O[(a/h)^{-1}] = O(\varepsilon_d^{-1}), \tag{2.23}$$

where $\varepsilon_d = a/h$. Thus the amplitude-wavelength scalings are quite distinct for the two theories, with (2.23) implying the existence of much longer waves of a given amplitude than would occur in the shallow-water theory. This important distinction was used by Koop and Butler (1981) to establish the validity of each theory in an experimental study of solitary internal waves.

2.4 Finite-depth equation for intermediate wavelengths

For the finite-depth case (2.4), the dispersion law (for small kh and d/H) can be written as (Joseph, 1977)

$$\frac{c(k) - c_0}{c_0} = \beta(k \coth kH - \frac{1}{H}), \tag{2.24}$$

which is transcendental in the wavenumber k. The corresponding evolution equation takes the form (Koop and Butler, 1981)

$$\eta_t + c_0 \eta_x + c_1 \eta\eta_x + c_2 \frac{\partial^2}{\partial x^2} \int_{-\infty}^{\infty} \eta(x',t) \left(\coth \frac{\pi}{2H}(x-x') - \operatorname{sgn} \frac{(x-x')}{H}\right) dx' = 0, \tag{2.25}$$

where $c_2 = c_0 \beta/2H$.

The solitary wave solution of (2.25) has the following form (Joseph, 1977):

$$\eta = a\left(\cosh^2 \frac{(x-ct)}{L} + \tan^2 \frac{H}{L} \sinh^2 \frac{(x-ct)}{L}\right)^{-1}, \tag{2.26}$$

where

$$c = c_0 - 2c_2\left(1 - \frac{2H}{L} \cot \frac{2H}{L}\right), \tag{2.27}$$

$$L = -8 c_2/c_1 \, a \cot \frac{H}{L}. \tag{2.28}$$

Note from (2.28) that the amplitude-wavelength scaling (aL^2) is now transcendental, in contrast to the algebraic relations in the previous two theories.

In conclusion it is worth remarking that all the evolution equations described above have soliton as well as solitary wave solutions. The various references for these solutions are given in Koop and Butler (1981).

3. MORE GENERAL THEORIES

3.1 Stratified shear flows

Nonlinear internal waves that occur in thermocline regions in the ocean or near atmospheric fronts will be significantly influenced by mean flows with vertical shear. A first assessement of the influence of a mean shear $U(z)$ on nonlinear internal waves in shallow water was made by Lee and Beardsley (1974). They studied both two-layer and continuously stratified cases, but in the latter they did not consider the possibility of singular modes (i.e., modes having a critical level $z = z_0$ where the phase speed is equal to the mean flow velocity: $c = U(z_0)$). These as well as the regular modes were studied by Maslowe and Redekopp (1980); also these authors derived a generalization of the BDO equation which includes a shear flow. The finite-depth theory for nonlinear waves in a shear flow has been developed by Tung et al. (1981), who in particular showed that when critical levels are present in the domain, wave-mean flow interaction may take place, which would prevent the formation of waves of permanent form.

In the above studies the governing evolution equations have constant coefficients. In many situations of interest, however, the basic shear flow on which the waves travel is slowly varying in both time and horizontal coordinates. This is especially true if the mean flow is due to the tides, for example, ebbing and flowing over a variable bottom topography. Grimshaw (1981a) developed a theory for variable coefficient KdV and BDO equations for long nonlinear waves on a shear flow which, in addition to the usual dependence on the vertical coordinate, is horizontally and temporally variable. Slowly varying solitary wave solutions for the variable coefficient KdV equation can be found by using the method developed in Grimshaw (1979). For the BDO equation, the slowly varying solitary wave solution has been constructed by Grimshaw (1981b), who shows that behind the solitary wave a shelf develops. (See also Grimshaw (1983) for a review of the theory of slowly varying solitary waves).

3.2 Channel of arbitrary cross-section

To properly describe long nonlinear internal waves in lakes, Farmer (1978) has suggested that changes in topography in the direction of propagation could be important. To study this problem, Grimshaw (1978) derived a set of evolution equations for a two-layer fluid in a channel of arbitrary cross-section. In particular, expressions for the phase speed of the waves were derived in terms of geometric properties of the cross-section. In a regime where dispersion balanced weak nonlinearity, the soliton theoretical phase speed of 0.24 m s^{-1} which takes into account the decreasing width and depth of Babine Lake, British Columbia, was found to agree very favourable with the observed phase speed of 0.2 m s^{-1} reported by Farmer (1978).

3.3 Dissipation

The attenutation of solitary internal waves has been studied via two mechanisms:

(i) energy dissipation in the viscous boundary layer along a solid surface and within the thermocline or interfacial regime (e.g., see Koop and Butler, 1981; Grimshaw, 1981a), and

(ii) radiation damping via downward propagating internal waves into a weakly stratified lower layer (e.g., see Maslowe and Redekopp, 1980; Grimshaw, 1981b).

The first mechanism is far from negligible in experimental studies of internal solitary waves. For example, Koop and Butler (1981) found that in the majority of their experiments, viscous stresses produce a 50 per cent attenuation in the internal wave amplitude as the disturbance propagates one length of the tank. Radiation damping is expected to be quite significant when a shallow-water solitary wave travels away from a two-layer shelf region into the deep water where the lower layer is stratified (see Figure 6).

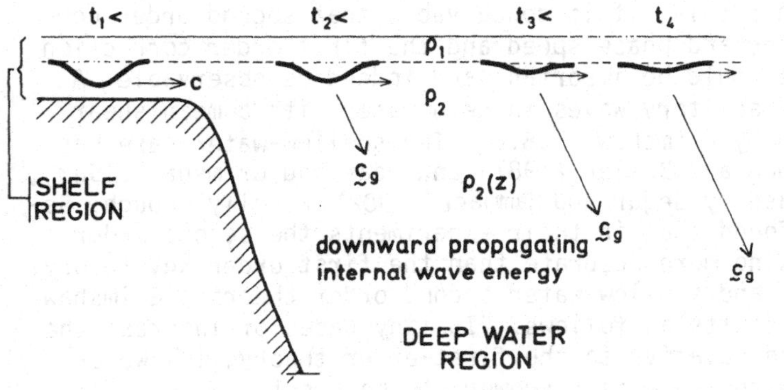

Figure 6

Radiation damping of a solitary wave travelling into the deep ocean.

3.4 Fission

Over many continental shelf regions in the world, groups of soliton-like internal waves are often seen propagating toward the shore (e.g., see Figure 5). These results suggest that the waves originate from the deep ocean. Assuming that in the deep ocean a solitary wave is incident upon a given shelf, then such soliton groups could be produced by fission (Djordjevic and Redekopp, 1978), which is illustrated in Figure 7. Roughly speaking, when a deep-water solitary wave reaches the shallow-shelf, it suddenly finds itself in an environment in which it can no longer exist as a solitary wave. Therefore, the localized wave profile at the edge of the shelf serves as an initial condition $\eta(x,0)$ which then evolves into a group of solitons, in a manner like that illustrated in Figure 3.

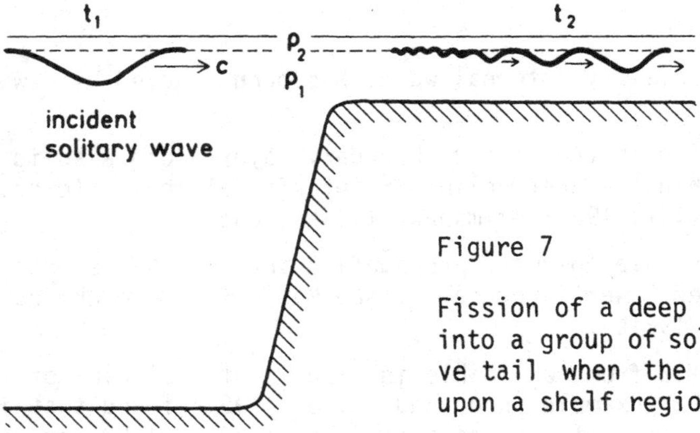

Figure 7

Fission of a deep water solitary wave into a group of solitons and dispersive tail when the former is incident upon a shelf region.

3.5 Second-order theories

Many of the solitary waves observed in the ocean, atmosphere, in lakes and in laboratory experiments, have quite large amplitudes, characterized by $\varepsilon = 0.5 - 0.6$ (see (1.1)). Since all the above theories involve expansions to first order in ε only, it is conceivable that second order corrections to the amplitude and phase speed and the first order correction to the wavelength scale could be important and indeed be observable. A second-order theory for solitary waves in deep water with a mean shear flow has been developed by Grimshaw (1981c). The shallow-water case has been investigated by Koop and Butler (1981) and Gear and Grimshaw (1983), and the finite-depth case by Segur and Hammack (1982). Funnily enough, Segur and Hammak (1982) found that in their experiments the second-order finite-depth theory was no more accurate than the first order KdV theory. For both the deep water and shallow-water second order theories Grimshaw (1983) summarized the results as follows: "In many cases of interest the phase speed is decreased relative to the first-order theory, the wavelength is increased and the wave is narrower at the crest".

3.6 Other effects

All the above theories are for nonrotating fluids. For time scales of a few hours or less, the Coriolis force can be safely neglected. This seems to be the case for many oceanic and lake observations of internal surges and solitary waves. In a numerical study of nonlinear waves in large lakes, Mørk et al. (1980) included the Coriolis force, which was found to produce a pronounced transverse tilting of the interface of a long-period (nonlinear) internal seiche. A similar cross-lake tilt was found by Oman (1982) in a numerical rotating nonlinear model of internal seiches in Lake of Zurich. Solitary waves as such, however, were not studied in either of these papers.

Another restriction of the above theories is that the stratified fluids considered are incompressible. Recently, with the atmosphere in mind, Grimshaw (1980/81) derived evolution equations for long nonlinear internal waves in a compressible fluid. Both KdV and BDO equations are derived, for both dry and moist atmospheres. Generally the effects of compressibility are small, but measurable, and are manifested mainly in the nonlinear term of the evolution equation.

4. OBSERVATIONS

4.1 Continental shelf regions

Among the first observations of long large-amplitude internal waves which have been interpreted as solitary waves are those of Halpern (1971) in Massachusetts Bay (see also Lee and Beardsley (1974) and Haury et al. (1979). Solitary internal waves have also been observed off La Jolla by Winant (1974), in the New York Bight, Indian Ocean and Carribean by Apel et al. (1975), in the Andaman Sea (near northern Sumatra) by Osborne and Burch (1980), in the Davis Strait in northern Canada (Mysak and Hodgins, 1981), in the Sulu Sea west of the Philipines (J.R. Holbrook, pers. comm., 1981), and on the Nova Scotia shelf (Sandstrom and Elliott, 1982). Many of the above authors argue that these waves which are often seen in satellite imagery, could be generated by the scattering of barotropic tidal energy into internal wave energy by large bottom topography changes (e.g., near the edge of the shelf) or by tidal flow through inter-island passages. However, the possibility of fission-generated waves on the shelf (Fig. 7) appears to have been overlooked, and should be regarded as a serious candidate.

4.2 Fjords and straits

Due to the large tidal currents at the narrow and shallow entrances to the Strait of Georgia, B.C., and the B.C. coastal fjords, large-amplitude internal waves are often generated in these regions. Paricularly fine examples in the Strait of Georgia are given in Gargett (1976), and in Knight Inlet by Farmer and Smith (1978). Similar waves have also been reported in the Strait of Gibraltar by Ziegenbein (1970), and are also probably common in the Scandinavian fjords.

4.3 Atmosphere

Atmosphere observation of solitary waves are less prolific than in the ocean and in lakes, but recently Christie et al. (1978, 1979) have documented a large number of solitary waves propagating on the noctunal inversion. Also, very large amplitude atmospheric solitary waves have been reported by Clarke et al. (1981) in association with the phenomenon of the Morning Glory of the Gulf of Carpentaria, in northern Australia.

4.4 Laboratory experiments

One of the first laboratory demonstrations of internal solitary waves was carried out by Davis and Acrivos (1967) who investigated the propagation of such waves in a thin layer of stratified fluid which was contained between two deep layers of different density. One form of the deep-water solitary wave theory was also developed in this paper, independent of that worked out at the same time by Benjamin (1967).

Maxworthy (1979) used a simple laboratory model to illustrate how barotropic tidal flow over a three-dimensional obstacle (e.g., a model of a sill in a fjord entrance) could generate a train of solitary waves. In another set of experiments Maxworthy (1980) generated nonlinear internal waves from the gravitational collapse of a mixed region (see also Kao and Pao, 1980).

An experimental study of solitary wave shape and amplitude-wavelength scale in a two-fluid system was conducted by Koop and Butler (1981). In particular, a careful comparison between a shallow- and a deep water configuration was made. In Segur and Hammack (1982), the KdV theory and the finite-depth theories were tested experimentally also using a two-fluid system. They found that the KdV equation predicts the shapes of the measured solitons with remarkable accuracy, much better than does the finite-depth equation.

4.5 Lakes

During the past decade or so there have been many observations of large-amplitude internal waves in lakes, and accordingly, our discussion here is not intended to be a catalogue of all such observations, but rather a brief discussion of the typical generation and evolution of such waves, together with a specific case study from the Lake of Zurich. What is observed in many long lakes is that following a strong gust of alongshore winds, the thermocline at one end of the lake is depressed and an internal surge is formed. Initially, the surge steepens owing to nonlinear effects, but as it propagates down the lake it evolves with a train of shorter period waves which often tend to have the appearance of a group of solitons or solitary waves (if spaced sufficiently far enough apart). Observations which show at least part of this pattern of development have been made by Thorpe et al. (1972) in Loch Ness, by Hunkins and Fliegel (1973) in Seneca Lake, New York State, by Farmer (1978) in Babine Lake, British Columbia, by Mørk et al. (1980) in Lake Mjøsa, Norway and by Mortimer and Horn (1982) in the Lake of Zurich. In very long lakes (e.g., Babine Lake), the waves tend to disappear at the far end of the lake because of dissipation or dispersion. However, in some of the shorter lakes, the surges are seen to travel back and forth along the lake several times, either due to being in part resonance with the wind (Thorpe,

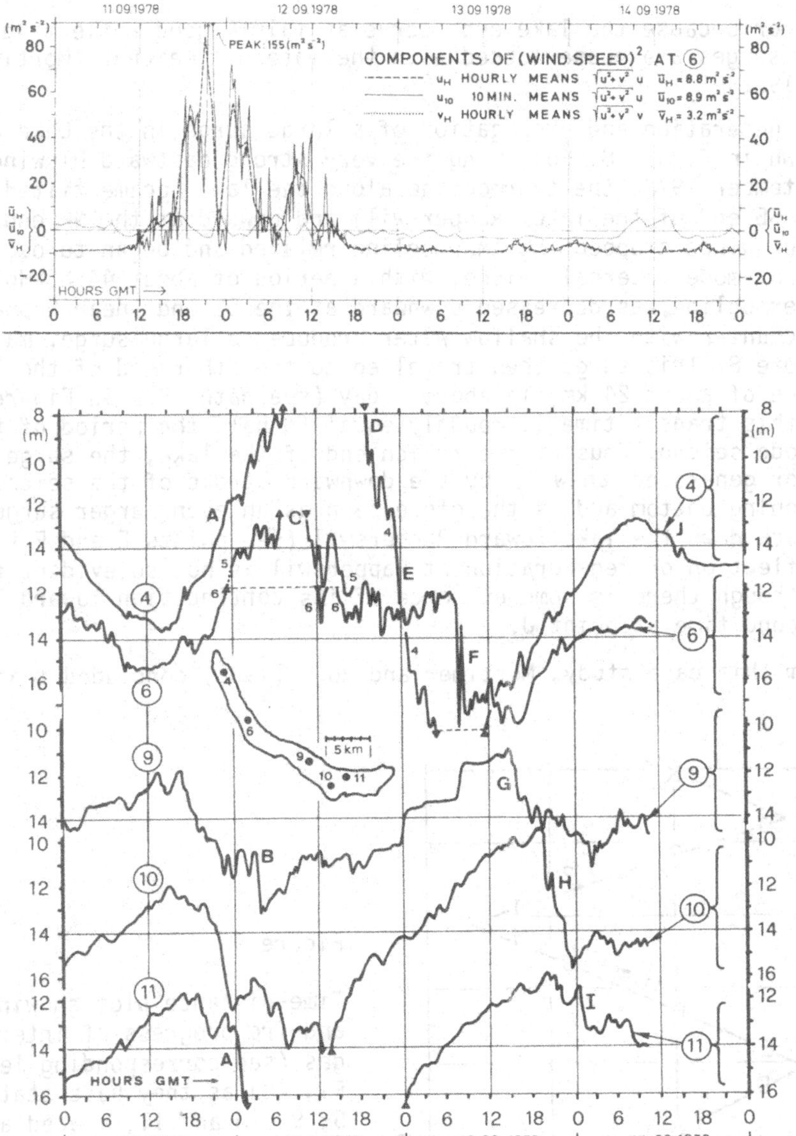

Figure 8

Depth variation of the 10^0C isotherm at mooring stations 4, 6, 9, 10 and 11 from September 11-14, 1978, in the Lake of Zurich. Mooring station numbers, placed near the corresponding traces, are circled. Letters A to J refer to the passages of internal surges past the indicated stations. The upper panel displays the u(EW) and v(NS) components of wind speed squared, 3 m above the lake surface at station 6. (From Mortimer and Horn, 1982).

1974), or because the lake ends serve as fairly good reflecting walls and/or the surges are phase-locked with the internal seiches (Mortimer and Horn, 1982).

The generation and propagation of a large surge in the Lake of Zurich is shown in Figure 8. Following the very strong eastward blowing winds on 11 September 1978, the thermocline along the lake became tilted, downward at the SE end of the lake (Rapperswil) and upward at the NW end (Zurich). When the winds stopped the thermocline relaxed and began to oscillate as a gravest-mode internal seiche, with a period of about 44 h. However, as the thermocline was depressed downward at the SE end (near Rapperswil) its encounter with the shallow water produced a large surge, marked by A in Figure 8. This surge then travelled to the other end of the lake (a distance of about 24 km) in about 1 day (see path ABCD in Figure 9). However, this transit time is roughly equal to half the period of the gravest-mode seiche. Thus at the Zurich end of the lake, the surge was amplified (or generated anew ?) by the downward stroke of the seiche against the sloping bottom and is therefore seen as an even larger surge travelling back down the lake toward Rapperswil (see points E and F in Figure 8). Its reflection or regeneration at Rapperswil is not so evident from the data, though there is some evidence of its continuation toward Zurich for the second time at point J.

From this case study, Mortimer and Horn (1982) concluded that the Zu-

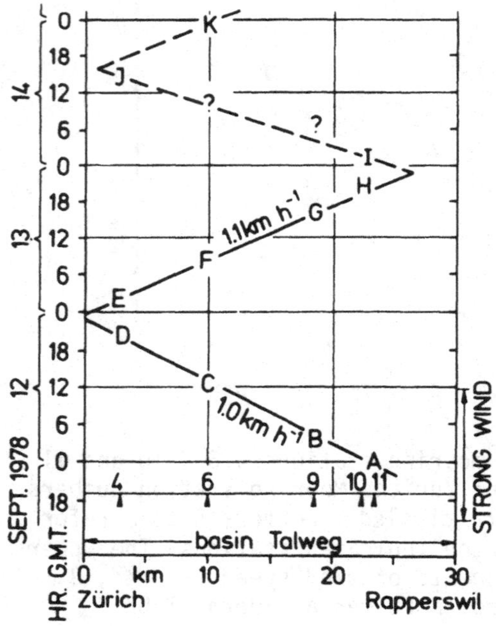

Figure 9

Time-distance plot showing the to-and-fro progress of internal surges (see corresponding letters in Fig. 8) as they pass stations 4, 6, 9, 10 and 11, spaced along the basin Talweg (constituting the distance scale). The question marks refer to doubtful signals. Straight lines serve to indicate approximate surge speed. (From Mortimer and Horn, 1982).

rich end of the lake is potentially a stronger surge generator than the Rapperswil end, but it is still an open question as to whether the surges are merely reflected at each end of the lake or generated anew by the downwelling stroke of the thermocline, approximately in phase with the gravest mode seiche. Another open question is why the speed of surge propagation (about 24 km day^{-1}) is *less* than the average long-wave internal wave speed c_0 (about 30 km day^{-1}). For solitary wave models (perhaps not applicable here), nonlinearity tends to give a phase speed larger than c_0. Perhaps the discrepancy is due to second-order effects (which decrease the phase speed), or due to the fact that at the Rapperswil end of the lake the stratification, if modelling by a two-layer fluid, has $h_1 = h_2$, in which case $c_1 = 0$ and nonlinearity disappears in the KdV equation. Thus another type of theory has to be considered, such as that developed in Segur and Hammack (1983). Also, perhaps fission-type processes are important as the surge approaches Rapperswil since there is a sudden rise in the bottom topography about 6-8 km from this end of the lake. Clearly, there is tremendous scope for future theoretical work related to the Lake of Zurich internal surges.

5. SUMMARY

In this chapter I have attempted to give a very brief overview of recent theoretical developments in nonlinear internal waves, with special focus on long solitary waves in shallow, deep and finite-depth fluids. Also, a number of nonlinear wave observations have been reported and a case study of a large-amplitude surge in the Lake of Zurich has been discussed. However, due to lack of space some important nonlinear wave problems have been omitted, such as Kelvin-Helmholtz instabilities and billow formation (e.g., see Thorpe et al., 1977).

On reviewing the literature of lake observations of nonlinear internal waves, it again became clear to me that there has been relatively little attempt to compare observation and theory carefully. Also, many observations remain unexplained, such as those outlined at the end of the previous section. Thus it is evident that there are many challenging lake problems involving nonlinear internal waves yet to be tackled, both theoretically and observationally.

ACKNOWLEDGEMENTS

These lecture notes were prepared during a sabbatical visit (1982-83) to the limnology group of the Laboratory of Hydraulics, Hydrology and Glaciology (VAW), Swiss Federal Institute of Technology, Zurich. The author is grateful to Dr. K. Hutter and his group for their kind hospitality. The author also wishes to thank Dr. R. Grimshaw for carefully reading a first draft of this paper. This work was supported by the VAW and the Izaak Walton Killam Memorial Fund for Advanced Studies.

REFERENCES

Ablowitz, M. and H. Segur, 1981. Solitons and the Inverse Scattering Transform. Soc. Industr. Appl. Math., 425 pp.

Apel, J.R., H.M. Byrne, J.R. Proni and R.L. Charnell, 1975. Observations of oceanic internal and surface waves from the earth resources technology satellite. J. Geophys. Res., 80, pp. 865-881.

Benjamin, T.B., 1966. Internal waves of finite amplitude and permanent form. J. Fluid Mech., 25, pp. 241-270.

Benjamin, T.B., 1967. Internal waves of permanent form in fluids of great depth. J. Fluid Mech., 29, pp. 559-592.

Cairns, J.L., 1967. Asymmetry of internal tidal waves in shallow coastal waters. J. Geophys. Res., 72., pp. 3563-3565.

Christie, D.R., K.J. Muirhead and A.L. Hales, 1978. On solitary waves in the atmosphere. J. Atmos. Sci., 35, pp. 805-825.

Christie, D.R., K.J. Muirhead and A.L. Hales, 1979. Intrusive density flows in the lower troposphere: a source of atmospheric solitons. J. Geophys. Res., 84, pp. 4959-4970.

Clarke, R.H., R.K. Smith and D.G. Reid, 1981. The Morning Glory of the Gulf of Carpentaria: an atmospheric undular bore. Mon. Weather Rev., 109, pp. 1726-1750.

Davis, R.E. and A. Acrivos, 1967. Solitary internal waves in deep water. J. Fluid Mech., 29, pp. 593-607.

De Vries, G., 1894. Bijdrage tot de kennis der lange golven. Doctoral dissertation, Univ. of Amsterdam.

Djordjevic, V.D. and L.D. Redekopp, 1978. The fission and disintegration of internal solitary waves moving over two-dimensional topography. J. Phys. Oceanogr., 8, pp. 1016-1024.

Farmer, D.M., 1978. Observations of long nonlinear internal waves in a lake. J. Phys. Oceanogr. 8, pp. 63-73.

Farmer, D.M. and J.D. Smith, 1978. Nonlinear internal waves in a fjord. In: "Hydrodynamics of Estuaries and Fjords", edit. by J.C.J. Nihoul, pp. 465-493, Elsevier.

Gardiner, C.S., J.M. Greene, M.D. Kruskal and R.M. Miura, 1967. Method for solving the Korteweg-de Vries equation. Phys. Rev. Lett., 19, pp. 1095-1097.

Gargett, A.E., 1976. Generation of internal waves in the Strait of Georgia, British Columbia. Deep-Sea Res., 23, pp. 17—32.

Gear, J.A. and R. Grimshaw, 1983. A second-order theory for solitary waves in shallow fluids. Phys. Fluids, 26, pp. 14-29.

Grimshaw, R., 1978. Long nonlinear internal waves in channels of arbitrary cross-section. J. Fluid Mech., 86, pp. 415—431.

Grimshaw, R., 1979. Slowly varying solitary waves I. Korteweg-de Vries equation. Proc. Roy. Soc. London, A, 368, pp. 359-375.

Grimshaw, R., 1980/81. Solitary waves in a compressible fluid. Pure Appl. Geophys., 119, pp. 780-797.

Grimshaw, R., 1981a. Evolution equations for long, nonlinear internal waves in stratified shear flows. Studies Appl. Math., 65, pp. 159-188.

Grimshaw, R., 1981b. Slowly varying solitary waves in deep fluids. Proc. Roy. Soc. London, A, 376, pp. 319-332.

Grimshaw, R., 1981c. A second-order theory for solitary waves in deep fluids. Phys. Fluid, 24, pp. 1611-1618.

Grimshaw, R., 1983. Solitary waves in density stratified fluids. In: "Nonlinear Deformation Waves". IUTAM Symposium, Tallinn, 1982, ed. U. Nigul, J. Engelbrecht, pp. 431-447, Springer.

Halpern, D., 1971. Observations of short-period internal waves in Massachusetts Bay. J. Marine Res., 29, pp. 116-132.

Haury, L.R., M.G. Briscoe and M.H. Orr, 1979. Tidally generated internal wave packets in Massachusetts Bay. Nature, 278, pp. 312-317.

Holyer, J.Y., 1979. Large amplitude interfacial waves. J. Fluid Mech., 93, pp. 433-448.

Hunkins, K. and M. Fliegel, 1973. Internal modular surges in Seneca Lake: a natural occurrence of solitons. J. Geophys. Res., 78, pp. 539-548.

Joseph, R.I., 1977. Solitary waves in a finite-depth fluid. J. Phys., A, 10, pp. L 225-227.

Kao, T.W. and H-P. Pao, 1980. Wake collapse in the thermocline and internal solitary waves. J. Fluid Mech., 97, pp. 117-127.

Keulegan, G.H., 1953. Characteristics of internal solitary waves. J. Res. Nat. Bureau of Standards, 51, pp. 133-140.

Koop, C.G. and G. Butler, 1981. An investigation of internal solitary waves in a two-fluid system. J. Fluid Mech., 112, pp. 225-251.

Korteweg, D.J. and G. de Vries, 1895. On the change of form of long waves advancing in a rectangular channel, and on a new type of long stationary waves. Phil. Mag., 39, pp. 422-443.

Kruskal, M.D., 1978. The birth of the soliton. In: "Nonlinear Evolution Equations Solvable by the Spectral Transform", edit. by F. Calogero, pp. 1-8, Pitman.

Kubota, T., D.R.S. Ko and L.D. Dobbs, 1978. Weakly-nonlinear, long internal gravity waves in stratified fluids of finite depth. AIAA J. Hydronautics, 12, pp. 157-165.

LeBlond, P.H. and L.A. Mysak, 1978. Waves in the Ocean. Elsevier, 602 pp.

Lee, C-Y. and R.C. Beardsley, 1974. The generation of long nonlinear internal waves in a weakly stratified shear flow. J. Geophys. Res., 79, pp. 453-462.

Locke, L., 1980. Waves beneath the sea. The Lamp, summer 1980, pp. 21-23.

Long, R.R., 1956. Solitary waves in one- and two-fluid systems. Tellus, 8, pp. 460-471.

Maslowe, S.A. and L.G. Redekopp, 1980. Long nonlinear waves in stratified shear flows. J. Fluid Mech., 101, pp. 321-348.

Maxworthy, T., 1979. A note on the internal solitary waves produced by tidal flow over a three-dimensional ridge. J. Geophys. Res., 84, pp. 338-346.

Maxworthy, T., 1980. On the formation of nonlinear internal waves from the gravitational collapse of mixed regions in two and three dimensions. J. Fluid Mech., 96, pp. 47-64.

Miles, J.W., 1981. The Korteweg-de Vries equation: a historical essay. J. Fluid Mech., 106, pp. 131-147.

Miura, R.M., 1976. The Korteweg-de Vries equation: a survey of results. SIAM Review, 18, pp. 412-459.

Mørk, G., B. Bjevik and S. Holte, 1980. Long internal waves in lakes. In: "Proceedings of Second International Symposium on Stratified Flows", Vol. 2, pp. 988-997, Trondheim.

Mortimer, C.H. and W. Horn, 1982. Internal wave dynamics and their implications for plankton biology in the Lake of Zurich. Vierteljahresschrift der Naturforschenden Gesellschaft, Zürich, 127, pp. 299-318.

Mysak, L.A. and D.O. Hodgins, 1981. Preliminary study of internal solitary waves in Davis Strait. Report prepared for Aquitane Co. of Canada Ltd., 21 pp.

Oman, G., 1982. Das Verhalten des geschichteten Zürichsees unter äusseren Windlasten. Mitteilung der Versuchsanstalt für Wasserbau, Hydrologie und Glaziologie, ETH, Zürich, 185 pp.

Ono, H., 1975. Algebraic solitary waves in stratified fluids. J. Phys. Soc. Japan, 39, pp. 1082-1091.

Osborne, A.R. and T.L Burch, 1980. Internal solitons in the Andaman Sea. Science, 208, pp. 451-460.

Pullin, D.I. and R.H.J. Grimshaw, 1983a. Nonlinear interfacial progressive waves near a boundary in a Boussinesq fluid. Phys. Fluid, 26, pp. 897-905.

Pullin, D.I. and R.H.J. Grimshaw, 1983b. Interfacial progressive gravity waves in a two-layer shear flow. Phys. Fluids, in press.

Russell, J.S., 1844. Report on waves. Rep. Meet. Brit. Assoc. Adv. Sci., 14 th, York, pp. 311-390, John Murray.

Sandstrom, H. and J.A. Elliott, 1982. Internal tide and solitons on the Scotian shelf - a nutrient pump at work. Unpublished manuscript.

Segur, H. and J.L. Hammack, 1982. Soliton models of long internal waves. J. Fluid Mech., 118, pp. 285-304.

Segur, H. and J.L. Hammack, 1983. Long internal waves of moderate amplitude. III: Layers of equal depth. Unpublished preprint.

Thorpe, S.A., 1968. On the shape of progressive internal waves. Phil. Trans. Roy. Soc. London, A, 263, pp. 563-614.

Thorpe, S.A., 1974. Near-resonant forcing in a shallow two-layer fluid: a model for the internal surge in Loch Ness. J. Fluid Mech., 69, pp. 509-527.

Thorpe, S.A., A. Hall and I. Crofts, 1972. The internal surge in Loch Ness. Nature, 237, pp. 96-98.

Thorpe, S.A. A.J. Hall, C. Taylor and J. Allen, 1977. Billows in Loch Ness. Deep-Sea Res., 24, pp. 371-379.

Tung, K.K., D.R.S. Ko and J.J. Chang, 1981. Weakly nonlinear internal waves in shear. Studies Appl. Math., 65, pp. 189-221.

Whitham, G.B., 1967. Non-linear dispersive waves. Proc. Roy. Soc. London, A, 299, pp. 6-25.

Winant, C.D., 1974. Internal surges in coastal waters. J. Geophys. Res., 79, pp. 4523-4526.

Ziegenbein, J., 1970. Spatial observations of short, internal waves in the Strait of Gibraltar. Deep-Sea Res., 17, pp. 867-875.

Hydrodynamics of Lakes: CISM Lectures
edited by K. Hutter, 1984
Springer Verlag Wien-New York

VERTICAL STRUCTURE OF CURRENT IN HOMOGENEOUS AND STRATIFIED WATERS

N.S. Heaps
Institute of Oceanographic Sciences,
Bidston Observatory,
Merseyside, L43 7RA, England.

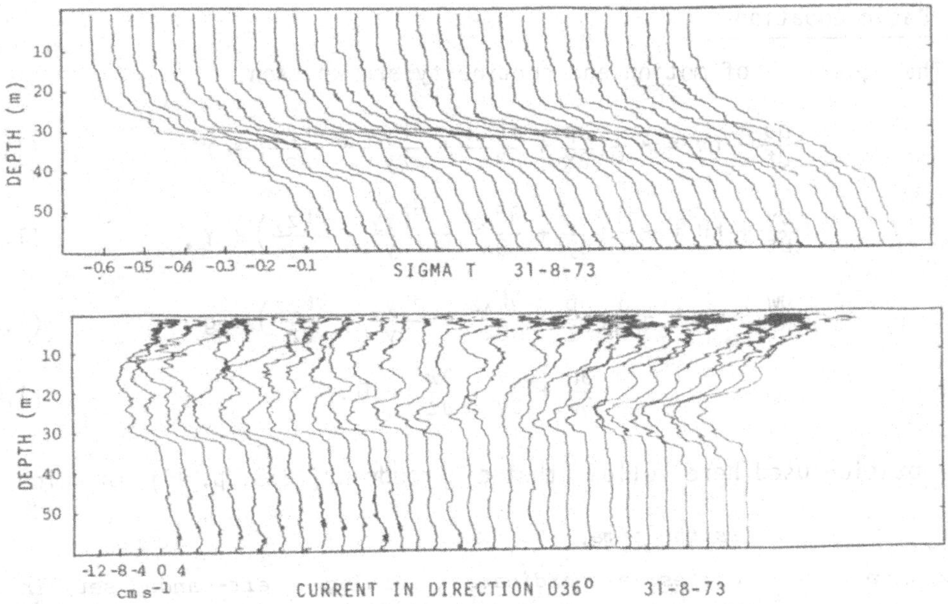

Vertical profiles of density (σ_t) and current at a position in Loch Ness, measured at regular intervals during an eight-hour period on 31 August 1973. Each successive density profile is displaced by 0.05 σ_t units to the right, and each velocity profile by 4 cm s^{-1}. From "Turbulence and Mixing in a Scottish Loch", by S.A. Thorpe (Philosophical Transactions of the Royal Society of London, A, Vol. 286, p. 137, 1977). Reproduced by kind permission of the Royal Society.

1. HYDRODYNAMIC EQUATIONS WITH APPLICATION TO A NARROW LAKE UNDER STEADY WIND

In this section the hydrodynamic equations are formulated, mainly in order to state basic principles and introduce a notation. Simple solutions are then developed for water movements in a narrow rectangular basin subjected to steady wind directed along its length. Vertical structures of current are derived for both one- and two-layered systems representing, respectively, a lake during conditions of winter homogeneity and summer stratification. In spite of their simplicity, for the most part achieved by linearization, the use of constant coefficients of eddy viscosity and the neglect of the Coriolis force, the solutions illustrate some important facts about the dynamics of wind-driven flows in a long narrow lake. Perhaps the main interest of the analysis lies in the actual construction of closed solutions, satisfying appropriate boundary conditions, for lake circulation.

1.1 Basic equations

The equations of motion and continuity are written

$$\frac{Du}{Dt} - fv = -\frac{1}{\rho}\left(\frac{\partial p}{\partial x} + \frac{\partial F_{xx}}{\partial x} + \frac{\partial F_{yx}}{\partial y} + \frac{\partial F_{zx}}{\partial z}\right) + X, \tag{1.1}$$

$$\frac{Dv}{Dt} + fu = -\frac{1}{\rho}\left(\frac{\partial p}{\partial y} + \frac{\partial F_{xy}}{\partial x} + \frac{\partial F_{yy}}{\partial y} + \frac{\partial F_{zy}}{\partial z}\right) + Y, \tag{1.2}$$

$$\frac{Dw}{Dt} = -\frac{1}{\rho}\left(\frac{\partial p}{\partial z} + \frac{\partial F_{xz}}{\partial x} + \frac{\partial F_{yz}}{\partial y} + \frac{\partial F_{zz}}{\partial z}\right) + g, \tag{1.3}$$

$$\frac{\partial u}{\partial x} + \frac{\partial v}{\partial y} + \frac{\partial w}{\partial z} = 0. \tag{1.4}$$

The notation used here follows that of Proudman (1953, p. 96) in which

- t — *is the time,*
- x, y, z — *Cartesian coordinates, forming a left-handed set, in which x, y are measured in the horizontal plane of the undisturbed water surface and z is depth below that surface,*
- u, v, w — *the components of current at depth z in the directions of increasing x, y, z respectively,*
- $F_{xx}, F_{yx}, \ldots$ — *components of internal stress due primarily to tubulence, where F_{rs} is the stress in the s-direction acting over a plane at right angles to the r-direction on the*

r - 0 side of that plane,

ρ *the density of the water,*

p *the pressure in the water,*

f *the geostrophic coefficient, equal to* $2\omega \sin \phi$ *where* ω *denotes the angular speed of the Earth's rotation and* ϕ *the latitude,*

X, Y *the tide-generating forces (body forces per unit mass) in the x, y directions,*

g *the acceleration due to gravity.*

Further,

$$\frac{D}{Dt} = \frac{\partial}{\partial t} + u \frac{\partial}{\partial x} + v \frac{\partial}{\partial y} + w \frac{\partial}{\partial z} . \tag{1.5}$$

Assuming that the hydrostatic law of pressure is satisfied, (1.3) reduces to

$$\frac{\partial p}{\partial z} = \rho g . \tag{1.6}$$

Equation (1.4) expresses continuity of volume for the water within a differential cuboid lying within the whole body of water. Continuity for the water within a vertical column extending from the lake surface to the lake bed gives

$$\frac{\partial \zeta}{\partial t} + \frac{\partial}{\partial x}[(h+\zeta)\bar{u}] + \frac{\partial}{\partial y}[(h+\zeta)\bar{v}] = 0 , \tag{1.7}$$

where h denotes the undisturbed depth, ζ the elevation of the water surface ($h+\zeta$ is the total depth at any time) and $\bar{u}$, $\bar{v}$ components of depth-mean current defined by

$$\bar{u} = \frac{1}{h+\zeta}\int_{-\zeta}^{h} u\, dz , \qquad \bar{v} = \frac{1}{h+\zeta}\int_{-\zeta}^{h} v\, dz . \tag{1.8}$$

Equations (1.4) and (1.7) are each derived from first principles by Proudman (1953, p. 17) but (1.7) may be deduced from (1.4) by vertical integration through the total depth, inserting kinematic boundary conditions at the lake surface and the lake bed.

1.2 Wind-induced steady circulation in a narrow homogeneous lake

Consider a rectangular basin of length ℓ and depth h, subjected to a steady longitudinal wind stress F_s. The stress F_s has a prescribed distribution along the length of the basin. Atmospheric pressure p_a is uniform and constant over the water surface. Take axes Oxyz with O in the undisturbed lake surface (at one end); Ox is directed along the length and Oz vertically downwards as shown in figure 1.

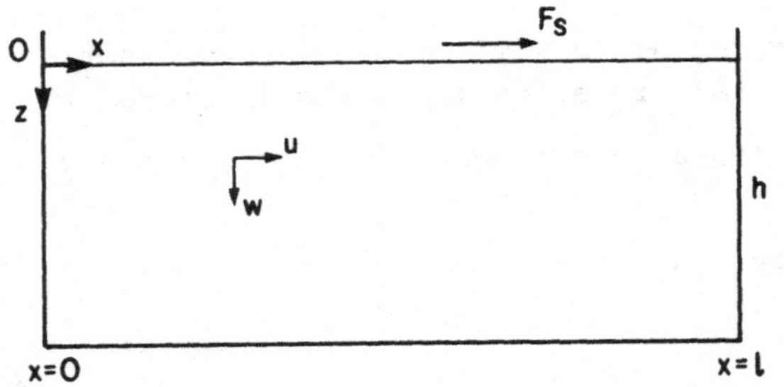

Figure 1

Rectangular basin. Definitions of wind force, velocity components and Cartesian coordinates x, y, z.

Assume that motion is two-dimensional in the vertical x - z plane; ignore geostrophic acceleration. Then, retaining only vertical shears and ignoring the tide-generating forces, the hydrodynamic equations (1.1), (1.4), (1.6) and (1.7) in linearized form (appropriate to small motion) reduce to

$$\frac{\partial p}{\partial x} + \frac{\partial F_{zx}}{\partial z} = 0 , \tag{1.9}$$

$$\frac{\partial u}{\partial x} + \frac{\partial w}{\partial z} = 0 , \tag{1.10}$$

$$\frac{\partial p}{\partial z} = \rho g , \tag{1.11}$$

$$\frac{\partial}{\partial x}\int_0^h u dz = 0 . \tag{1.12}$$

Introducing a vertical eddy viscosity, μ, we may write

$$F_{zx} = - \rho\mu \frac{\partial u}{\partial z} . \tag{1.13}$$

With homogeneous water, $\rho =$ constant. Therefore, satisfying the surface condition

$$p = p_a \quad \text{at} \quad z = -\zeta \tag{1.14}$$

equation (1.11) may be integrated to give

$$p = p_a + \rho g (z+\zeta), \tag{1.15}$$

so that

$$\frac{\partial p}{\partial x} = \rho g \frac{\partial \zeta}{\partial x} . \tag{1.16}$$

Hence, from (1.9),

$$\frac{\partial F_{zx}}{\partial z} = -\rho g \frac{\partial \zeta}{\partial x}$$

which may be integrated to yield

$$F_{zx} = -\rho g \frac{\partial \zeta}{\partial x} z + A. \qquad (1.17)$$

Introducing (1.13),

$$-\rho\mu \frac{\partial u}{\partial z} = -\rho g \frac{\partial \zeta}{\partial x} z + A.$$

Then, assuming that eddy viscosity μ is a constant, it follows that

$$-\rho\mu u = -\frac{1}{2} \rho g \frac{\partial \zeta}{\partial x} z^2 + Az + B. \qquad (1.18)$$

The constants of integration A and B may be determined by satisfying a bottom boundary condition and invoking continuity. Thus, assuming that bottom current is zero:

$$u = 0 \quad \text{at} \quad z = h, \qquad (1.19)$$

gives

$$0 = -\frac{1}{2} \rho g \frac{\partial \zeta}{\partial x} h^2 + Ah + B \qquad (1.20)$$

and using this to eliminate B from (1.18) we get

$$\rho\mu u = \frac{1}{2} \rho g \frac{\partial \zeta}{\partial x} (z^2 - h^2) + A(h - z). \qquad (1.21)$$

To find A, note that continuity (1.12) implies

$$\int_0^h u dz = \text{constant},$$

which since there is no horizontal flow across the ends of the basin:

$$u = 0 \quad \text{at} \quad x = 0 \quad \text{and} \quad x = \ell, \qquad (1.22)$$

reduces to

$$\int_0^h u dz = 0. \qquad (1.23)$$

Combining (1.21) and (1.23) yields

$$A = \frac{2}{3} \rho g h \frac{\partial \zeta}{\partial x}. \qquad (1.24)$$

Therefore, from (1.17),

$$F_{zx} = \frac{1}{3} \rho g \frac{\partial \zeta}{\partial x} (2h - 3z) \qquad (1.25)$$

and from (1.21),

$$u = \frac{g}{6\mu} \frac{\partial\zeta}{\partial x}(h-z)(h-3z). \tag{1.26}$$

To find $\partial\zeta/\partial x$ the surface condition

$$F_{zx} = F_s \quad \text{at} \quad z = 0 \tag{1.27}$$

is inserted in (1.25) to give

$$F_s = \frac{2}{3} \rho g h \frac{\partial\zeta}{\partial x}, \tag{1.28}$$

whence

$$\frac{\partial\zeta}{\partial x} = \frac{3F_s}{2\rho g h}. \tag{1.29}$$

A return to (1.25) and (1.26) then produces the further results

$$F_{zx} = \frac{F_s}{2h}(2h-3z) \tag{1.30}$$

and

$$u = \frac{F_s}{4\rho\mu h}(h-z)(h-3z). \tag{1.31}$$

To find the vertical current w, from (1.10) we have

$$w = -\int_h^z \frac{\partial u}{\partial x}\, dz, \tag{1.32}$$

where the lower limit is chosen to satisfy the bottom boundary condition

$$w = 0 \quad \text{at} \quad z = h. \tag{1.33}$$

Inserting (1.31) into (1.32) then gives the result

$$w = -\frac{F_s'}{4\rho\mu h} z(z-h)^2, \tag{1.34}$$

where

$$F_s' = \frac{dF_s}{dx}. \tag{1.35}$$

Along streamlines of the flow, in the vertical plane of the basin,

$$\frac{dz}{dx} = \frac{w}{u} = -\frac{F_s'}{F_s}\frac{z(h-z)}{h-3z}. \tag{1.36}$$

Integrating this equation:

$$- \int \frac{3z-h}{z(z-h)} \, dz = \int \frac{F_s'}{F_s} \, dx \, ,$$

whence

$$- \log \{ z(z-h)^2 \} = \log (F_s/C)$$

and therefore

$$z(z-h)^2 F_s = C \, . \tag{1.37}$$

Equation (1.37) gives the equation of a streamline; varying the constant of integration C yields the whole family of lines.

In the above solution, note that while the vertically-integrated flow across each end of the basin is zero (equation (1.23)), currents at various levels through the depth at each end are only zero if $F_s = 0$ at $x = 0$ and $x = \ell$ (equation (1.31)). Therefore, to satisfy completely the no-flow conditions at each end we need to define F_s such that

$$F_s = 0 \quad \text{at} \quad x = 0 \quad \text{and} \quad x = \ell \, . \tag{1.38}$$

Practically, these conditions do not limit the generality with which the distribution of the wind stress F_s can be prescribed along the length of the basin.

The vertical structures of stress and current, obtained from (1.30), (1.31) and (1.34) and given non-dimensionally by

$$\hat{F}_{zx} = F_{zx}/F_s = 1 - 3\xi/2 \, , \tag{1.39}$$

$$\hat{u} = u / (\frac{h F_s}{4\rho\mu}) = (1-\xi)(1-3\xi) \, , \tag{1.40}$$

$$\hat{w} = w / (\frac{h^2 F_s'}{4\rho\mu}) = -\xi(1-\xi)^2 \, , \tag{1.41}$$

where $\xi = z/h$, are plotted in figure 2. Note the wind-driven flow near the surface and the counter current at greater depths - driven by pressures, opposing the wind, associated with the gradient of surface elevation (equation (1.29)). That gradient is upward in the wind direction; the return currents produced by it are so-called "slope" or "gradient" currents. Examination of equation (1.21) shows that the wind-drift current decreases linearly with depth while the gradient current decreases quadratically with depth.

It is evident from (1.39) that the bottom stress comes out to half the wind stress. Also it may be deduced from (1.40) and (1.41) that the coefficient of eddy viscosity, μ, must increase at least in proportion to h^2 for u and w to remain bounded as h becomes large. This behaviour is consi-

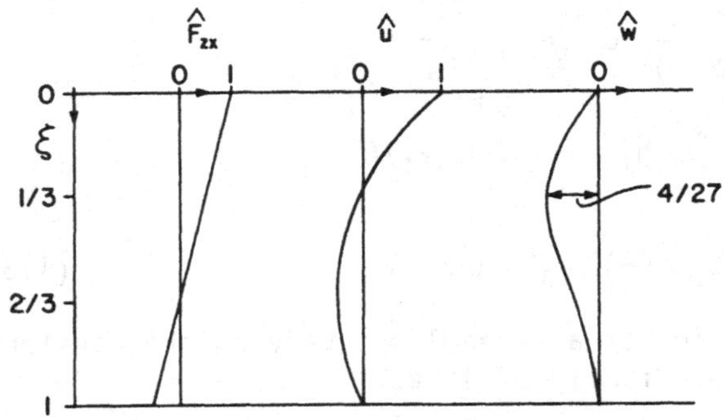

Figure 2

Vertical structure of stress and current, as given by (1.39), (1.40) and (1.41).

stent with the fact that the magnitudes of coefficients of eddy viscosity usually increase with the size of the region considered (Proudman, 1953, p. 104). Note that fundamentally, and consistent with its dimensions, eddy viscosity may be expressed as the product of a typical length scale and a typical velocity for the motion with which it is associated. Thereby the extent of the region occupied by the motion, and the intensity of the turbulence are represented in the definition. Generally, eddy viscosity varies through space and time from one flow region to another (Csanady, 1976, 1979, 1980).

The general character of the streamlines in the vertical x - z plane, corresponding to a wind stress distribution which reaches a maximum midway between the ends of the basin, is sketched out in figure 3. Manifestly, the wind maintains a single circulation cell. A number of cells are produced by a wind stress which attains maximum and minimum values alternately in its variation along the length of the basin.

Integrating (1.29), satisfying continuity of volume for the entire basin:

$$\int_0^{\ell} \zeta \, dx = 0 \, , \tag{1.42}$$

yields elevation of the water surface in the form

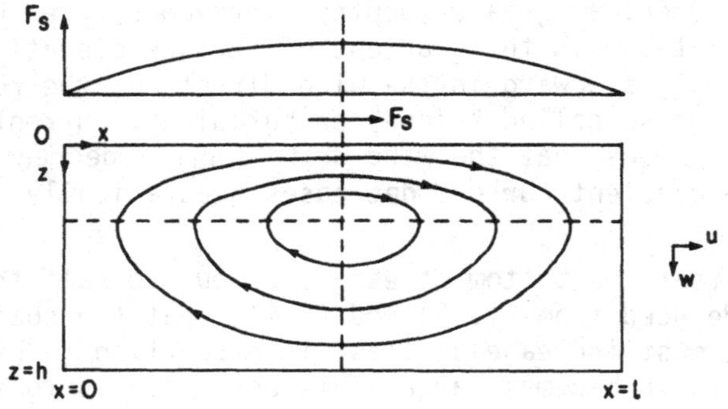

Figure 3

Streamlines.

$$\zeta = \frac{3}{2\rho gh}\left[\int_0^x F_s(x')\,dx' - \frac{1}{\ell}\int_0^\ell dx \int_0^x F_s(x')\,dx'\right], \tag{1.43}$$

where x' is a running variable for x. When F_s is uniformly distributed, (1.43) reduces to

$$\zeta = \frac{3F_s}{4\rho gh}(2x-\ell). \tag{1.44}$$

Note that the integration of (1.9) and (1.11) to yield u, given here by (1.31), has also been carried out by Bye (1965) using however a formulation for μ based on the concept of Prandtl's mixing length (Proudman, 1953, pp. 102-103).

1.3 Influence of bottom slip on the wind-induced circulation

The preceding solution for circulation in a homogeneous narrow lake assumes zero slip at the bottom, expressed by (1.19). Alternatively a slip condition may be used relating the force of friction on the bottom to the bottom current:

$$F_{zx} = k\rho u \quad \text{at} \quad z = h. \tag{1.45}$$

Equation (1.45), in which k is a constant, may be regarded as a convenient theoretical alternative to the well-known quadratic law of bottom stress (Proudman, 1953, p. 136). The quadratic law may be linearized when there are strong background tidal currents (Bowden, 1953), but such a procedure is not realistic when considering water movements in lakes where the tides are weak. Employing a simple physical argument, the use of (1.45) accounts for the supposed presence of a thin boundary layer adjacent to the bottom through which the current u diminishes linearly with increasing depth from $(u)_{z=h}$ to zero. In such circumstances, if h_b denotes the thickness of the layer and μ_b its vertical eddy viscosity, assumed to be a constant, then from (1.13):

$$F_{zx} = -\rho\mu_b(-u/h_b) \quad \text{at} \quad z = h,$$

so that the coefficient of friction k in (1.45) is given by $k = \mu_b / h_b$. Realistically, however, the bottom boundary layer has a more complicated structure, described by Bowden (1978).

Repeating the analysis of Section 1.2, but replacing (1.19) by (1.45), we obtain

$$\frac{\partial\zeta}{\partial x} = \frac{3}{2(1+\delta)}\,\frac{F_s}{\rho gh}, \tag{1.46}$$

$$\hat{F}_{zx} = \frac{1+\delta-3\xi/2}{1+\delta}, \tag{1.47}$$

$$\hat{u} = \frac{(1-\xi)(1-3\xi+4\delta)-2\delta}{1+\delta}, \tag{1.48}$$

$$\hat{w} = -\frac{\xi(1-\xi)(1-\xi+2\delta)}{1+\delta}, \tag{1.49}$$

where

$$\delta = \frac{1}{\beta+2}, \quad \beta = \frac{kh}{\mu} \tag{1.50}$$

are non-dimensional parameters. These results replace those given earlier in (1.29), (1.39), (1.40) and (1.41). The earlier results, corresponding to zero bottom current, may be derived from the present ones by setting $\delta = 0$ corresponding to the limiting case: $k \to \infty$, $\beta \to \infty$.

The ranges of k, β and δ are

$$0 \le k < \infty, \quad 0 \le \beta < \infty, \quad 1/2 \ge \delta > 0. \tag{1.51}$$

At one extreme ($k \to \infty$, $\beta \to \infty$, $\delta = 0$) the condition of zero slip is satisfied at the bottom, while at the other extreme ($k = 0$, $\beta = 0$, $\delta = 1/2$) the condition of free slip holds there. In going from zero slip to free slip, we observe:

a) the gradient of surface elevation $\partial\zeta/\partial x$ diminishes from $(3/2)(F_s/\rho gh)$ to $F_s/\rho gh$ (equation (1.46));

b) the bottom stress diminishes to zero from its original value equal to half the wind stress, the vertical profile of stress remains linear (equation (1.47));

c) horizontal current increases by one third at the surface while the maximum of its vertical profile moves to the bottom and doubles its value (equation (1.48));

d) vertical current between the surface and the bottom becomes larger but maintains the same type of profile; vertical gradient of that current at the bottom becomes non-zero (equation (1.49)).

1.4 Wind-induced steady circulation in a narrow lake stratified in two layers

Next suppose that the rectangular basin has two horizontal layers of water, each homogeneous, differing in density (figure 4a). The upper layer, of depth h_1 and density ρ_1, represents the epilimnion and the lower layer, of depth h_2 and density ρ_2, the hypolimnion in a thermally stratified lake. The interface between the layers, assumed to be impermeable, represents the thermocline. Physically $\rho_1 < \rho_2$ and, in accordance with previous notation, $h_1 + h_2 = h$; ρ_1, ρ_2, h_1 and h_2 are constants.

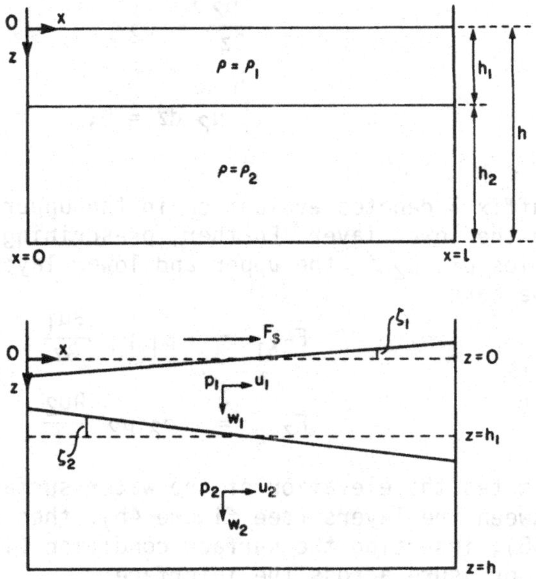

a) The water at rest.

b) The water in motion.

Figure 4 Two-layered basin.

Again suppose that a steady longitudinal wind stress F_s acts over the water surface. Then, making the same basic assumptions as before and employing the same notation, the following equations may be formulated governing the wind-induced circulation in the vertical x - z plane. Referring back to equations (1.9), (1.10), (1.11) and (1.23), it follows that for the upper layer:

$$\frac{\partial p_1}{\partial x} + \frac{\partial F_{zx_1}}{\partial z} = 0, \tag{1.52}$$

$$\frac{\partial u_1}{\partial x} + \frac{\partial w_1}{\partial z} = 0, \tag{1.53}$$

$$\frac{\partial p_1}{\partial z} = \rho_1 g, \tag{1.54}$$

$$\int_0^{h_1} u_1 \, dz = 0 \tag{1.55}$$

and for the lower layer:

$$\frac{\partial p_2}{\partial x} + \frac{\partial F_{zx_2}}{\partial z} = 0, \tag{1.56}$$

$$\frac{\partial u_2}{\partial x} + \frac{\partial w_2}{\partial z} = 0, \tag{1.57}$$

$$\frac{\partial p_2}{\partial z} = \rho_2 g, \tag{1.58}$$

$$\int_{h_1}^{h} u_2 \, dz = 0, \tag{1.59}$$

where suffix 1 denotes evaluation in the upper layer and suffix 2 evaluation in the lower layer. Further, prescribing constant vertical eddy viscosities μ_1, μ_2 in the upper and lower layers respectively, as in (1.13) we have

$$F_{zx_1} = -\rho_1 \mu_1 \frac{\partial u_1}{\partial z}, \tag{1.60}$$

$$F_{zx_2} = -\rho_2 \mu_2 \frac{\partial u_2}{\partial z}. \tag{1.61}$$

If ζ_1 denotes the elevation of the water surface and ζ_2 that of the interface between the layers (see figure 4b), then the integration of (1.54) and (1.58), inserting the surface condition (1.14) and satisfying continuity of pressure across the interface:

$$p_1 = p_2 \quad \text{at} \quad z = h_1 - \zeta_2 \tag{1.62}$$

yields

$$\left.\begin{aligned} p_1 &= p_a + \rho_1 g(z+\zeta_1), \\ p_2 &= p_a + \rho_1 g(h_1+\zeta_1-\zeta_2) + \rho_2 g(z-h_1+\zeta_2). \end{aligned}\right\} \tag{1.63}$$

Therefore, from (1.52) and (1.56):

$$\frac{\partial F_{zx_1}}{\partial z} = -\rho_1 g \frac{\partial \zeta_1}{\partial x}, \tag{1.64}$$

$$\frac{\partial F_{zx_2}}{\partial z} = -\rho_1 g \frac{\partial \zeta_1}{\partial x} - (\rho_2 - \rho_1) g \frac{\partial \zeta_2}{\partial x}. \tag{1.65}$$

Integrating these equations with respect to z, using (1.60) and (1.61), satisfying the continuity conditions (1.55) and (1.59), ensuring continuity of stress across the interface:

$$F_{zx_1} = F_{zx_2}, \quad \text{at} \quad z = h_1 \tag{1.66}$$

and enforcing a slip condition at the bottom:

$$F_{zx_2} = k_2 \rho_2 u_2, \quad \text{at} \quad z = h, \tag{1.67}$$

where k_2 is a constant, yields

$$F_{zx_1} = g\rho_1 h_1 (1-\eta) G_1 + \frac{2}{3}(1+\delta_2) g\rho_1 h_2 G_2 , \tag{1.68}$$

$$F_{zx_2} = g\rho_1 h_2 \left[\frac{2}{3}(1+\delta_2) - \nu\right] G_2 , \tag{1.69}$$

$$u_1 = \frac{gh_1^2}{6\mu_1}(3\eta^2 - 6\eta + 2) G_1 + \frac{g h_1 h_2}{3\mu_1}(1+\delta_2)(1-2\eta) G_2 , \tag{1.70}$$

$$u_2 = \frac{gh_2^2}{6\mu_2}\left(\frac{\rho_1}{\rho_2}\right)\left[3\nu^2 - 4(1+\delta_2)\nu + 1 + 2\delta_2\right] G_2 , \tag{1.71}$$

in which

$$G_1 = \frac{\partial\zeta_1}{\partial x}, \qquad G_2 = \frac{\partial\zeta_1}{\partial x} + \left(\frac{\rho_2-\rho_1}{\rho_1}\right)\frac{\partial\zeta_2}{\partial x} \tag{1.72}$$

$$\delta_2 = \frac{1}{\beta_2 + 2}, \qquad \beta_2 = \frac{k_2 h_2}{\mu_2} \tag{1.73}$$

and

$$\eta = \frac{z}{h_1}, \qquad \nu = \frac{z - h_1}{h_2} . \tag{1.74}$$

Here, η, ν are normalized vertical coordinates for the upper and lower layers respectively; the range of depth in the upper layer is covered by $0 \le \eta \le 1$ and in the lower layer by $0 \le \nu \le 1$.

To determine G_1 and G_2 in (1.68) - (1.71), and thence the gradients $\partial\zeta_1/\partial x$ and $\partial\zeta_1/\partial x$, the surface condition:

$$F_{zx_1} = F_s , \quad \text{at} \quad z = 0 \tag{1.75}$$

and a slip condition at the interface:

$$F_{zx_1} = k_1 \rho_1 (u_1 - u_2), \quad \text{at} \quad z = h_1 , \tag{1.76}$$

where k_1 is a constant, are satisfied. As a result we get

$$G_1 = \left[1 + \frac{2}{3}(1+\delta_2)\lambda\right]\frac{F_s}{g\rho_1 h_1}, \qquad G_2 = -\frac{\lambda F_s}{g\rho_1 h_2} , \tag{1.77}$$

where

$$\lambda = \frac{\beta_1}{\frac{4}{3}(1+\delta_2)(3+\beta_1) + (1+2\delta_2)\alpha\beta_1} , \tag{1.78}$$

and

$$\alpha = \frac{\rho_1 \mu_1 h_2}{\rho_2 \mu_2 h_1}, \qquad \beta_1 = \frac{k_1 h_1}{\mu_1} . \tag{1.79}$$

Substituting the expressions for G_1, G_2 given by (1.77) into (1.68) - (1.71) gives vertical structures of stress and current, in non-dimensional form, as follows:

$$\hat{F}_{zx_1} = F_{zx_1}/F_s = 1 - \eta - \frac{2}{3}\lambda(1+\delta_2)\eta \,, \qquad (1.80)$$

$$\hat{F}_{zx_2} = F_{zx_2}/F_s = \lambda\left(\nu - \frac{2}{3}(1+\delta_2)\right) , \qquad (1.81)$$

$$\hat{u}_1 = u_1/\left(\frac{h_1 F_s}{6\rho_1\mu_1}\right) = 3\eta^2 - 6\eta + 2 + \frac{2}{3}\lambda(1+\delta_2)(3\eta^2 - 1) \,, \qquad (1.82)$$

$$\hat{u}_2 = u_2/\left(\frac{h_2 F_s}{6\rho_2\mu_2}\right) = -\lambda\left(3\nu^2 - 4(1+\delta_2)\nu + 1 + 2\delta_2\right) . \qquad (1.83)$$

Using these forms for u_1 and u_2, the vertical currents may be obtained by integrating (1.53) and (1.57) with respect to z, inserting the interfacial condition:

$$w_1 = w_2 , \quad \text{at} \quad z = h_1 \qquad (1.84)$$

and the bottom condition:

$$w_2 = 0 , \quad \text{at} \quad z = h . \qquad (1.85)$$

The integrations yield

$$\hat{w}_1 = w_1/\left(\frac{h_1^2 F_s'}{6\rho_1\mu_1}\right) = \eta(1-\eta)\left(\eta - 2 + \frac{2}{3}\lambda(1+\delta_2)(\eta+1)\right) , \qquad (1.86)$$

$$\hat{w}_2 = w_2/\left(\frac{h_2^2 F_s'}{6\rho_2\mu_2}\right) = \lambda\nu\left(\nu^2 - 2(1+\delta_2)\nu + 1 + 2\delta_2\right) . \qquad (1.87)$$

The vertical structures of stress and current given by (1.80) - (1.83) and (1.86), (1.87) are sketched out in figure 5 for the case of free slip at the bottom ($k_2 = \beta_2 = 0$, $\delta_2 = 1/2$). Since $0 \le \lambda < 3/4(1+\delta_2)$ from (1.78), it follows that $0 \le \lambda < 1/2$ in this case. The particular profiles shown in the figure correspond to $\lambda = 0.3$.

A study of figure 5 indicates that, for the distribution of wind stress F_s shown in figure 3, there is a wind-induced circulation in the upper layer (similar in character to that depicted in figure 3) together with an *opposite* circulation in the lower layer. The latter circulation is driven by the frictional stress exerted by the upper layer on the lower layer across the interface; that stress acts in the opposite direction to the wind. When $k_1 = 0$ and there is free slip at the interface with zero stress there (equation (1.76)), then $\lambda = 0$ from (1.78) and therefore $u_2 = w_2 = 0$

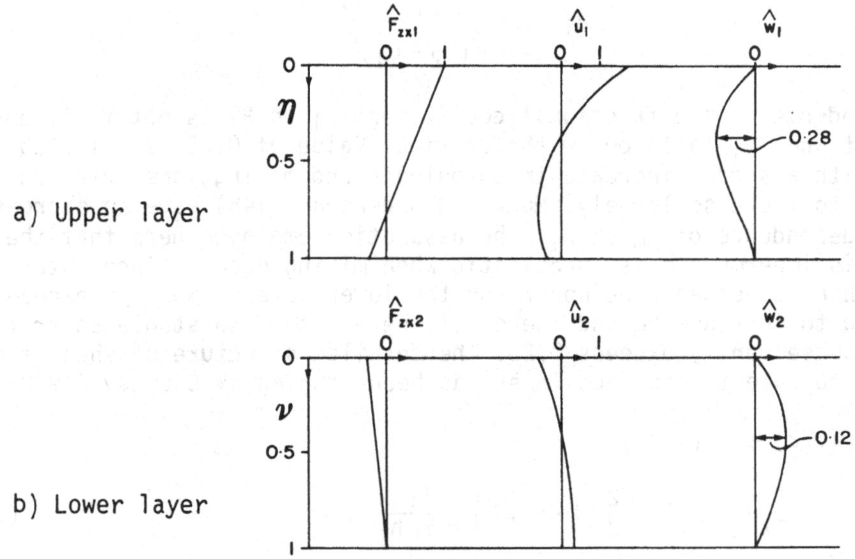

Figure 5 Vertical structure of stress and current for a two-layered basin.

from (1.83) and (1.87) indicating no motion in the lower layer. Clearly, λ is an important non-dimensional parameter representing the influence of frictional conditions at the interface on the wind-induced motion in the basin; λ controls the degree to which horizontal momentum is transmitted across the interface from the upper to the lower layer.

The use of the slip condition (1.76) accounts for the supposed presence of a shear layer spanning the thermocline of thickness h_m say, through which the current u changes linearly from u_1 to u_2 as the depth increases. If this thin layer has a constant vertical eddy viscosity μ_m, we may write

$$F_{zx_1} = -\rho_1 \mu_m \left(\frac{u_2 - u_1}{h_m}\right), \quad \text{at} \quad z = h_1 \tag{1.88}$$

so that, from (1.76), $k_1 = \mu_m/h_m$. Shear instability in the thermocline, which must influence the values of μ_m and h_m (and hence k_1), has been discussed by Mortimer (1961, 1974). Turbulence will either increase or subside at thermocline level according to whether the Richardson number

$$R_i = g \frac{\partial\rho}{\partial z} / \rho\left(\frac{\partial u}{\partial z}\right)^2 \tag{1.89}$$

there is less or greater than approximately 0.25. In terms of the present model:

$$R_i = \frac{g\, h_m (\rho_2 - \rho_1)}{\rho_1 (u_2 - u_1)^2} . \tag{1.90}$$

The dependence of the frictional coefficient k_1 on R_i is not fully understood but when R_i falls below the critical value of 0.25, and instability occurs with a sudden increase in turbulence and mixing, the friction is expected to increase largely. Munk and Anderson (1948) gave an expression for the dependence of μ_m on R_i. The assumption employed here that the interface is impermeable is unrealistic when mixing occurs since water is then exchanged between the upper and the lower layers. Such an exchange will tend to increase h_m and there will be a return to stable shear flow when R_i subsequently exceeds 0.25. The detailed structure of shear flow within a turbulent interface layer has been studied by Csanady (1978).

From (1.72) and (1.77):

$$\frac{\partial \zeta_1}{\partial x} = \left(1 + \frac{2}{3}(1+\delta_2)\,\lambda\right) \frac{F_s}{g\,\rho_1\, h_1} , \tag{1.91}$$

$$\frac{\partial \zeta_2}{\partial x} = -\left(\frac{\rho_1}{\rho_2 - \rho_1}\right) \left(1 + \frac{2}{3}(1 + \delta_2)\lambda + \frac{\lambda\, h_1}{h_2}\right) \frac{F_s}{g\,\rho_1\, h_1} . \tag{1.92}$$

Therefore the surface gradient $\partial\zeta_1/\partial x$ is upward in the wind direction and the interfacial gradient downward in that direction. In practice $\rho_1/(\rho_2-\rho_1) \sim 1000$ and therefore $\partial\zeta_2/\partial x$ is much larger than $\partial\zeta_1/\partial x$. While in reality ζ_1 normally attains values of the order of a centimetre or so, under strong winds ζ_2 may reach heights of several metres and the thermocline may cut the free surface at the windward end of a narrow lake (Mortimer, 1952). The present linear analysis is based on the assumption of small displacements and cannot cover the latter situation. Csanady (1982, pp. 99-104) has given a theoretical treatment to account for the surfacing of the thermocline along the shore of a large lake such as Lake Ontario.

1.5 General response of a narrow lake to wind

Mortimer (1952), in an investigation of water movements in Windermere, distinguished two main phases in the response of a long narrow stratified lake to wind: i) quasi-steady states of motion, such as have been considered here, and (ii) surface and internal seiches which develop after the wind has dropped. A theory for the *total* response of a narrow two-layered lake to a wind pulse has been given by Heaps and Ramsbottom (1966) and applied to Windermere; internal friction at the interface between the layers was assumed to be permanently zero. Internal seiches in narrow stratified lakes have been studied by Mortimer (1953). A theory for surface and internal seiches, in a narrow lake uniformly stratified in three layers, has been given by Heaps (1961).

2. EKMAN'S THEORY AND ITS APPLICATION

Consideration is now given to the vertical structure of wind-induced motion in broad seas and lakes, taking the effects of the Earth's rotation into account. The water is assumed to be homogeneous. Ekman's basic solution for wind currents in the open sea is presented; the generalisation of that theory to account for steady-state and time-dependent motions in real basins is then described. The depth variation of vertical eddy viscosity (assumed constant in Ekman's theory) is discussed, referring to recent research.

2.1 Ekman spiral

Consider the currents produced in a homogeneous sea of infinite extent by a steady wind stress F_S, uniformly distributed over the water surface, acting in the y-direction. Ignoring tide-generating forces and recognizing that the profile of current through every vertical must be the same with horizontal derivatives all zero, the equations (1.1) and (1.2) reduce to

$$-fv = -\frac{1}{\rho}\frac{\partial F_{zx}}{\partial z}, \tag{2.1}$$

$$fu = -\frac{1}{\rho}\frac{\partial F_{zy}}{\partial z}. \tag{2.2}$$

Taking

$$F_{zx} = -\rho\mu\frac{\partial u}{\partial z}, \quad F_{zy} = -\rho\mu\frac{\partial v}{\partial z}, \tag{2.3}$$

where μ is a constant eddy viscosity, it follows that

$$-fv = \mu\frac{d^2u}{dz^2}, \tag{2.4}$$

$$fu = \mu\frac{d^2v}{dz^2}. \tag{2.5}$$

Combining (2.4) and (2.5) yields the complex equation

$$\frac{d^2w'}{dz^2} = \frac{(1+i)^2\pi^2}{D^2}w' \tag{2.6}$$

where (the prime does not indicate differentiation)

$$w' = u + iv \qquad (2.7)$$

and

$$D = \pi\sqrt{\frac{2\mu}{f}}. \qquad (2.8)$$

Satisfying the wind stress boundary condition at the sea surface:

$$F_{zx} + iF_{zy} = -\rho\mu \frac{\partial w'}{\partial z} = iF_s, \quad \text{at} \quad z = 0 \qquad (2.9)$$

and ensuring that current tends to zero as the depth increases to infinity:

$$w' \to 0, \quad \text{as} \quad z \to \infty \qquad (2.10)$$

the solution of (2.6) is

$$w' = (1+i)\frac{DF_s}{2\pi\rho\mu} e^{-(1+i)\pi z/D}, \qquad (2.11)$$

which may be written alternatively in the form

$$w' = \frac{\sqrt{2}\,\pi F_s}{\rho f D} e^{-\pi z/D} e^{i(\pi/4-\pi z/D)}. \qquad (2.12)$$

Therefore the surface current is at 45^0 to the right of the wind (to the left of the wind in the southern hemisphere) and is given by

$$w_s' = \frac{\sqrt{2}\,\pi F_s}{\rho f D} e^{i\pi/4}. \qquad (2.13)$$

As depth z below the surface increases, the current decreases exponentially (as $e^{-\pi z/D}$) and its angle of deflection to the right of the wind, namely $\pi/2-(\pi/4-\pi z/D) = \pi/4+\pi z/D$, increases. This current system is called the Ekman spiral after Ekman (1905) who discovered it (figure 6). At depth $z = D$, the current is $e^{-\pi} = 1/23$ of its surface value and has turned through 180^0. The depth D may be taken as a measure of how far downwards the wind effect penetrates and, accordingly, is known as the "frictional depth". Since

$$-\rho\mu\frac{\partial w'}{\partial z} = \left(\frac{\rho\mu\pi\sqrt{2}}{D}\right) e^{i\pi/4} w'$$

it follows that, at any depth z, the direction of the frictional stress F_z say, is always turned contra solem through 45^0 from the direction of the current, w_z say, at the same depth (figure 7). The total transport of water across the spiral is

$$\int_0^\infty w'\,dz = \frac{F_s}{f\rho}, \qquad (2.14)$$

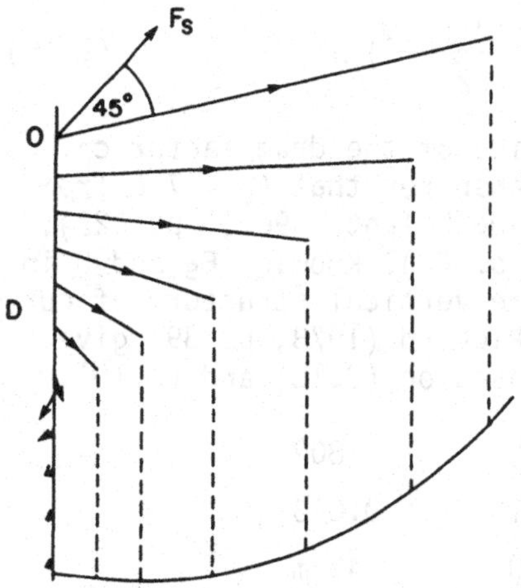

Figure 6 Ekman spiral.

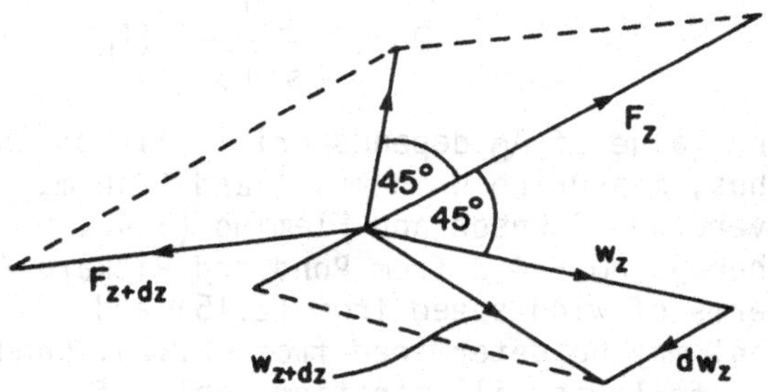

Figure 7 Forces and currents on the upper and lower faces of a slab between z and z+dz in the Ekman spiral.

manifestly directed perpendicular to the wind stress and to the right of it (compare (2.9)).

Expressing the wind stress in terms of the square of the wind speed:

$$F_S = \rho_A c_D W^2 \tag{2.15}$$

where W denotes the wind speed, ρ_A the density of the air and c_D a drag coefficient, it follows from (2.13) that

$$|w_S'| = \left(\frac{\sqrt{2}\,\pi\,\rho_A\,c_D}{\rho}\right)\frac{W^2}{fD} = \left(\frac{\pi\,\rho_A\,c_D}{\sqrt{2}\,\omega\,\rho}\right)\frac{W^2}{D\sin\phi}\,. \tag{2.16}$$

Therefore, the ratio of the surface current to the wind speed, the so-called wind factor (α', say), is

$$\alpha' = \frac{|w_S'|}{W} = \left(\frac{\pi\,\rho_A\,c_D}{\sqrt{2}\,\omega\,\rho}\right)\frac{W}{D\sin\phi}\,. \tag{2.17}$$

On the basis of observations in various seas and oceans prior to 1939, Ekman derived the empirical relation

$$\alpha' = \frac{0.0127}{\sqrt{\sin\phi}}\,. \tag{2.18}$$

Comparing (2.17) and (2.18) yields

$$D = \frac{A_0 W}{\sqrt{\sin\phi}} \quad \left(A_0 = \frac{\pi \rho_A c_D}{0.0127 \times \sqrt{2}\,\omega\rho}\right). \tag{2.19}$$

The value of A_0 depends critically on the choice of the drag factor c_D. Thus, measuring W in $m\,s^{-1}$ and D in m, it transpires that $A_0 = 7.6$ from Sverdrup, Johnson and Fleming (1946, p. 494) and Defant (1961, p. 422), whereas $A_0 = 4.3$ from Pond and Pickard (1978, p. 88). Knowing F_s and D in terms of wind speed from (2.15) and (2.19), the vertical structure of current may be determined from (2.12). Pond and Pickard (1978, p. 89) give the following illuminating table of values, based on (2.18) and (2.19):

	ϕ =	10^0	45^0	80^0
	α'	0.030	0.015	0.013
$W = 10\ m\,s^{-1}$:	D =	100	50	45 m
$W = 20\ m\,s^{-1}$:	D =	200	100	90 m

Combining (2.8) and (2.19) gives a formula for the eddy viscosity in the Ekman spiral in terms of the wind speed:

$$\mu = (\omega A_0^2/\pi^2)\, W^2 . \tag{2.20}$$

According to Defant (1961, pp. 422-423) this formula with W in $m\,s^{-1}$ holds for $W > 6\,m\,s^{-1}$ with $\rho\omega A_0^2/\pi^2 = 4.3$, while μ is proportional to W^3 for $W < 6\,m\,s^{-1}$. It follows from (2.15) and (2.20) that μ is directly proportional to F_s.

Observations of the angle of deflection of the surface current from the wind direction have shown considerable deviations from the classical 45^0 result. The Ekman spiral is not a clearly observed distribution of current. The reasons may have to do with the highly idealized nature of the Ekman theory outlined above, specifically in its assumption of

(i) no land boundaries,
(ii) infinitely deep water,
(iii) eddy viscosity uniform through the vertical,
(iv) steady state motion,
(v) homogeneous water.

Repeating the analysis of equations (2.1)-(2.12) but considering a sea of finite depth h and inserting a zero slip condition at the sea bed:

$$u = v = 0 \quad \text{at} \quad z = h \tag{2.21}$$

instead of satisfying (2.10), yields

$$w' = (1+i)\,\frac{DF_s}{2\pi\rho\mu}\;\frac{\sinh\{(1+i)\,a\,(1-\xi)\}}{\cosh\{(1+i)\,a\}} \tag{2.22}$$

where

$$a = \pi h/D, \qquad \xi = z/h . \tag{2.23}$$

Vertical structures of current from (2.22) are illustrated in figure 8 for various values of h/D, showing the important influence of bottom friction in a shallow sea. As h decreases the angle of deflection of the current from the wind direction decreases and the effect of the Earth's rotation becomes smaller. Thus, for small h,

$$w' \sim i\, h\, F_S(1-\xi)/\rho\, \mu$$

from (2.22), indicating currents in the wind direction, independent of ω, decreasing linearly with depth.

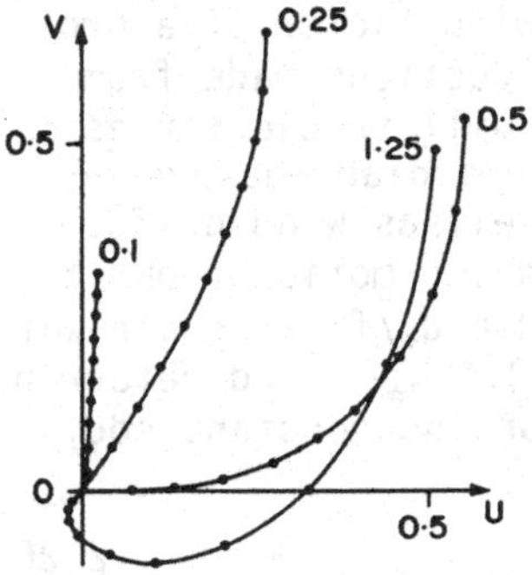

Figure 8

Vertical structure of wind-induced current for h/D = 0.1, 0.25, 0.5, 1.25: marked in steps of 0.1 h from the surface down to the bottom.

Under a steady current, friction at the sea bed generates an Ekman spiral directly above the bottom. As the current decreases towards the bottom from the mainstream flow, it rotates contra solem - in the opposite sense to the wind-driven near-surface Ekman spiral (Defant, 1961, p. 408; Pond and Pickard, 1978, p. 95). The vertical structure of current in a deep sea therefore has an Ekman layer near the surface of frictional depth D say, and one near the bottom of frictional depth D' say. When the total depth h becomes small and the layers overlap ($h < D + D'$) the respective spirals tend to cancel each other, producing the results shown in figure 8.

2.2 Distributions of vertical eddy viscosity

Ekman's problem of wind-induced flow defined by (2.1)-(2.10) has been solved by Madsen (1977) replacing the constant eddy viscosity by one which increases linearly with depth from a value of zero at the surface $z = 0$. The law of increase follows Prandtl's theory for a turbulent boundary layer (Defant, 1961, pp. 387-390; Sverdrup, Johnson and Fleming, 1946, pp. 479-480):

$$\mu = \kappa\, u_* \, z, \tag{2.24}$$

where u_* is the friction velocity defined by

$$u_\star = \sqrt{\frac{F_s}{\rho}} \qquad (2.25)$$

and $\kappa = 0.4$ is von Karman's constant. Evaluating surface current at $z = z_0$, where z_0 is a roughness length of order 0.15 cm, a logarithmic singularity in velocity at $z = 0$ is avoided. Thus (2.24) and the current structure associated with it effectively apply for $z \geq z_0$.

The results obtained by Madsen, compared with those from Ekman's theory, show a more rapid decrease and rotation of the current vector with depth: due to a steep logarithmic fall of the velocity component in the wind direction, near to the surface and downwards from it. For a range of wind speeds, Madsen's model gives an angle of deflection of the surface current to the right of the wind stress of about 10° and a wind factor of approximately 0.03. These values agree essentially with deductions made from surface drift experiments and observations of oil spill tradjectories (Pearce and Cooper, 1981, figure 1), though such Lagrangian measurements will include Stokes drift due to surface waves as well as wind drift. Madsen's model shows that the depth of the wind-induced motion (down to a level of practically no current) is approximately $0.4\, u_\star/f$: a length which corresponds to D in the Ekman theory. Setting $D = 0.4\, u_\star/f$ and referring back to (2.8) yields an expression for the corresponding constant eddy viscosity in the Ekman theory:

$$\mu = 0.008\, u_\star^2/f\,. \qquad (2.26)$$

Svensson (1979) employed a turbulence model (using $k-\varepsilon$ closure condition) to determine the vertical structure of current in the surface Ekman layer. A vertical eddy viscosity distribution of the form shown in figure 9 was obtained, showing a linear increase in viscosity near the sea surface as in Madsen's model, turning however into a decrease lower down. The depth of penetration of the motion was found to be approximately $1.0\, u_\star/f$. The use of a constant eddy viscosity:

$$\mu = 0.026\, u_\star^2/f \qquad (2.27)$$

was found to give velocity and shear stress distributions in good agreement with those obtained from the turbulence model - except that close to the surface (within $u_\star/f$ of it) a constant eddy viscosity gives a linear variation in velocity in the wind direction whereas the turbulence model gives a logarithmic variation. An eddy viscosity decaying with

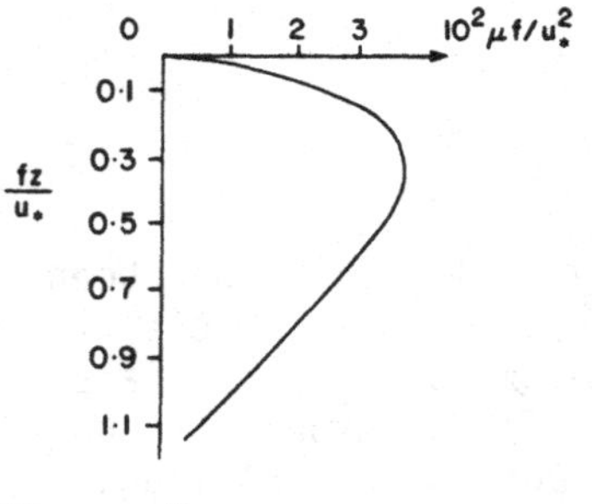

Figure 9

Eddy viscosity distribution from a turbulence model, using $\kappa-\varepsilon$ closure condition.

depth reflects the expected condition of diminishing turbulence intensity as distance below the surface increases. Theories prescribing exponential decays in eddy viscosity with increasing depth have been given by Dobroklonskiy (1969), Lai and Rao (1976) and Witten and Thomas (1976) while, for shallow water, an eddy viscosity proportional to height z_B above the bottom has been considered by Thomas (1975) and one proportional to $z_B^{4/3}$ by Fjeldstad (1929). In various theories, therefore, eddy viscosity either increases or decreases as the free surface is approached from below. While a decrease seems most likely on the weight of evidence so far available, there appears to be little or no observational evidence to confirm this for a full range of wind speeds. Clearly, a knowledge of the behaviour of μ near the sea surface is important in the theoretical estimation of near-surface currents.

On the basis of work by Csanady (1976, 1978, 1979; 1982, p. 13) the surface Ekman layer might be considered to consist of (a) a relatively thin "wall" layer in the immediate vicinity of the free surface through which vertical eddy viscosity increases linearly with depth from a small value μ_S at the surface and (b) an outer layer through which the eddy viscosity remains constant. Thus:

$$\begin{aligned} \mu &= \kappa u_* z + \mu_S , & 0 \le z \le H/\kappa R_e , \\ &= u_* H/R_e + \mu_S , & H/\kappa R_e \le z \le H , \end{aligned} \tag{2.28}$$

where H denotes the depth of the Ekman layer and R_e a Reynolds number lying between 12 and 20. Taking $R_e = 12$ gives $H/\kappa R_e = 0.2H$, indicating a wall layer of thickness equal to one fifth of the total depth of the Ekman layer. Then, with $H = 0.4 u_*/f$ say, the eddy viscosity in the outer layer (evaluated as $u_* H/R_e$) comes to $0.033\, u_*^2/f$. Pearce and Cooper (1981) have used (2.28) for computations of wind-induced flow in shallow water of depth H (figure 10) employing a slip condition at the sea bed. Bowden, Fairbairn and Hughes (1959) proposed a similar distribution of vertical eddy viscosity to account for the presence of a wall layer adjacent to the sea bed in tidal flow; the corresponding theory was worked out by Heaps and Jones (1981). Generally the profile of eddy viscosity is modified by a wall layer at the sea surface and a wall layer at the sea bed. These layers have distinct length and velocity scales when the corresponding Ekman layers do not overlap. At the bottom, the stress, and hence u_*, depends on the motion of the water.

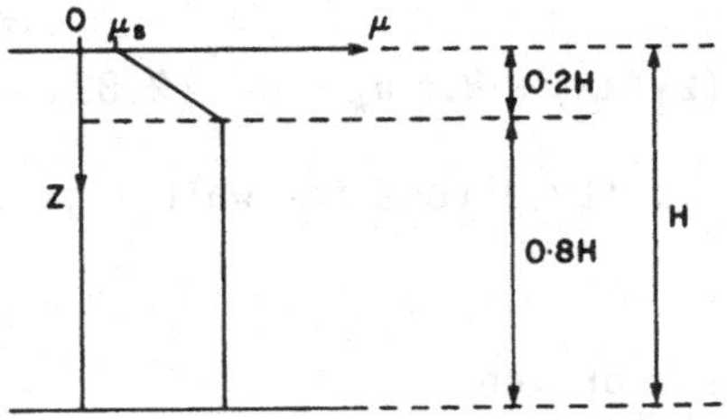

Figure 10

Eddy viscosity distribution with a surface wall layer.

Stress is assumed to be constant through a wall layer, implying a logarithmic variation in velocity through it. Thus, ignoring the effect of the Earth's rotation in a surface wall layer and adopting a variation of μ through it given by (2.24), for a wind stress F_S in the y-direction as before we have

$$-\rho(\kappa u_* z)\frac{\partial u}{\partial z} = 0, \qquad -\rho(\kappa u_* z)\frac{\partial v}{\partial z} = F_S = \rho u_*^2$$

whence

$$\frac{\partial u}{\partial z} = 0\,, \qquad \frac{\partial v}{\partial z} = -\frac{u_*}{\kappa z}$$

and therefore

$$u = u_S\,, \qquad v = v_S - \frac{u_*}{\kappa}\log\left(\frac{z}{z_0}\right). \tag{2.29}$$

Here, u_S, v_S denote the components of current at the free surface (the top of the wall layer), taken at $z = z_0$. At the bottom of the wall layer, $z = z_1$ say: $u = u_1$ and $v = v_1$ where u_1, v_1 are the surface currents from the Ekman theory given by (2.13), since the wind stress is transmitted without change through the wall layer to that part of the Ekman layer, below, where μ is a constant. Therefore, from (2.29),

$$u_S = u_1, \qquad v_S = v_1 + (u_*/\kappa)\log(z_1/z_0)\,. \tag{2.30}$$

This shows that surface current obtained with a wall layer may be obtained from surface current derived from the pure Ekman theory by adding a logarithmic contribution in the wind direction. Such an addition increases the magnitude of the surface current and reduces its angle of deflection from the wind direction.

Instead of (2.29), Csanady (1982, p. 23) effectively writes

$$u = u_S\,, \qquad v = v_S - \frac{u_*}{\kappa}\log\left(\frac{z}{z_0}\right) - 8.5\,u_* \tag{2.31}$$

so that the surface is located at $z = z_0\, e^{-8.5\,\kappa}$. It then follows that

$$u_S = u_1, \qquad v_S = v_1 + (u_*/\kappa)\log(z_1/z_0) + 8.5\,u_* \tag{2.32}$$

and therefore when $z_0 \sim z_1$, the main jump in velocity across the wall layer is close to $8.5\,u_*$.

2.3 Extension of Ekman's theory to determine wind-driven currents in a shallow sea or lake: steady state

Retaining dependence on all three space coordinates x, y and z, for steady-state motion (1.1) and (1.2) are reduced to

$$- fv = - \frac{1}{\rho}\left(\frac{\partial p}{\partial x} + \frac{\partial F_{zx}}{\partial z}\right), \tag{2.33}$$

$$fu = - \frac{1}{\rho}\left(\frac{\partial p}{\partial y} + \frac{\partial F_{zy}}{\partial z}\right). \tag{2.34}$$

Again horizontal shears and the tide-generating forces are ignored and nonlinear terms are omitted. Since the water is homogeneous, ρ constant, the hydrostatic law (1.15) may be assumed. Therefore, introducing an eddy viscosity μ as in (2.3), the above equations become

$$- fv = - g \frac{\partial \zeta}{\partial x} + \frac{\partial}{\partial z}\left(\mu \frac{\partial u}{\partial z}\right), \tag{2.35}$$

$$fu = - g \frac{\partial \zeta}{\partial y} + \frac{\partial}{\partial z}\left(\mu \frac{\partial v}{\partial z}\right). \tag{2.36}$$

Taking $w' = u + iv$ as before and writing

$$\frac{\partial \zeta}{\partial n} = \frac{\partial \zeta}{\partial x} + i \frac{\partial \zeta}{\partial y}, \tag{2.37}$$

equations (2.35) and (2.36) may be combined to give

$$\frac{\partial}{\partial z}\left(\mu \frac{\partial w'}{\partial z}\right) - i f w' = g \frac{\partial \zeta}{\partial n}. \tag{2.38}$$

If the wind stress has components τ_{sx}, τ_{sy} in the x and y directions respectively, and

$$\tau_s = \tau_{sx} + i \tau_{sy} \tag{2.39}$$

then the surface boundary condition on w', prescribing wind stress, is

$$- \rho \mu \frac{\partial w'}{\partial z} = \tau_s, \quad \text{at} \quad z = 0. \tag{2.40}$$

Also, if h denotes the depth of water, varying generally with position (x,y), then a bottom boundary condition on w', prescribing zero slip, is

$$w' = 0, \quad \text{at} \quad z = h. \tag{2.41}$$

Assuming a constant eddy viscosity, a closed analytic solution of (2.38), satisfying (2.40) and (2.41), takes the form (Welander, 1957):

$$w' = P(z) \tau_s + Q(z) \frac{\partial \zeta}{\partial n}, \tag{2.42}$$

where P, Q are simple hyperbolic functions of z in the complex domain. Integrating (2.42) through the depth yields

$$W' = \int_0^h w'\,dz = \frac{A}{\rho f}\tau_s + \frac{gh}{f}B\frac{\partial\zeta}{\partial n}, \qquad (2.43)$$

where A, B are functions only of h/D, D being the frictional depth given by (2.8). Then writing

$$W' = U + iV, \quad A = A_1 + iA_2, \quad B = B_1 + iB_2, \qquad (2.44)$$

where

$$U = \int_0^h u\,dz, \qquad V = \int_0^h v\,dz \qquad (2.45)$$

and A_1, A_2, B_1, B_2 are real, it follows from (2.43) that

$$U = \frac{1}{\rho f}(A_1\tau_{sx} - A_2\tau_{sy}) + \frac{gh}{f}\left(B_1\frac{\partial\zeta}{\partial x} - B_2\frac{\partial\zeta}{\partial y}\right), \qquad (2.46)$$

$$V = \frac{1}{\rho f}(A_2\tau_{sx} + A_1\tau_{sy}) + \frac{gh}{f}\left(B_2\frac{\partial\zeta}{\partial x} + B_1\frac{\partial\zeta}{\partial y}\right). \qquad (2.47)$$

In the steady state, (1.7) reduces to

$$\frac{\partial U}{\partial x} + \frac{\partial V}{\partial y} = 0 \qquad (2.48)$$

and therefore a stream function ψ may be introduced such that

$$U = \frac{\partial\psi}{\partial y}, \qquad V = -\frac{\partial\psi}{\partial x}. \qquad (2.49)$$

Hence, inserting (2.49) into (2.46) and (2.47):

$$\frac{\partial\psi}{\partial y} = \frac{1}{\rho f}(A_1\tau_{sx} - A_2\tau_{sy}) + \frac{gh}{f}\left(B_1\frac{\partial\zeta}{\partial x} - B_2\frac{\partial\zeta}{\partial y}\right), \qquad (2.50)$$

$$-\frac{\partial\psi}{\partial x} = \frac{1}{\rho f}(A_2\tau_{sx} + A_1\tau_{sy}) + \frac{gh}{f}\left(B_2\frac{\partial\zeta}{\partial x} + B_1\frac{\partial\zeta}{\partial y}\right). \qquad (2.51)$$

Solving for $\partial\zeta/\partial x$ and $\partial\zeta/\partial y$ gives

$$\frac{\partial\zeta}{\partial x} = \frac{f}{g}\left(\beta_1\frac{\partial\psi}{\partial y} - \beta_2\frac{\partial\psi}{\partial x}\right) - \frac{1}{\rho g}(\beta_3\tau_{sx} + \beta_4\tau_{sy}), \qquad (2.52)$$

$$\frac{\partial\zeta}{\partial y} = -\frac{f}{g}\left(\beta_1\frac{\partial\psi}{\partial x} + \beta_2\frac{\partial\psi}{\partial y}\right) + \frac{1}{\rho g}(\beta_4\tau_{sx} - \beta_3\tau_{sy}), \qquad (2.53)$$

where

$$\beta_1 = B_1/\left((B_1^2 + B_2^2)\,h\right), \qquad \beta_2 = B_2/\left((B_1^2 + B_2^2)\,h\right),$$
$$\beta_3 = \beta_1 A_1 + \beta_2 A_2, \qquad \beta_4 = \beta_2 A_1 - \beta_1 A_2. \tag{2.54}$$

Then eliminating ζ from (2.52) and (2.53) yields a second-order elliptic equation for ψ of the form

$$\nabla^2\psi + \gamma_1 \frac{\partial\psi}{\partial x} + \gamma_2 \frac{\partial\psi}{\partial y} = \gamma_3 \tag{2.55}$$

where

$$\gamma_1 = \frac{1}{\beta_1}\left(\frac{\partial\beta_1}{\partial x} - \frac{\partial\beta_2}{\partial y}\right), \qquad \gamma_2 = \frac{1}{\beta_1}\left(\frac{\partial\beta_1}{\partial y} + \frac{\partial\beta_2}{\partial x}\right),$$

$$\gamma_3 = \frac{1}{\rho f \beta_1}\left[\beta_3\left(\frac{\partial\tau_{sx}}{\partial y} - \frac{\partial\tau_{sy}}{\partial x}\right) + \beta_4\left(\frac{\partial\tau_{sx}}{\partial x} + \frac{\partial\tau_{sy}}{\partial y}\right) + \right. \tag{2.56}$$
$$\left. + \left(\frac{\partial\beta_4}{\partial x} + \frac{\partial\beta_3}{\partial y}\right)\tau_{sx} + \left(\frac{\partial\beta_4}{\partial y} - \frac{\partial\beta_3}{\partial x}\right)\tau_{sy}\right].$$

Equation (2.55) may be solved numerically on a finite-difference grid covering the lake or sea area under consideration, with ψ given on the circumscribing lateral boundary by

$$\psi = \int_0^s \left(U\frac{dy}{ds} - V\frac{dx}{ds}\right) ds, \tag{2.57}$$

where s is distance measured along the boundary from some fixed position on it; U, V are prescribed functions of s defining flow on the boundary. For a completely closed lake, along the entire periphery the condition of zero normal flow requires

$$\psi = 0\,. \tag{2.58}$$

Having evaluated ψ over the area involved, the horizontal fields of $\partial\zeta/\partial x$ and $\partial\zeta/\partial y$ may be determined from (2.52), (2.53) and thence the vertical structure of horizontal current at any position from (2.42). The main aim of the analysis is to find $\partial\zeta/\partial n$ in (2.42) and thereby the system of gradient currents set up by boundary influence. Equation (2.42) compounds wind currents and gradient currents to yield the total three-dimensional structure of horizontal current. Vertical currents may be deduced subsequently from a vertically-integrated form of (1.4), namely

$$w = \frac{\partial}{\partial x}\int_z^h u\,dz + \frac{\partial}{\partial y}\int_z^h v\,dz\,, \tag{2.59}$$

a form derived, for example, by Heaps (1972).

The above method, first proposed by Welander (1957), has been used to calculate steady wind-driven currents in Lake Ontario (Bonham-Carter and Thomas, 1973) and in Lake Erie (Gedney and Lick, 1972). Consistent with the development described above, eddy viscosity was assumed to be a constant. However, the use of the method when vertical eddy viscosity varies linearly from zero at the bottom to a maximum $\kappa u_* h$ at the surface has been investigated by Thomas (1975). The case in which vertical eddy viscosity decreases exponentially with depth from a prescribed constant maximum at the surface has been investigated by Witten and Thomas (1976).

A further development of the model can be envisaged with the no-slip boundary condition (2.41) replaced by the more general slip condition:

$$\tau_b = -\rho\mu \frac{\partial w'}{\partial z} = k\rho w', \quad \text{at} \quad z = h, \tag{2.60}$$

where k is a constant and τ_b denotes the bottom stress with x, y components: τ_{bx}, τ_{by}, i.e.

$$\tau_b = \tau_{bx} + i\,\tau_{by}\,. \tag{2.61}$$

If, instead of (2.60), it could be assumed that

$$\tau_b = -\rho\mu \frac{\partial w'}{\partial z} = \frac{K\rho W'}{h}, \quad \text{at} \quad z = h, \tag{2.62}$$

where K is a constant, then the vertically-integrated form of (2.38):

$$gh \frac{\partial \zeta}{\partial n} = \left(\frac{\tau_s - \tau_b}{\rho}\right) - i f W' \tag{2.63}$$

would become

$$gh \frac{\partial \zeta}{\partial n} = \frac{\tau_s}{\rho} - \left(\frac{K}{h} + if\right) W' \tag{2.64}$$

and eliminating W' from (2.64) and (2.43) would give $\partial\zeta/\partial n$ for substitution into (2.42). The functions P(z), Q(z) would, in these circumstances, result from the solution of (2.38) subject to conditions (2.40) and (2.62). However the validity of (2.62) is questionable since bottom current $(w')_{z=h}$ is not generally proportional to the depth-mean current W'/h.

2.4 Ekman's approach applied to determine time-dependent wind-driven currents in a shallow sea or lake

Retaining dependence on the time t, equations (1.1) and (1.2) are reduced to

$$\frac{\partial u}{\partial t} - fv = -\frac{1}{\rho}\left(\frac{\partial p}{\partial x} + \frac{\partial F_{zx}}{\partial z}\right), \tag{2.65}$$

$$\frac{\partial v}{\partial t} + fu = -\frac{1}{\rho}\left(\frac{\partial p}{\partial y} + \frac{\partial F_{zy}}{\partial z}\right). \tag{2.66}$$

The same basic assumptions are made as before, namely that horizontal shears, tide-generating forces and nonlinear terms may be neglected. Introducing the hydrostatic law (1.15) and an eddy viscosity μ, the equations become

$$\frac{\partial u}{\partial t} - fv = -g\frac{\partial \zeta}{\partial x} + \frac{\partial}{\partial z}\left(\mu\frac{\partial u}{\partial z}\right), \tag{2.67}$$

$$\frac{\partial v}{\partial t} + fu = -g\frac{\partial \zeta}{\partial y} + \frac{\partial}{\partial z}\left(\mu\frac{\partial v}{\partial z}\right), \tag{2.68}$$

which are identical to (2.35) and (2.36) but with the time derivatives $\partial u/\partial t$, $\partial v/\partial t$ now included. Combining (2.67) and (2.68), in the same way that (2.35) and (2.36) were previously combined, yields

$$\frac{\partial w'}{\partial t} + ifw' = -g\frac{\partial \zeta}{\partial n} + \frac{\partial}{\partial z}\left(\mu\frac{\partial w'}{\partial z}\right). \tag{2.69}$$

Welander (1957), assuming a constant eddy viscosity, considered two distinct solutions of this equation relating to the vertical structure of current. The first gives the dynamic response

$$w' = w_1'(z,t) \tag{2.70}$$

to unit wind stress in the x-direction, suddenly created at $t = 0$ and maintained; w_1' satisfies

$$\frac{\partial w'}{\partial t} + ifw' = \mu\frac{\partial^2 w'}{\partial z^2} \tag{2.71}$$

subject to the conditions:

$$\begin{aligned} -\rho\mu\frac{\partial w'}{\partial z} &= 1, \quad \text{at} \quad z = 0,\\ w' &= 0, \quad \text{at} \quad z = h,\\ w' &= 0, \quad \text{at} \quad t = 0. \end{aligned} \tag{2.72}$$

These conditions prescribe the wind stress, zero bottom slip and an initial state of rest, respectively. The second solution gives the response

$$w' = w_2'(z,t) \tag{2.73}$$

to unit surface slope in the x-direction, suddenly created at $t = 0$ and maintained; w_2' satisfies

$$\frac{\partial w'}{\partial t} + ifw' = -g + \mu\frac{\partial^2 w'}{\partial z^2} \tag{2.74}$$

subject to:

$$-\rho\mu \frac{\partial w'}{\partial z} = 0, \quad \text{at} \quad z = 0,$$
$$w' = 0, \quad \text{at} \quad z = h, \qquad (2.75)$$
$$w' = 0, \quad \text{at} \quad t = 0.$$

Analytical expressions for w_1' and w_2', in the form of infinite series, are known from the work of Fjeldstad (1930) and Hidaka (1933).

Summing contributions to w' through time, coming from the succession of differential wind stress and surface slope increments which make up the continuous variations of those quantities, gives

$$w' = \int_0^t \left[w_1'(z,t') \frac{\partial \tau_s}{\partial t}(t-t') + w_2'(z,t') \frac{\partial^2 \zeta}{\partial t \, \partial n}(t-t') \right] dt', \qquad (2.76)$$

where it is assumed that τ_s and $\partial\zeta/\partial n$, defined earlier by (2.39) and (2.37), start from zero values at $t = 0$. Integrating by parts in (2.76) produces the alternative result:

$$w' = \int_0^t \left[\tau_s(t-t') \frac{\partial w_1'}{\partial t}(z,t') + \frac{\partial \zeta}{\partial n}(t-t') \frac{\partial w_2'}{\partial t}(z,t') \right] dt'. \qquad (2.77)$$

The vertical integration of this yields

$$W' = U + iV$$
$$= \int_0^t \left[\tau_s(t-t') \frac{\partial W_1'}{\partial t}(t') + \frac{\partial \zeta}{\partial n}(t-t') \frac{\partial W_2'}{\partial t}(t') \right] dt', \qquad (2.78)$$

where

$$W_1' = \int_0^h w_1' \, dz, \qquad W_2' = \int_0^h w_2' \, dz. \qquad (2.79)$$

Equation (1.7) written in the form:

$$\frac{\partial \zeta}{\partial t} + \frac{\partial U}{\partial x} + \frac{\partial V}{\partial y} = 0 \qquad (2.80)$$

with U and V given by (2.78), then provides an integro-differential equation for ζ which, in principle, may be solved numerically proceeding step-by-step through time. At each step, a boundary condition on $\partial\zeta/\partial n$ has to be satisfied by specifying the normal component of W' on the boundary (that normal component is zero at a coastal boundary). The vertical structure of horizontal current may be determined through time from (2.77) as the calculations advance.

The method described above was suggested by Welander (1957) but, as far as I know, has never been put into practice. Another approach due to Jelesnianski (1970), used by him for modelling hurricane surges on the east coast of the United States, solves the vertically-integrated form of (2.69):

$$\frac{\partial W'}{\partial t} + ifW' = -gh\frac{\partial \zeta}{\partial n} + \frac{\tau_s - \tau_b}{\rho}, \tag{2.81}$$

along with the continuity equation (2.80), by performing two-dimensional initial-value computations on a horizontal finite-difference grid covering the sea area under consideration. (For the methodology of such a two-dimensional model, see Ramming and Kowalik (1980, pp. 112-168)). These calculations yield the changing distribution of ζ and W' over the area through time. Then, knowing $\partial\zeta/\partial n$, the vertical structure of horizontal current w' at any place and time may be determined from the convolution integral (2.77). However a central problem is the evaluation of the bottom stress τ_b in (2.81). Jelesnianski took

$$\tau_b = -\rho\mu\left(\frac{\partial w'}{\partial z}\right)_{z=h}$$

and, inserting w' from (2.77), obtained

$$\tau_b = \frac{2\mu}{h^2}\left[\int_0^t \Big(\tau_s(t-t')\,K_1(t') - \rho g h\frac{\partial\zeta}{\partial n}(t-t')\,K_2(t')\Big)\,e^{-ift'}\,dt'\right] \tag{2.82}$$

where

$$K_1(t) = \sum_{n=0}^{\infty} (-1)^n \pi(n+1/2)\exp\left(-((n+1/2)\pi)^2\frac{\mu t}{h^2}\right), \tag{2.83}$$

$$K_2(t) = \sum_{n=0}^{\infty} \exp\left(-((n+1/2)\pi)^2\frac{\mu t}{h^2}\right) \tag{2.84}$$

for large $\mu t/h^2$. Other series expansions for K_1 and K_2 hold for small $\mu t/h^2$. For the numerical computations, Jelesnianski used the approximations:

$$K_1 = 0.8835\left(\frac{\pi}{2}\right)\left[e^{-\pi^2\mu t/4h^2} - e^{-9\pi^2\mu t/4h^2}\right], \tag{2.85}$$

$$K_2 = 1.2337\,e^{-\pi^2\mu t/4h^2} \tag{2.86}$$

and developed a recursion formula by which the integral in (2.82) could be evaluated economically from one time level to the next in the two-dimensional model. A complete system of computation depending on equations

(2.77), (2.80), (2.81), (2.82), (2.85) and (2.86) was thereby established for the determination of variations of sea level and three-dimensional current structure through time. Effectively, the system consists of a one-dimensional model through the vertical and a two dimensional model in the horizontal, coupled together through bottom friction. The method has been applied very effectively by Forristall (1974, 1980) and Forristall, Hamilton and Cardone (1977) to evaluate the structure of storm-generated currents in the Gulf of Mexico. In that work, deviations from the original formulation have included the use of a slip condition at the sea bed in place of the zero slip condition ((2.72), (2.75)) and the use of a two-layered current system with different eddy viscosities in the upper and lower layers in place of the single-layered system with the same eddy viscosity throughout.

Manisfestly the main concern of the foregoing approach lies in the treatment of bottom friction. However if τ_b can be expressed in terms of the depth-mean current, as in (2.62), then the problem becomes easier since in those circumstances (2.80) and (2.81) may be solved directly for ζ and W' without appeal to (2.77). The forms for w_1' and w_2' are then different to before, since a new bottom boundary condition is involved in (2.72) and (2.75). Moreover, with the one-dimensional dynamics through the vertical (equations (2.69) - (2.77)) then effectively decoupled from the two-dimensional dynamics in the horizontal (equations (2.80), (2.81)), the former may be solved numerically permitting the use of a general variation of eddy viscosity through the depth.

An alternative approach to that of Jelesnianski, outlined above, has been suggested by Kielmann and Kowalik (1980). They solve (2.69) as a difference equation in t and a differential equation in z, and derive an expression for τ_b at time $t+\Delta t$ in terms of τ_s at time $t+\Delta t$, $\partial\zeta/\partial n$ at time $t+\Delta t/2$ and W' at time t. This expression may then be used in developing the two-dimensional solution for ζ and W'. The associated vertical current structure is obtained by referring back to the solution of the difference-differential equation.

For the prediction of wind-induced water levels in Lake Erie, Platzman (1963) adopted another approach to the solution of the dynamical equations (2.69), (2.80). This is now described. Assuming constant vertical eddy viscosity, (2.69) may be put into the form:

$$\frac{\partial^2 w'}{\partial z^2} = \frac{1}{\mu}\left(if + \frac{\partial}{\partial t}\right)w' + \frac{g}{\mu}\frac{\partial\zeta}{\partial n}. \tag{2.87}$$

Solving this, satisfying the surface wind-stress condition (2.40) and the bottom boundary condition of zero slip (2.41), yields

$$w' = \frac{\sinh\sigma(1-\xi)}{\sigma\cosh\sigma}\,\frac{h\tau_s}{\rho\mu} + \frac{1}{\sigma^2}\left(\frac{\cosh\sigma\xi}{\cosh\sigma} - 1\right)\frac{gh^2}{\mu}\frac{\partial\zeta}{\partial n}, \tag{2.88}$$

where $\xi = z/h$ and

$$\sigma^2 = \sigma_0^2 + \lambda\ , \tag{2.89}$$

in which

$$\sigma_0^2 = \frac{ifh^2}{\mu}\ , \qquad \lambda = \frac{h^2 \partial}{\mu\, \partial t}\ . \tag{2.90}$$

Then, integrating (2.88) through the depth, from the surface $z = 0$ to the bottom $z = h$, gives

$$\frac{\mu}{h^2}\left(\sigma^2 + G(\sigma)\right) W' = -gh\frac{\partial\zeta}{\partial n} + \left(1 + H(\sigma)\right)\frac{\tau_s}{\rho}\ , \tag{2.91}$$

where

$$G(\sigma) = \frac{\sigma \tanh \sigma}{1 - \sigma^{-1} \tanh \sigma}\ , \tag{2.92}$$

$$H(\sigma) = \frac{\sigma^{-1} \tanh \sigma - \operatorname{sech} \sigma}{1 - \sigma^{-1} \tanh \sigma}\ . \tag{2.93}$$

Making Taylor series approximations of the form

$$\begin{aligned} G(\sigma) &= G_0(\sigma_0) + \lambda\, G_1(\sigma_0)\ , \\ H(\sigma) &= H_0(\sigma_0) + \lambda\, H_1(\sigma_0)\ , \end{aligned} \tag{2.94}$$

in (2.91), it follows that

$$\frac{\partial W'}{\partial t} + ifAW' = -gh\frac{\partial\zeta}{\partial n}B + \left(C + \frac{J}{if}\frac{\partial}{\partial t}\right)\frac{\tau_s}{\rho}\ , \tag{2.95}$$

where

$$\begin{aligned} A &= \frac{1 + \sigma_0^{-2} G_0}{1 + G_1}\ , & B &= \frac{1}{1 + G_1}\ , \\ C &= \frac{1 + H_0}{1 + G_1}\ , & J &= \frac{\sigma_0^2 H_1}{1 + G_1}\ . \end{aligned} \tag{2.96}$$

It turns out that J is small and therefore, for practical purposes, (2.95) may be written

$$\frac{\partial W'}{\partial t} + ifAW' = -gh\frac{\partial\zeta}{\partial n}B + C\frac{\tau_s}{\rho}\ . \tag{2.97}$$

In particular, for large depths ($|\sigma_0| \gg 1$), $A = B = C = 1$ and (2.97) reduces to

$$\frac{\partial W'}{\partial t} + ifW' = -gh\frac{\partial\zeta}{\partial n} + \frac{\tau_s}{\rho}\ , \tag{2.98}$$

while for small depths ($|\sigma_0| \ll 1$, relating to a shallow sea or lake),

$$A = 5/2\sigma_0^2 + 43/42, \qquad B = 5/6, \qquad C = 5/4$$

and (2.97) becomes

$$\frac{\partial W'}{\partial t} + i\left(\frac{43}{42} f\right) W' = - gh \frac{\partial \zeta}{\partial n} + \frac{\tau_s - \hat{\tau}_b}{\rho}, \tag{2.99}$$

where

$$\hat{\tau}_b = \frac{5\rho\mu}{2h^2} W' - \frac{1}{6} \rho g h \frac{\partial \zeta}{\partial n} - \frac{\tau_s}{4}. \tag{2.100}$$

Equation (2.99) is a vertically-integrated form of (2.69). Comparing it with (2.81), the effective bottom friction $\hat{\tau}_b$ is seen to depend on the depth-mean current W'/h with a coefficient inversely proportional to the depth. Manifestly, bottom friction reduces the effective pressure gradient and augments the effective wind stress; the Coriolis force is affected only slightly.

Equations (2.99) and (2.80) constitute a two-dimensional model for determining ζ and W', incorporating the effects of bottom stress. Having computed the variations of $\partial\zeta/\partial n$ from this system, in the horizontal and through time, vertical current structure may be determined correspondingly from (2.77). Following Platzman's application of this theory to Lake Erie, Jelesnianski (1967) has applied the method to the computation of hurricane surges: employing a slip condition at the sea bed (see (2.60)) instead of the no-slip condition (2.41).

3. SPECTRAL APPROACH

Solution of the hydrodynamic equations for time-dependent motion in a shallow sea or lake may be interestingly accomplished by expanding variables in a series of orthogonal functions through the depth. The coefficients in those expansions, varying in the horizontal and through time, may be evaluated numerically from a two-dimensional system of partial differential equations. The application of this vertical spectral approach to determine wind-induced motion is now described. Treatments are given relating to homogeneous and stratified water. Thus, again, vertical current structure is found within the framework of a calculation evaluating horizontal fields of motion.

3.1 Wind-induced motion: homogeneous water

Referring back to (2.67), (2.68), (2.80) and (2.45), the basic hydrodynamic equations are taken as

$$\frac{\partial u}{\partial t} - fv = - g \frac{\partial \zeta}{\partial x} + \frac{\partial}{\partial z}\left(\mu \frac{\partial u}{\partial z}\right), \tag{3.1}$$

$$\frac{\partial v}{\partial t} + fu = -g\frac{\partial \zeta}{\partial y} + \frac{\partial}{\partial z}(\mu \frac{\partial v}{\partial z}), \tag{3.2}$$

$$\frac{\partial \zeta}{\partial t} + \frac{\partial}{\partial x}\int_0^h u\,dz + \frac{\partial}{\partial y}\int_0^h v\,dz = 0, \tag{3.3}$$

where generally it is assumed that $\mu = \mu(x,y,z)$.

Solutions are sought for u, v and ζ subject to a prescribed surface wind stress:

$$-\rho\mu\frac{\partial u}{\partial z} = \tau_{sx}, \quad -\rho\mu\frac{\partial v}{\partial z} = \tau_{sy}, \quad \text{at} \quad z = 0 \tag{3.4}$$

and a bottom-stress condition:

$$-\rho\mu\frac{\partial u}{\partial z} = \tau_{bx}, \quad -\rho\mu\frac{\partial v}{\partial z} = \tau_{by}, \quad \text{at} \quad z = h. \tag{3.5}$$

Surface and bottom conditions have already been formulated in (2.40) and (2.60) using complex notation.

With a view to eliminating the z-coordinate from (3.1) and (3.2), by the vertical integration of each equation from the surface $z = 0$ to the bottom $z = h$, consider the evaluation of the integrals

$$I_x = \int_0^h Z\frac{\partial}{\partial z}(\mu\frac{\partial u}{\partial z})\,dz, \quad I_y = \int_0^h Z\frac{\partial}{\partial z}(\mu\frac{\partial v}{\partial z})\,dz \tag{3.6}$$

where $Z = Z(x,y,z)$ is an as yet unknown kernal function. Integrating by parts twice in each integral of (3.6), using (3.4) and (3.5), yields

$$\begin{aligned} I_x &= Z_0\frac{\tau_{sx}}{\rho} - Z_h\frac{\tau_{bx}}{\rho} + \mu_0 Z_0' u_0 - \mu_h Z_h' u_h - \lambda h\hat{u}, \\ I_y &= Z_0\frac{\tau_{sy}}{\rho} - Z_h\frac{\tau_{by}}{\rho} + \mu_0 Z_0' v_0 - \mu_h Z_h' v_h - \lambda h\hat{v}, \end{aligned} \tag{3.7}$$

where

$$\hat{u} = \frac{1}{h}\int_0^h Zu\,dz, \quad \hat{v} = \frac{1}{h}\int_0^h Zv\,dz \tag{3.8}$$

and $Z' = \partial Z/\partial z$; suffix 0 denotes evaluation at $z = 0$ and suffix h evaluation at $z = h$. In thus deriving (3.7), it is necessary to assume that

$$\frac{\partial}{\partial z}(\mu Z') = -\lambda Z, \tag{3.9}$$

where λ is independent of z.

Taking: $$Z_0 = 1, \quad Z_0' = 0, \quad \mu_h Z_h' + k Z_h = 0 \tag{3.10}$$

and assuming linear bottom slip (see (2.60)):

$$\tau_{bx} = k\rho u_h , \quad \tau_{by} = k\rho v_h , \tag{3.11}$$

where k is a constant, it follows from (3.7) that

$$I_x = \frac{\tau_{sx}}{\rho} - \lambda h \hat{u} , \quad I_y = \frac{\tau_{sy}}{\rho} - \lambda h \hat{v} . \tag{3.12}$$

Therefore, multiplying (3.1) and (3.2) by Z, integrating from $z = 0$ to $z = h$, and then dividing by h, yields

$$\frac{\partial \hat{u}}{\partial t} + \lambda \hat{u} - f\hat{v} = -g a \frac{\partial \zeta}{\partial x} + \frac{\tau_{sx}}{\rho h} , \tag{3.13}$$

$$\frac{\partial \hat{v}}{\partial t} + \lambda \hat{v} + f\hat{u} = -g a \frac{\partial \zeta}{\partial y} + \frac{\tau_{sy}}{\rho h} \tag{3.14}$$

where

$$a = \frac{1}{h}\int_0^h Z\, dz . \tag{3.15}$$

Equations (3.13), (3.14) are vertically-integrated forms of (3.1), (3.2) involving $\hat{u}$, $\hat{v}$, ζ instead of u, v, ζ. Note that u, v are functions of x,y,z,t while $\hat{u}, \hat{v}, \zeta$ are functions of x, y, t only. Now,

$$\lambda = \lambda_r, \quad Z = Z_r, \quad (r = 1, 2, \ldots, \infty), \tag{3.16}$$

where λ_r denote the ascending eigenvalues and Z_r the corresponding eigenfunctions associated with the differential equation (3.9) when solved in the range $0 \le z \le h$ subject to the end conditions (3.10). Hence, from (3.13) and (3.14):

$$\frac{\partial \hat{u}_r}{\partial t} + \lambda_r \hat{u}_r - f\hat{v}_r = -g a_r \frac{\partial \zeta}{\partial x} + \frac{\tau_{sx}}{\rho h} , \tag{3.17}$$

$$\frac{\partial \hat{v}_r}{\partial t} + \lambda_r \hat{v}_r + f\hat{u}_r = -g a_r \frac{\partial \zeta}{\partial y} + \frac{\tau_{sy}}{\rho h} \tag{3.18}$$

where, from (3.8)

$$\hat{u}_r = \frac{1}{h}\int_0^h Z_r u\, dz , \quad \hat{v}_r = \frac{1}{h}\int_0^h Z_r v\, dz . \tag{3.19}$$

and, from (3.15),

$$a_r = \frac{1}{h}\int_0^h Z_r\, dz \,. \tag{3.20}$$

Inserting (3.16) in (3.9):

$$\frac{\partial}{\partial z}(\mu Z_r') = -\lambda_r Z_r \,, \qquad \frac{\partial}{\partial z}(\mu Z_s') = -\lambda_s Z_s \tag{3.21}$$

where r, s denote any two positive integers. Hence

$$(\lambda_r - \lambda_s) Z_r Z_s = \frac{\partial}{\partial z}\Big(\mu (Z_r Z_s' - Z_s Z_r')\Big) \tag{3.22}$$

so that

$$(\lambda_r - \lambda_s)\int_0^h Z_r Z_s\, dz = \mu_h (Z_r Z_s' - Z_s Z_r')_h - \mu_0 (Z_r Z_s' - Z_s Z_r')_0 \,. \tag{3.23}$$

Incorporating the conditions of (3.10) then yields

$$\int_0^h Z_r Z_s\, dz = 0 \quad (r \neq s) \,. \tag{3.24}$$

This shows that the eigenfunctions form an orthogonal set. As a consequence, writing

$$u = \sum_{r=1}^{\infty} A_r Z_r \,, \qquad v = \sum_{r=1}^{\infty} B_r Z_r \tag{3.25}$$

where A_r, B_r are independent of z, and substituting these series for u and v into (3.19), we obtain

$$\hat{u}_r = A_r/\phi_r \,, \qquad \hat{v}_r = B_r/\phi_r \tag{3.26}$$

where

$$\phi_r = h \Big/ \int_0^h Z_r^2\, dz \,. \tag{3.27}$$

Therefore, from (3.26) and (3.25):

$$u = \sum_{r=1}^{\infty} \phi_r \hat{u}_r Z_r \,, \qquad v = \sum_{r=1}^{\infty} \phi_r \hat{v}_r Z_r \,. \tag{3.28}$$

The substitution of u, v from (3.28) into (3.3) gives the equation of continuity in terms of $\hat{u}_r$, $\hat{v}_r$:

$$\frac{\partial \zeta}{\partial t} + \sum_{r=1}^{\infty}\left(\frac{\partial}{\partial x}(h\, a_r \phi_r \hat{u}_r) + \frac{\partial}{\partial y}(h\, a_r \phi_r \hat{v}_r)\right) = 0 \,. \tag{3.29}$$

For practical purposes infinite series are truncated to include only the first M terms. Then, from (3.17), (3.18) and (3.29):

$$\frac{\partial\zeta}{\partial t} = -\sum_{r=1}^{M}\left(\frac{\partial}{\partial x}(h\,a_r\,\phi_r\,\hat{u}_r) + \frac{\partial}{\partial y}(h\,a_r\,\phi_r\,\hat{v}_r)\right), \tag{3.30}$$

$$\frac{\partial\hat{u}_r}{\partial t} = -\lambda_r\,\hat{u}_r + f\,\hat{v}_r - g\,a_r\,\frac{\partial\zeta}{\partial x} + \frac{\tau_{sx}}{\rho h}, \quad (r=1,2,\ldots,M), \tag{3.31}$$

$$\frac{\partial\hat{v}_r}{\partial t} = -\lambda_r\,\hat{v}_r - f\,\hat{u}_r - g\,a_r\,\frac{\partial\zeta}{\partial y} + \frac{\tau_{sy}}{\rho h}, \quad (r=1,2,\ldots,M), \tag{3.32}$$

These $2M+1$ equations constitute a two-dimensional set, without z dependence, which may be solved numerically for ζ, $\hat{u}_r$, $\hat{v}_r$ $(r=1,2,\ldots,M)$ on a horizontal finite-difference grid covering the sea or lake area under consideration. The horizontal fields of ζ, $\hat{u}_r$, $\hat{v}_r$ are advanced through time in a series of consecutive time increments Δt, values at time $t+\Delta t$ being deduced from those at time t applying (3.30), (3.31) and (3.32) in succession. Assuming that motion is generated by wind stress from an initial state of rest ($u=v=\zeta=0$ at $t=0$) the solution is developed from zero values, namely

$$\hat{u}_r = \hat{v}_r = \zeta = 0, \quad \text{at} \quad t=0. \tag{3.33}$$

Throughout the development, appropriate dynamical conditions have to be satisfied along the lateral boundaries of the area. Thus, along coastal boundaries, zero normal flow ($u\cos\Theta + v\sin\Theta = 0$) requires

$$\hat{u}_r\cos\Theta + \hat{v}_r\sin\Theta = 0 \tag{3.34}$$

where Θ denotes the inclination of the normal to the axis of x.

As the solution of (3.30) - (3.32) evolves through time, yielding the changing horizontal fields of $\hat{u}_r$ and $\hat{v}_r$, the vertical structure of horizontal current at any place and time may, on the basis of (3.28), be deduced from

$$u = \sum_{r=1}^{M}\phi_r\,\hat{u}_r\,Z_r\,, \qquad v = \sum_{r=1}^{M}\phi_r\,\hat{v}_r\,Z_r\,. \tag{3.35}$$

This structure may be regarded as consisting of a series of current modes through the vertical of progressively increasing order, denoted by $r = 1, 2, \ldots, M$. The x and y directed currents in the r-mode are, respectively, $\phi_r\,\hat{u}_r\,Z_r$ and $\phi_r\,\hat{v}_r\,Z_r$. Since

$$\begin{gathered} Z_r = Z_r(x,y,z) \\ \hat{u}_r,\ \hat{v}_r = \hat{u}_r,\ \hat{v}_r\,(x,y,t)\,, \qquad \phi_r = \phi_r(x,y) \end{gathered} \tag{3.36}$$

it is evident that the eigenfunction Z_r determines the *shape* of the current profile through the vertical in the r-mode, since only Z_r has z-dependence.

Heaps (1972) applied the preceding theory to calculate the changing three-dimensional system of currents set up in a closed rectangular basin by a uniform longitudinal wind stress of 15 dyn cm^{-2} suddenly created over the water surface at $t = 0$ (figure 11). Eddy viscosity μ was assumed

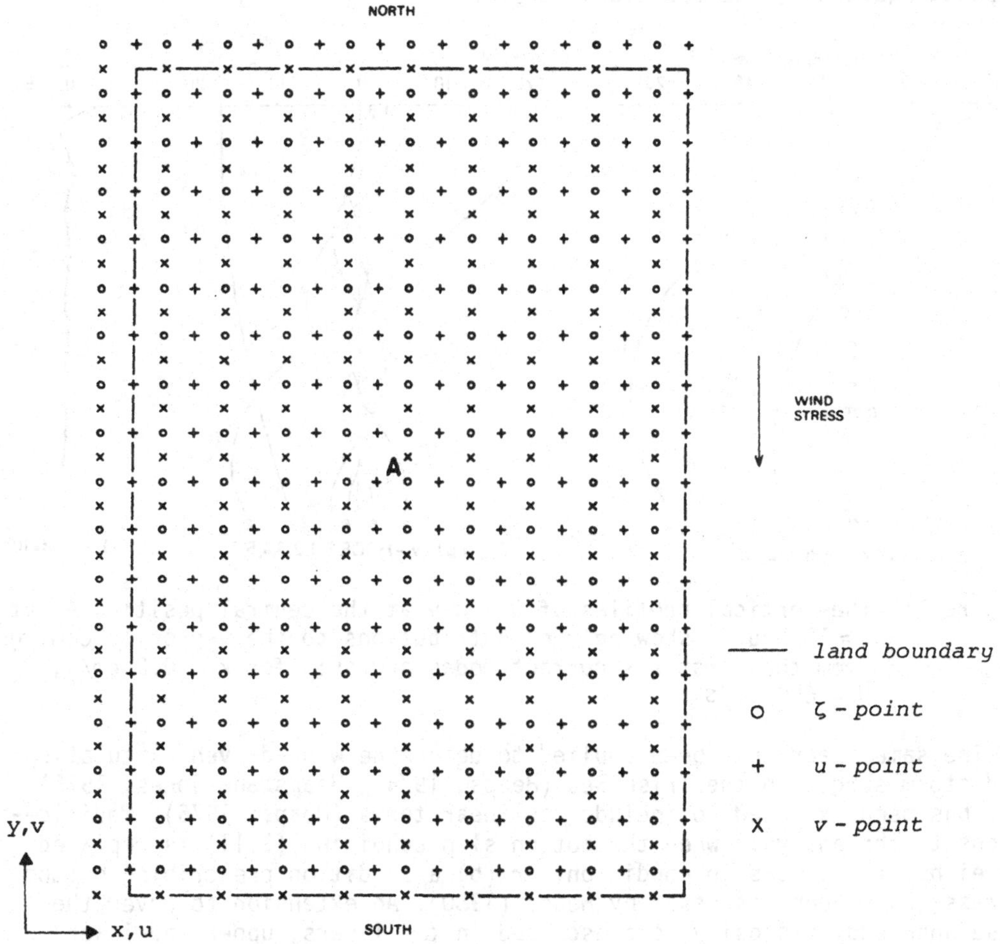

Figure 11 Closed rectangular sea basin with dimensions and rotation representative of the North Sea. The basin is subjected to a uniform longitudinal wind stress from the north and the resulting elevations of the sea surface (ζ) and current components (u,v) at any depth are evaluated at the grid points shown.

to be a constant; the Z_r then come out as simple cosine functions. Figure 12 shows the calculated vertical profiles of u and v at the central position of the basin 15 hours after the onset of the wind. The first ten current modes were taken into account (M = 10) and the series for u and v were completed to infinity approximately, beyond the first ten terms, by adding certain closed expressions Δu, Δv respectively. The contributions to the v-profile coming from each of the first six modes and Δv are shown in the figure. The variations through the vertical shown in figure 12 are plotted against ξ, the fractional depth.

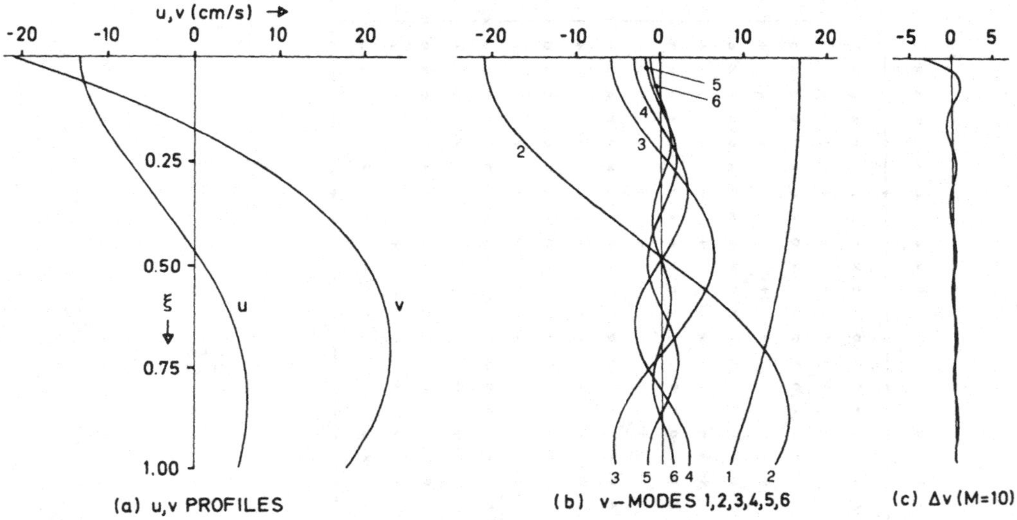

Figure 12 The vertical profiles of u and v at the central position A at t = 15 hours, showing the contributions to the v-profile coming from the first six current modes and Δv: for k = 0.2 cm/s, N = 650 cm/s.

The same theory has been applied to determine wind-driven circulations and storm surges in the Irish Sea (Heaps, 1974, Heaps and Jones, 1975) and has been extended to include nonlinear terms (Heaps, 1976). Modifications to the analysis when the botton slip condition (3.11) is replaced by either (a) a no-slip condition, or (b) a condition prescribing bottom stress, have been discussed by Heaps (1980). An extension to cover the case when eddy viscosity is prescribed in *two* layers, upper and lower, has been applied to compute tides in the Irish Sea with a bed friction layer (Heaps, 1981; Heaps and Jones, 1981).

The use of the Galerkin method with horizontal components of current expanded in terms of an arbitrary set of basis functions through the vertical, generalizes the method described above. Making allowance for an arbitrary variation of eddy viscosity is then facilited. Basis sets consisting respectively of B-splines, cosines, Chebyshev polynomials, Legendre polynomials and eigenfunctions (themselves expressed in terms of an arbitrary set of basis functions, chosen to be B-splines) have been employed by Davies (1977a,b ; 1983) and Davies and Owen (1979). The hydrodynamical equations in fully nonlinear form have been treated in this manner (Davies, 1980). Applications to determine tides, storm surges and meteorologically-induced circulations on the north-west European continental shelf have been described by Davies (1981a,b; 1982a) and Davies and Furnes (1980).

3.2 Wind-induced motion: continuously stratified water of constant depth

Accounting for variable density, the hydrodanamic equations are written in the form

$$\frac{\partial u}{\partial t} - fv = -\frac{1}{\rho}\left(\frac{\partial p}{\partial x} + \frac{\partial F_{zx}}{\partial z}\right), \tag{3.37}$$

$$\frac{\partial v}{\partial t} + fu = -\frac{1}{\rho}\left(\frac{\partial p}{\partial y} + \frac{\partial F_{zy}}{\partial z}\right), \tag{3.38}$$

$$\frac{\partial u}{\partial x} + \frac{\partial v}{\partial y} + \frac{\partial w}{\partial z} = 0, \tag{3.39}$$

$$\frac{\partial p}{\partial z} = \rho g, \tag{3.40}$$

$$\frac{\partial \rho}{\partial t} + w\frac{\partial \rho}{\partial z} = 0. \tag{3.41}$$

Here, equations (3.37) - (3.40) come from (1.1), (1.2), (1.4) and (1.6) respectively, while (3.41) is the adiabaticity equation $D\rho/Dt = 0$ with terms involving products of the motion ignored.

Suppose that, in equilibrium, the water is continuously stratified through the depth with density given by

$$\rho = \rho_*(z), \quad 0 \le z \le h. \tag{3.42}$$

The density at the water surface in these conditions is denoted by ρ_0, so that

$$\rho_0 = \rho_*(0). \tag{3.43}$$

During the motion:

$$\rho = \rho_*(z) + \rho_1(x, y, z, t) \tag{3.44}$$

and correspondingly, from (3.40)

$$p = p_a + p_* + p_1 \, , \tag{3.45}$$

where

$$p_* = g \int_0^z \rho_* \, dz \tag{3.46}$$

and

$$p_1 = g \int_0^z \rho_1 \, dz + g \rho_0 \zeta \, . \tag{3.47}$$

Here, ρ_*, p_* denote the static and ρ_1, p_1 the dynamic parts of the density and pressure. As before, p_a denotes atmospheric pressure on the sea surface, regarded as a constant, and ζ the elevation of that surface above its mean level $z = 0$.

Replacing ρ by ρ_0 according to the Boussinesq approximation, writing

$$F_{zx} = - \rho_0 \mu \frac{\partial u}{\partial z} \, , \qquad F_{zy} = - \rho_0 \mu \frac{\partial v}{\partial z} \tag{3.48}$$

and introducing (3.45), the equations (3.37) and (3.38) reduce to

$$\frac{\partial u}{\partial t} - fv = - \frac{1}{\rho_0} \frac{\partial p_1}{\partial x} + \frac{\partial}{\partial z} (\mu \frac{\partial u}{\partial z}) \, , \tag{3.49}$$

$$\frac{\partial v}{\partial t} + fu = - \frac{1}{\rho_0} \frac{\partial p_1}{\partial y} + \frac{\partial}{\partial z} (\mu \frac{\partial v}{\partial z}) \, . \tag{3.50}$$

Further, differentiating with respect to t in (3.40) and then substituting for $\partial \rho / \partial t$ from (3.41) yields

$$\frac{\partial}{\partial z} (\frac{\partial p}{\partial t}) = - g \frac{\partial \rho}{\partial z} w \, . \tag{3.51}$$

Ignoring products of the motion, taking account of (3.44) - (3.47), this relation reduces to the linearized form:

$$\frac{\partial}{\partial z} (\frac{\partial p_1}{\partial t}) = - g \frac{d \rho_*}{dz} w \, . \tag{3.52}$$

From (3.47)

$$\frac{\partial p_1}{\partial t} = g \rho_0 \frac{\partial \zeta}{\partial t} \, , \quad \text{at} \quad z = 0 \tag{3.53}$$

and therefore because of the kinematic condition

$$w = - \frac{\partial \zeta}{\partial t} \, , \quad \text{at} \quad z = 0 \tag{3.54}$$

we obtain the surface boundary condition:

$$\frac{\partial p_1}{\partial t} + g\rho_0 w = 0\,, \quad \text{at} \quad z = 0\,. \tag{3.55}$$

A bottom boundary condition, exactly true for a flat bottom, is

$$w = 0\,, \quad \text{at} \quad z = h\,. \tag{3.56}$$

Adopting the approach described by Simons (1980, p. 47), solutions of the equations (3.39), (3.49), (3.50) and (3.52), subject to (3.55) and (3.56), are sought of the form:

$$\begin{aligned} w &= \hat{w}(x,y,t)\, Z(z)\,, \\ u &= \hat{u}(x,y,t)\, Z'(z)\,, \\ v &= \hat{v}(x,y,t)\, Z'(z)\,, \\ p_1 &= g\rho_0 \hat{\xi}(x,y,t)\, Z'(z)\,, \end{aligned} \tag{3.57}$$

where $Z' = dZ/dz$. Note: in separating the vertical and horizontal variations, it follows from the equation of continuity (3.39) and the equations of motion (3.49) and (3.50) that if $w \propto Z$ then $u, v, p_1 \propto Z'$; the use of ρ_0 instead of ρ_* in p_1 follows from the Boussinesq approximation. Substituting (3.57) into (3.52), (3.55) and (3.56) yields

$$\frac{\partial \hat{\xi}}{\partial t} Z'' + \frac{1}{\rho_0} \frac{d\rho_*}{dz} \hat{w} Z = 0, \tag{3.58}$$

$$\frac{\partial \hat{\xi}}{\partial t} Z'(0) + \hat{w}\, Z(0) = 0, \tag{3.59}$$

$$Z(h) = 0\,. \tag{3.60}$$

Separating the variables in (3.58) leads to

$$\frac{\partial \hat{\xi}/\partial t}{\hat{w}} = -\frac{1}{\rho_0} \frac{d\rho_*}{dz} \left(\frac{Z}{Z''}\right) = \frac{c^2}{g}\,, \tag{3.61}$$

where c is the separation constant having the dimensions of a velocity. Therefore,

$$\hat{w} = \frac{g}{c^2} \frac{\partial \hat{\xi}}{\partial t} \tag{3.62}$$

and

$$Z'' = -\left(N^2/c^2 \right) Z, \tag{3.63}$$

where

$$N = \left(\frac{g}{\rho_0} \frac{d\rho_*}{dz}\right)^{1/2} \tag{3.64}$$

approximates the Brunt-Väisälä frequency (Gill, 1982, p. 129; Hutter (eq. 3.16)). Because of (3.62), the results of (3.59) and (3.60) may be written

$$Z'(o) + \frac{g}{c^2} Z(0) = 0, \qquad Z(h) = 0 . \tag{3.65}$$

Solving the eigenvalue problem defined by (3.63) and (3.65) yields

$$c = c_r, \qquad Z = Z_r \qquad (r = 1, 2, \ldots, \infty), \tag{3.66}$$

where c_r denotes an eigenvalue and Z_r the corresponding eigenfunction. Each Z_r defines a vertical profile of current: a so-called normal mode, dependent on the vertical structure of density. Generally, the motion may be regarded as consisting of a linear combination of such modes at any place and time. Therefore, on the basis of (3.57), taking account of (3.62) and (3.66), a solution of the hydrodynamic equations is sought of the form:

$$\begin{aligned} w &= \sum_{r=1}^{\infty} \frac{g}{c_r^2} \frac{\partial \hat{\zeta}_r}{\partial t}(x, y, t)\, Z_r(z) , \\ u &= \sum_{r=1}^{\infty} \hat{u}_r (x, y, t)\, Z_r'(z) , \\ v &= \sum_{r=1}^{\infty} \hat{v}_r (x, y, t)\, Z_r'(z) , \\ p_1 &= \sum_{r=1}^{\infty} g\, \rho_0\, \hat{\zeta}_r (x, y, t)\, Z_r'(z) . \end{aligned} \tag{3.67}$$

For the consistency of (3.47) and (3.67) at $z = 0$:

$$\zeta = \sum_{r=1}^{\infty} \hat{\zeta}_r (x, y, t)\, Z_r'(0) . \tag{3.68}$$

Next, equations for the unknown functions $\hat{u}_r$, $\hat{v}_r$, $\hat{\zeta}_r$ in (3.67) are derived by invoking (3.39), (3.49) and (3.50). First, however, the orthogonality of the functions Z_r' is established. Thus, (3.66) in (3.63) gives

$$\frac{Z_r''}{N^2} = - \frac{Z_r}{c_r^2}, \qquad \frac{Z_s''}{N^2} = - \frac{Z_s}{c_s^2} \tag{3.69}$$

where r, s are positive integers. Therefore,

$$\frac{d}{dz}\left(\frac{Z_r''}{N^2}\right) = - \frac{Z_r'}{c_r^2}, \qquad \frac{d}{dz}\left(\frac{Z_s''}{N^2}\right) = - \frac{Z_s'}{c_s^2} \tag{3.70}$$

and hence

$$\left(\frac{1}{c_r{}^2} - \frac{1}{c_s{}^2}\right) Z_r' \, Z_s' = \frac{d}{dz}\left(\frac{Z_r' \, Z_s'' - Z_s' \, Z_r''}{N^2}\right) \tag{3.71}$$

so that

$$\left(\frac{1}{c_r{}^2} - \frac{1}{c_s{}^2}\right) \int_0^h Z_r' \, Z_s' \, dz = \left[\frac{Z_r' \, Z_s'' - Z_s' \, Z_r''}{N^2}\right]_0^h . \tag{3.72}$$

Now, from (3.63)

$$Z_r''(h) = -(N^2/c_r^2)\, Z_r(h), \qquad Z_s''(h) = -(N^2/c_s^2)\, Z_s(h) \tag{3.73}$$

and therefore, from (3.65)

$$Z_r''(h) = Z_s''(h) = 0. \tag{3.74}$$

Also, from (3.63) and (3.65)

$$\begin{aligned} Z_r''(0) &= -\frac{N^2}{c_r{}^2} Z_r(0), \qquad & Z_s''(0) &= -\frac{N^2}{c_s{}^2} Z_s(0), \\ Z_r'(0) &= -\frac{g}{c_r{}^2} Z_r(0), \qquad & Z_s'(0) &= -\frac{g}{c_s{}^2} Z_s(0). \end{aligned} \tag{3.75}$$

Introducing (3.74) and (3.75) into (3.72) gives

$$\int_0^h Z_r' \, Z_s' \, dz = 0 \qquad (r \neq s) \tag{3.76}$$

which shows that the Z_r' constitute an orthogonal set. Therefore, multiplying the equations for u, v and p_1 in (3.67) by Z_s' and then integrating each through the vertical yields

$$\begin{aligned} \int_0^h u \, Z_s' \, dz &= h \, \hat{u}_s / \chi_s , \\ \int_0^h v \, Z_s' \, dz &= h \, \hat{v}_s / \chi_s , \\ \int_0^h p_1 \, Z_s' \, dz &= g \rho_0 h \, \hat{\zeta}_s / \chi_s , \end{aligned} \tag{3.77}$$

where

$$\chi_s = h \Big/ \int_0^h Z_s'^2 \, dz . \tag{3.78}$$

Consider now the equation of continuity (3.39). Substituting from (3.67) for u, v and w gives

$$\sum_{r=1}^{\infty}\left(\frac{\partial \hat{u}_r}{\partial x} + \frac{\partial \hat{v}_r}{\partial y} + \frac{g}{c_r^2}\frac{\partial \hat{\zeta}_r}{\partial t}\right) Z_r' = 0 \, .$$

Then multiplying by Z_s' and integrating vertically from $z = 0$ to $z = h$, using the orthogonality condition (3.76), gives

$$\frac{\partial \hat{\zeta}_s}{\partial t} = -\frac{c_s^2}{g}\left(\frac{\partial \hat{u}_s}{\partial x} + \frac{\partial \hat{v}_s}{\partial y}\right). \tag{3.79}$$

Further, consider the equation of motion (3.49). Multiplying by Z_s' and integrating vertically yields

$$\int_0^h \frac{\partial u}{\partial t} Z_s' \, dz - f \int_0^h v \, Z_s' \, dz$$

$$= -\frac{1}{\rho_0}\int_0^h \frac{\partial p_1}{\partial x} Z_s' \, dz + \int_0^h Z_s' \frac{\partial}{\partial z}\left(\mu \frac{\partial u}{\partial z}\right) dz \, .$$

Incorporating the results of (3.77), this becomes

$$\frac{\partial \hat{u}_s}{\partial t} - f \hat{v}_s + g \frac{\partial \hat{\zeta}_s}{\partial x} = \frac{\chi_s}{h}\int_0^h Z_s' \frac{\partial}{\partial z}\left(\mu \frac{\partial u}{\partial z}\right) dz \, . \tag{3.80}$$

Similarly, working from (3.50) leads to the result:

$$\frac{\partial \hat{v}_s}{\partial t} + f \hat{u}_s + g \frac{\partial \hat{\zeta}_s}{\partial y} = \frac{\chi_s}{h}\int_0^h Z_s' \frac{\partial}{\partial z}\left(\mu \frac{\partial v}{\partial z}\right) dz \, . \tag{3.81}$$

Interest now centres on the evaluation of the integrals in (3.80) and (3.81), namely

$$I_u^{(s)} = \int_0^h Z_s' \frac{\partial}{\partial z}\left(\mu \frac{\partial u}{\partial z}\right) dz, \qquad I_v^{(s)} = \int_0^h Z_s' \frac{\partial}{\partial z}\left(\mu \frac{\partial v}{\partial z}\right) dz \, . \tag{3.82}$$

Integrating by parts twice gives

$$I_u^{(s)} = \left[Z_s' \, \mu \frac{\partial u}{\partial z} - \mu \, u \, Z_s'' \right]_0^h + \int_0^h u \frac{\partial}{\partial z}\left(\mu \, Z_s''\right) dz \, . \tag{3.83}$$

Since, consistent with (3.48), the x-components of surface and bottom stress may be expressed by

$$\tau_{sx} = -\left(\rho_0 \mu \frac{\partial u}{\partial z}\right)_{z=0}, \qquad \tau_{bx} = -\left(\rho_0 \mu \frac{\partial u}{\partial z}\right)_{z=h} \tag{3.84}$$

it follows that

$$I_u^{(s)} = Z_s'(0)\,\frac{\tau_{sx}}{\rho_0} - Z_s'(h)\,\frac{\tau_{bx}}{\rho_0} + \mu_0\, u_0\, Z_s''\,(0) - \mu_h\, u_h\, Z_s''\,(h) + \int_0^h u\,\frac{\partial}{\partial z}(\,\mu\, Z_s''\,)\,dz\,, \tag{3.85}$$

where suffices 0 and h, on μ and u, denote evaluation at $z = 0$ and $z = h$ respectively. Taking account of (3.69) and (3.74), equation (3.85) reduces to

$$I_u^{(s)} = Z_s'(0)\,\frac{\tau_{sx}}{\rho_0} - Z_s'(h)\,\frac{\tau_{bx}}{\rho_0} + \mu_0\, u_0\, Z_s''\,(0) - \frac{1}{c_s^2}\int_0^h u\,\frac{\partial}{\partial z}(\,\mu\, N^2\, Z_s)\,dz\,. \tag{3.86}$$

Assuming that

$$\mu = \Lambda\,/\,N^2, \tag{3.87}$$

where Λ is independent of z, and using (3.77), equation (3.86) simplifies further to

$$I_u^{(s)} = Z_s'(0)\,\frac{\tau_{sx}}{\rho_0} - Z_s'(h)\,\frac{\tau_{bx}}{\rho_0} + \mu_0\, u_0\, Z_s''(0) - \frac{\Lambda h}{X_s\, c_s^2}\,\hat{u}_s. \tag{3.88}$$

The assumption (3.87) that the vertical eddy viscosity is inversely proportional to the static stability has been employed in theoretical analyses by Fjeldstad (1964), Mork (1968, 1971) and McCreary (1981a,b). It takes some account of the reduction in eddy viscosity at thermocline level. From the point of view of the present theory, it introduces a compatibility between the frictional eigenvalue modes (such as are employed exclusively in Section 3.1) and the eigenvalue modes associated with the density structure (employed in the present section). The integral $I_v^{(s)}$ may be evaluated in the same way as $I_u^{(s)}$ leading to the result

$$I_v^{(s)} = Z_s'(0)\,\frac{\tau_{sy}}{\rho_0} - Z_s'(h)\,\frac{\tau_{by}}{\rho_0} + \mu_0\, v_0\, Z_s''(0) - \frac{\Lambda h}{X_s\, c_s^2}\,\hat{v}_s. \tag{3.89}$$

Introducing $I_u^{(s)}$ from (3.88) into (3.80), and $I_v^{(s)}$ from (3.89) into (3.81), gives

$$\frac{\partial \hat{u}_s}{\partial t} = -\frac{\Lambda}{c_s^2}\,\hat{u}_s + f\,\hat{v}_s - g\,\frac{\partial \hat{\zeta}_s}{\partial x} + \frac{X_s}{h}\Big(Z_s'(0)\,\frac{\tau_{sx}}{\rho_0} - Z_s'(h)\,\frac{\tau_{bx}}{\rho_0} + \mu_0\, u_0\, Z_s''(0)\Big) \tag{3.90}$$

$$\frac{\partial \hat{v}_s}{\partial t} = -\frac{\Lambda}{c_s^2}\hat{v}_s - f\hat{u}_s - g\frac{\partial \hat{\zeta}_s}{\partial y} +$$

$$+ \frac{\chi_s}{h}\left(Z_s'(0)\frac{\tau_{sy}}{\rho_0} - Z_s'(h)\frac{\tau_{by}}{\rho_0} + \mu_0 v_0 Z_s''(0)\right). \qquad (3.91)$$

Equations (3.79), (3.90) and (3.91) constitute a two-dimensional set for the numerical generation of $\hat{\zeta}_s$, $\hat{u}_s$, $\hat{v}_s$ ($s = 1, 2, \ldots, \infty$) through time and horizontal space, satisfying appropriate lateral boundary conditions and starting from an initial prescribed state of motion. Having thus solved for $\hat{\zeta}_s$, $\hat{u}_s$ and $\hat{v}_s$, the vertical structure of current may be determined from (3.67). Importantly, the wind stress (τ_{sx}, τ_{sy}) appears in (3.90) and (3.91) as the forcing agent producing the motion; that stress may be determined from the wind speed as, for example, in (2.15). The bottom stress (τ_{bx}, τ_{by}) also appears and may be determined from the linear law:

$$\tau_{bx} = k\rho u_h, \qquad \tau_{by} = k\rho v_h, \qquad (3.92)$$

where, from (3.67),

$$u_h = \sum_{r=1}^{\infty} \hat{u}_r Z_r'(h), \qquad v_h = \sum_{r=1}^{\infty} \hat{v}_r Z_r'(h). \qquad (3.93)$$

Also appearing is the surface current (u_0, v_0) which may be evaluated either in terms of the wind speed, using a wind factor (equation (2.17)) and an assumed angular deflection from the wind direction, or from (3.67):

$$u_0 = \sum_{r=1}^{\infty} \hat{u}_r Z_r'(0), \qquad v_0 = \sum_{r=1}^{\infty} \hat{v}_r Z_r'(0). \qquad (3.94)$$

The use of (3.93) and (3.94) couples together the vertical modes of different order.

The general nature of the vertical modes $Z_r(z)$: $r = 1, 2, \ldots, \infty$ is demonstrated easily by examining the case when $d\rho_*/dz$ is a constant, corresponding to a linear density variation through the vertical. Then, the general solution of (3.63) is

$$Z = A\cos az + B\sin az, \qquad (3.95)$$

where $a = N/c$ and A, B are constants of integration. Satisfying the surface and bottom conditions of (3.65) yields

$$-\frac{A}{B} = \frac{ac^2}{g} = \tan ah, \qquad (3.96)$$

whence

$$\tan ah = \frac{h\,N^2/g}{ah}. \qquad (3.97)$$

The smallness of hN^2/g (due to the smallness of the vertical gradient of density) means that, except when ah is small, the roots of (3.97) are given approximately by

$$\tan ah = 0$$

so that

$$ah = (r-1)\pi , \qquad r = 2, 3, \ldots, \infty . \tag{3.98}$$

Correspondingly, from (3.95) and (3.96)

$$Z_r = B_r \sin \frac{(r-1)\pi z}{h} \tag{3.99}$$

and

$$c_r = \frac{Nh}{(r-1)\pi}, \tag{3.100}$$

where B_r is a constant. The normal modes given by (3.99) are baroclinic with $w = 0$ at the surface $z = 0$. The associated current modes

$$Z_r' = \frac{(r-1)\pi}{h} B_r \cos \frac{(r-1)\pi z}{h} \tag{3.101}$$

oscillate with $r-1$ nodal points through the depth and have a zero depth-averaged value. When ah is small, approximately from (3.96):

$$\frac{ac^2}{g} = ah$$

giving the first eigenvalue

$$c_1 = (gh)^{1/2}. \tag{3.102}$$

Correspondingly from (3.95) and (3.96):

$$Z_1 = B_1(z-h) \tag{3.103}$$

and therefore

$$Z_1' = B_1 \tag{3.104}$$

where B_1 is a constant, indicating a barotropic mode with no current variation through the depth.

The development of spectral models to predict the vertical structure of current in a stratified sea or lake is in its early stages. Heaps (1983) and Heaps and Jones (1983) have formulated a three-layered model for the motion of a stratified shelf sea in which horizontal components of current are expanded through the vertical, within each layer, in terms of a set of eigenfunctions. Density and vertical eddy viscosity are prescribed in each layer; continuity of current and shear stress is satisfied across the internal interfaces. On the other hand, Davies (1982b) has shown how the Galerkin method may be applied in the vertical to compute the motion in a stratified sea with continuous vertical variations of eddy viscosity and density arbitrarily specified. Further developments may be expected during the next few years.

REFERENCES

Bonham-Carter, G. and Thomas, J.H., 1973. Numerical calculation of steady wind-driven currents in Lake Ontario and the Rochester embayment. Proc. 16th Conf. Great Lakes Res., Internat. Assoc. Great Lakes Res., pp. 640-662.

Bowden, K.F., 1953. Note on wind drift in a channel in the presence of tidal currents. Proc. Roy. Soc. A, 219, pp. 426-446.

Bowden, K.F., 1978. Physical problems of the benthic boundary layer. Geophysical Surveys, 3, pp. 255-296.

Bowden, K.F., Fairbairn, L.A. and Hughes, P., 1959. The distribution of shearing stresses in a tidal current. Geophys. J.R. astr. Soc., 2, pp. 288-305.

Bye, J.A.T., 1965. Wind-driven circulation in unstratified lakes. Limnol. Oceanogr., 10, pp. 451-458.

Csanady, G.T., 1976. Mean circulation in shallow seas. J. Geophys. Res., 81, pp. 5389-5399.

Csanady, G.T., 1978. Turbulent interface layers. J. Geophys. Res., 83, pp. 2329-2342.

Csanady, G.T., 1979. A developing turbulent surface shear layer model. J. Geophys. Res., 84, pp. 4944-4948.

Csanady, G.T., 1980. The evolution of a turbulent Ekman layer. J. Geophys. Res., 85, pp. 1537-1547.

Csanady, G.T., 1982. Circulation in the coastal ocean. Dordrecht, Holland: Reidel.

Davies, A.M., 1977a. The numerical solution of the three-dimensional hydrodynamic equations, using a B-spline representation of the vertical current profile. pp. 1-25 in, Bottom Turbulence, (ed. J.C.J. Nihoul). Amsterdam: Elsevier.

Davies, A.M., 1977b. Three-dimensional model with depth-varying eddy viscosity. pp. 27-48 in, Bottom Turbulence, (ed. J.C.J. Nihoul). Amsterdam: Elsevier.

Davies, A.M., 1980. Application of the Galerkin method to the formulation of a three-dimensional nonlinear hydrodynamic numerical sea model. Appl. Math. Modelling, 4, pp. 245-256.

Davies, A.M., 1981a. Three-dimensional modelling of surges. pp. 45-47 in, Floods due to High Winds and Tides, (ed. D.H. Peregrine). London: Academic.

Davies, A.M., 1981b. Three-dimensional hydrodynamic numerical models. Part I: A homogeneous ocean-shelf model. Part 2: A stratified model of the northern North Sea. pp. 270-426 in, The Norwegian Coastal Current, (ed. R. Saetre and M. Mork). University of Bergen.

Davies, A.M., 1982a. Meteorologically-induced circulation on the North-West European continental shelf: from a three-dimensional numerical model. Oceanol. Acta, 5, pp. 269-280.

Davies, A.M., 1982b. On computing the three-dimensional flow in a stratified sea using the Galerkin method. Appl. Math. Modelling, 6, pp. 347-362.

Davies, A.M., 1983. Formulation of a linear three-dimensional hydrodynamic sea model using a Galerkin-eigenfunction method. Int. J. Numer. Meth. Eng., 3, pp. 33-60.

Davies, A.M. and Furnes, G.K., 1980. Observed and computed M_2 tidal currents in the North Sea. J. Phys. Oceanogr., 10, pp. 237-257.

Davies, A.M. and Owen, A., 1979. Three-dimensional numerical sea model using the Galerkin method with a polynomial basis set. Appl. Math. Modelling, 3, pp. 421-428.

Defant, A., 1961. Physical Oceanography (Vol. 1). Oxford: Pergamon.

Dobroklonskiy, S.V., 1969. Drift currents in the sea with an exponentially decaying eddy viscosity coefficient. Oceanology, 9, pp. 19-25.

Ekman, V.W., 1905. On the influence of the Earth's rotation on ocean currents. Ark. Mat. Astr. Fys., 2(11), pp. 1-52.

Fjeldstad, J.E., 1929. Ein Beitrag zur Theorie der winderzeugten Meeresströmungen. Gerlands Bietr. Geophys., 23. pp. 237-247.

Fjeldstad, J.E., 1930. Ein Problem aus der Windstom-theorie. Z. Angew. Math. Mech., 10, pp. 121-137.

Fjeldstad, J.E., 1964. Internal waves of tidal origin. Part 1: Theory and analysis of observations. Geophys. Publr., 25(5), pp. 1-73.

Forristall, G.Z., 1974. Three-dimensional structure of storm-generated currents. J. Geophys. Res., 79, pp. 2721-2729.

Forristall, G.Z., 1980. A two-layer model for hurricane-driven currents on an irregular grid. J. Phys. Oceanogr., 9, pp. 1417-1438.

Forristall, G.Z., Hamilton, R.C. and Cardone, V.J., 1977. Continental shelf currents in tropical storm Delia: observations and theory. J. Phys. Oceanogr., 7, pp. 532-546.

Gedney, R.T. and Lick, W., 1972. Wind-driven currents in Lake Erie. J. Geophys. Res., 77, pp. 2714-2723.

Gill, A.E., 1982. Atmosphere-Ocean Dynamics. New York: Academic.

Heaps, N.S., 1961. Seiches in a narrow lake, uniformly stratified in three layers. Geophys. J. R. astr. Soc., 5, pp. 134-156.

Heaps, N.S., 1972. On the numerical solution of the three-dimensional hydrodynamical equations for tides and storm surges. Mém. Soc. r. sci. Liège, ser. 6, 2, pp. 143-180.

Heaps, N.S., 1974. Development of a three-dimensional numerical model of the Irish Sea. Rapp. P.-v. Réun. Cons. perm. int. Explor. Mer, 167, pp. 147-162.

Heaps, N.S., 1976. On formulating a non-linear numerical model in three-dimensions for tides and storm surges. pp. 368-387 in, Computing Methods in Applied Sciences, (ed. R. Glowinski and J.L. Lions). Berlin: Springer-Verlag.

Heaps, N.S., 1980. Spectral method for the numerical solution of the three-dimensional hydrodynamic equations for tides and surges. pp. 75-90 in, Mathematical Modelling of Estuarine Physics, (ed. J. Sündermann and K.-P.Holz). Berlin: Springer-Verlag.

Heaps, N.S., 1981. Three-dimensional model for tides and surges with vertical eddy viscosity prescribed in two-layers. I. Mathematical formulation. Geophys. J. R. astr. Soc., 64, pp. 291-302.

Heaps, N.S. 1983. Development of a three-layered spectral model for the motion of a stratified shelf sea. I. Basic equations. pp. 336-400 in, Physical Oceanography of Coastal and Shelf Seas, (ed. B. Johns). Amsterdam: Elsevier.

Heaps, N.S. and Jones, J.E., 1975. Storm surge computations for the Irish Sea using a three-dimensional numerical model. Mém. Soc. r. sci. Liège, Ser. 6, 7, pp. 289-333.

Heaps, N.S. and Jones, J.E., 1981. Three-dimensional model for tides and surges with vertical eddy viscosity prescribed in two layers. II. Irish Sea with bed friction layer. Geophys. J. R. astr. Soc., 64, pp. 303-320.

Heaps, N.S. and Jones, J.E., 1984. Development of a three - layered spectral model for the motion of a stratified sea. II. Experiments with a rectangular basin representing the Celtic Sea. pp. 401-465 in, Physical Oceanography of Coastal and Shelf Seas, (ed. B. Johns). Amsterdam: Elsevier.

Heaps, N.S. and Ramsbottom, A.E., 1966. Wind effects on the water in a narrow two-layered lake. Phil. Trans. Roy. Soc. A, 259, pp. 391-430.

Hidaka, K., 1933. Non-stationary ocean currents. Mem. Imp. Mar. Obs. Kobe, 5, pp. 141-266.

Jelesnianski, C.P., 1967. Numerical computations of storm surges with bottom stress. Mon. Weather Rev., 95, pp. 740-756.

Jelesnianski, C.P., 1970. Bottom stress time history in linearized equations of motion for storm surges. Mon. Weather Rev., 98, pp. 462-478.

Kielmann, J. and Kowalik, Z., 1980. A bottom stress formulation for storm surge problems. Oceanol. Acta, 3, pp. 51-58.

Lai, R.Y.S. and Rao, D.B., 1976. Wind drift currents in deep sea with variable eddy viscosity. Arch. Met. Geophys. Bioklim. A, 25, pp. 131-140.

Madsen, O.S., 1977. A realistic model of the wind-induced Ekman boundary layer. J. Phys. Oceanogr., 7, pp. 248-255.

McCreary, J.P., 1981a. A linear stratified ocean model of the equatorial undercurrent. Phil. Trans. Roy. Soc. A, 298, pp. 603-635.

McCreary, J.P., 1981b. A linear stratified ocean model of the coastal undercurrent. Phil. Trans. Roy. Soc. A, 302, pp. 385-413.

Mork, M., 1968. The response of a stratified sea to atmospheric forces. Geophysical Institute, University of Bergen.

Mork, M., 1971. On the time-dependent motion induced by wind and atmospheric pressure in a continuously stratified ocean of varying depth. Geophysical Institute, University of Bergen.

Mortimer, C.H., 1952. Water movements in lakes during summer stratification; evidence from the distribution of temperature in Windermere. Phil. Trans. Roy. Soc. B, 236, pp. 355-404.

Mortimer, C.H., 1953. The resonant response of stratified lakes to wind. Schweiz. Z. Hydrol., 15, pp. 94-151.

Mortimer, C.H., 1961. Motion in thermoclines. Verh. Internat. Verein. Limnol., 14, pp. 79-83.

Mortimer, C.H., 1974. Lake hydrodynamics. Mitt. Internat. Verein. Limnol., 20, pp. 124-197.

Munk, W.H. and Anderson, E.R., 1948. Notes on a theory of the thermocline. J. Mar. Res., 7, pp. 276-295.

Pearce, B.R. and Cooper, C.K., 1981. Numerical circulation model for wind induced flow. J. Hydraul. Div. Am. Soc. Civ. Engrs , 107 (HY3), pp. 285-302.

Platzman, G.W., 1963. The dynamical prediction of wind tides on Lake Erie. Meteorol. Monogr., 4 (26), 44 pp.

Pond, S. and Pickard, G.L., 1978. Introductory Dynamic Oceanography. Oxford: Pergamon.

Proudman, J., 1953. Dynamical Oceanography. London: Methuen.

Ramming, H.G. and Kowalik, Z., 1980. Numerical modelling of marine hydrodynamics. Amsterdam: Elsevier.

Simons, T.J., 1980. Circulation models of lakes and inland seas. Can. Bull. Fish. Aquat. SCI. 203, pp. 1-146.

Svensson, U., 1979. The structure of the turbulent Ekman layer. Tellus, 31, pp. 340-350.

Sverdrup, H.U., Johnson, M.M. and Fleming, R.H., 1946. The Oceans, their physics, chemistry and general biology. New York: Prentice Hall.

Thomas, J.H., 1975. A theory of steady wind-driven currents in shallow water with variable eddy viscosity. J. Phys. Oceanogr., 5, pp. 136-142.

Welander, P., 1957. Wind action on a shallow sea: some generalizations of Ekman's theory. Tellus, 9, pp. 45-52.

Witten, A.J. and Thomas, J.H., 1976. Steady wind-driven currents in a large lake with depth-dependent eddy viscosity. J. Phys. Oceanogr., 6, pp. 85.-92.

Hydrodynamics of Lakes: CISM Lectures
edited by K. Hutter, 1984
Springer Verlag Wien-New York

NUMERICAL MODELLING OF BAROTROPIC CIRCULATION PROCESSES

Jürgen Sündermann

Institute of Oceanography
University of Hamburg

Heimhuder Str. 71, 2000 Hamburg 13
Federal Republic of Germany

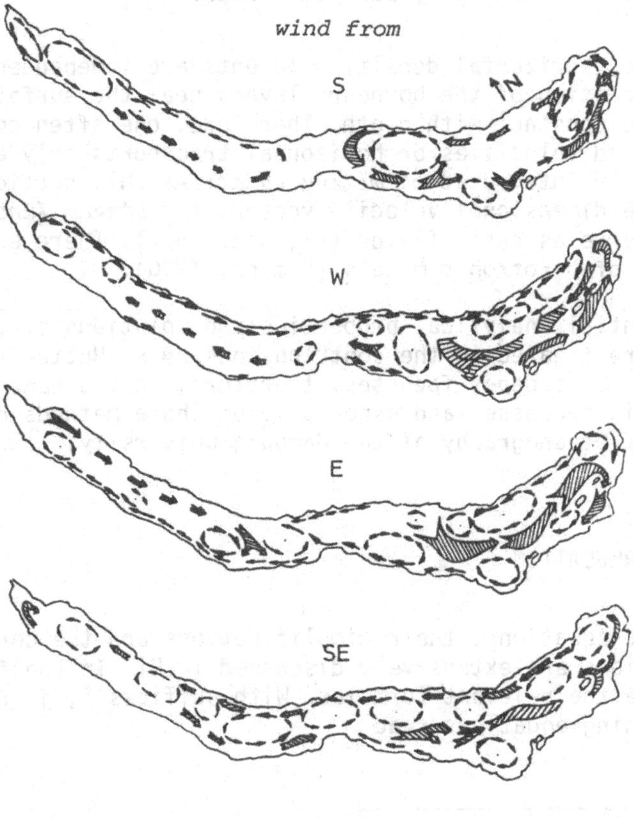

An artist's interpretation of the mean transport in Lake of Zurich subject to steady uniform wind from S, W, E, SE

1. INTRODUCTION

The dynamics of lakes has a hydrodynamic and a thermodynamic component. By neglection of the latter, assuming a thermally homogeneous water body, a *barotropic* model is obtained. It is characterized by vanishing horizontal density gradients; the surface slope alone determines the horizontal pressure gradients (H1, section 1)*). Vertical density gradients can be considered by a barotropic multi-layer model (H1, section 5).

Physical processes to be described by a barotropic model are

- the three-dimensional wind-driven circulation,
- up- and downwelling,
- surface waves at different scales,
- transport of dynamically passive tracers.

For these, horizontal density gradients are independent of depth and, with the exception of the boundary layers near the surface and the bottom, currents are constant with depth. Therefore, one often considers vertically averaged velocities or horizontal transports only and may thus use the vertically integrated *primitive equations* (H1, section 3.4). In lieu of the three-dimensional velocity vectors the *stream function* and the *vorticity* serve as basic fields (H1, section 4). There exist, therefore, two classes of barotropic models (Simons, 1980).

The essential analytical properties and solutions of the barotropic equations are treated in the contributions by K. Hutter and L.A. Mysak in this volume. This paper focusses, therefore, on the *numerical modelling* of barotropic processes and especially on those methods developed at the Institute of Oceanography of the Hamburg university.

2. BASIC FORMULATION

The basic equations, their simplifications and the corresponding boundary conditions are extensively discussed in H1. In the following we merely compile the relevant formulae. With suffices i, j running from 1 to 3 the governing equations read

*) *H1 refers to Hutter's article "Fundamental equations and approximations".*

- *3D equation of motion*

$$\frac{\partial v_i}{\partial t} + v_j \frac{\partial v_i}{\partial x_j} + \varepsilon_{ij} v_j = \frac{1}{\rho}(k_i - \frac{\partial p}{\partial x_i}) + \frac{\partial}{\partial x_j}(A \frac{\partial v_i}{\partial x_j}), \tag{2.1}$$

- *3D equation of continuity*

$$\frac{\partial v_j}{\partial x_j} = 0, \tag{2.2}$$

- *3D convection-diffusion equation*

$$\frac{\partial c}{\partial t} + v_j \frac{\partial c}{\partial x_j} = \frac{\partial}{\partial x_j}(D \frac{\partial c}{\partial x_j}) + s, \tag{2.3}$$

with the following notations:

- v_i components of the velocity vector,
- p pressure,
- c concentration of a tracer,
- ε_{ij} Coriolis tensor, $\varepsilon_{ij} = \left(\begin{array}{cc|c} 0 & -f & 0 \\ f & 0 & 0 \\ \hline 0 & 0 & 0 \end{array}\right)$
- ρ density,
- k_i components of the external forces,
- A eddy viscosity coefficient,
- D eddy diffusivity coefficient,
- s sources and sinks of concentration,
- t time,
- x_i position vector.

The following *boundary conditions* are imposed:

— Lateral solid boundaries:

$$v_i = 0, \tag{2.4}$$

$$n_j \frac{\partial c}{\partial x_j} = 0. \tag{2.5}$$

– Free surface (i, j = 1, 2):

$$\frac{\partial \zeta}{\partial t} + v_j \frac{\partial \zeta}{\partial x_j} - v_3 = 0 , \tag{2.6}$$

$$A \frac{\partial v_i}{\partial x_3} = \tau_s^{\,i} , \tag{2.7}$$

$$D \frac{\partial c}{\partial x_3} = Q_s . \tag{2.8}$$

– Bottom (i, j = 1, 2):

$$v_j \frac{\partial h}{\partial x_j} + v_3 = 0 , \tag{2.9}$$

$$A \frac{\partial v_i}{\partial x_3} = \tau_b^{\,i} , \tag{2.10}$$

$$n_j \frac{\partial c}{\partial x_j} = 0 .$$

with the following notations:

ζ water elevation,

h undisturbed water depth,

n_j components of the normal vector,

$\tau_s{}^i$ wind stress at the surface,

$\tau_b{}^i$ bottom stress,

Q_s concentration flux at the surface.

In the shallow water approximation equations are scaled such that the pressure is hydrostatic,

$$0 = \frac{1}{\rho} \left(k_3 - \frac{\partial p}{\partial x_3}\right) . \tag{2.11}$$

With $k_3 = -g$ and by integrating from the surface to the bottom, equation (2.11) yields the pressure in the form

$$p(z) = p^{atm}(\zeta) + g \int_{-z}^{\zeta} \rho \, dx_3 , \tag{2.12}$$

where the following notations are used:

g gravity acceleration,

$p^{atm}(\zeta)$ atmospheric pressure at the surface.

With the aid of (2.12) the horizontal pressure gradients can be expressed in terms of horizontal water elevation and density gradients ($i = 1, 2$), viz.

$$\frac{\partial p}{\partial x_i} = \frac{\partial p^{atm}(\zeta)}{\partial x_i} + g\,\frac{\partial \zeta}{\partial x_i} + g\int_{-z}^{\zeta} \frac{\partial \rho}{\partial x_i}\,dx_3\,. \tag{2.13}$$

The second term on the right hand side is the *barotropic* part of the pressure gradient, the third the *baroclinic* one. In barotropic models this latter is neglected. Furthermore, in case of lake circulation the horizontal gradients of the atmospheric pressure may also be ignored.

Instead of equations (2.1), (2.2) (2.3), often their *vertically integrated* counterparts are used. Restricting attention to the barotropic case these equations read ($i, j = 1, 2$)

$$\frac{\partial \overline{v_i}}{\partial t} + \bar{v}_j\,\frac{\partial \overline{v_i}}{\partial x_j} + \overline{\varepsilon_{ij}}\;\bar{v}_j = -\,g\,\frac{\partial \zeta}{\partial x_i} + \frac{\partial}{\partial x_j}\left(\bar{A}\,\frac{\partial \overline{v_i}}{\partial x_j}\right) - \tau_b^{\,i} + \tau_s^{\,i} \tag{2.14}$$

$$\frac{\partial \zeta}{\partial t} + \frac{\partial}{\partial x_j}\Big((h+\zeta)\,\overline{v_j}\Big) = 0 \tag{2.15}$$

$$\frac{\partial \bar{c}}{\partial t} + \overline{v_j}\,\frac{\partial \bar{c}}{\partial x_j} = \frac{\partial}{\partial x_j}\left(\bar{D}\,\frac{\partial \bar{c}}{\partial x_j}\right) + \bar{S} \tag{2.16}$$

in which $\bar{\varepsilon}_{ij}$ is the upper left 2×2-matrix of the ε_{ij} defined previously. Moreover, the vertically averaged velocities $\overline{v_i}$ are defined by ($i = 1, 2$):

$$\overline{v_i} = \frac{1}{h+\zeta}\int_{-h}^{\zeta} v_i\,dx_3\,,$$

and all other quantities with overbar are redefined appropriately considering mean conditions.

Equations (2.14), (2.15) are referred to as *primitive* equations. In lieu of them one may also use the so-called *divergence* and *vorticity* equations (Simons, 1980). This is based on Helmholtz' theorem stating that any vector field can be expressed as a sum of a solenoidal and a irrotational vector field. Then, two scalar functions, the *vector potential* ϕ and the *stream function* ψ, can be introduced, and the horizontal divergence and vorticity respectively, can be related to them. In this case instead of $\overline{v_i}$ ($i = 1, 2$) ϕ and ψ are calculated (H1, section 4).

3. NUMERICAL ALGORITHMS

For real lakes the basic equations cannot be solved analytically; solutions must be sought numerically. Thus, the time and the space domain have to be discretized, the differential equations must be replaced by algebraic equations for the unknown nodal values of the computational grid (Sündermann, 1979).

3.1 Time integration

With respect to time derivatives the systems (2.1), (2.2), (2.3) or (2.14), (2.15), (2.16) can be represented by the following matrix equations

$$\frac{d}{dt} q_i(x_i,t) = L_{ij}\, q_j(x_i,t)\,, \tag{3.1}$$

with the initial condition

$$q_i(x_i,0) = q_i^{\,o}(x_i)\,;$$

$q_i(x_i,t)$ denotes an unknown quantity, and L_{ij} is a quasi-linear differential operator. The discretization of the time domain by finite time steps Δt transforms (3.1) into a difference equation. Mostly forward differences are used, leading to an *explicit* procedure

$$q_i^{\,n+1} = L_{ij}^{*}(\Delta t) q_i^{\,n},$$

in which L_{ij}^{*} is a linear difference operator and n denotes the time step. L_{ij}^{*} should be a *convergent*, *stable* and *consistent* approximation of L_{ij}.

There are several stability criteria; they are based on the necessary condition of von Neumann

$$|\lambda_m| < 1 + O(\Delta t) \tag{3.2}$$

where λ_m is the m-th eigenvalue of L_{ij}^{*}.

Applying (3.2) to the basic equations (2.1), (2.2), (2.3) and assuming an instantaneous balance of the local time derivative with one of the other terms, a set of stability criteria is obtained which must all be fulfilled, if stability should be guaranteed.

They are $(i, j = 1, 2)$:

$$Cr \equiv \frac{\Delta t}{\Delta x_i}\left(|v_j\ v_j|^{1/2} + (2\,gh)^{1/2}\right) \leqq 1\,, \tag{3.3}$$

$$\frac{2\,A\,\Delta t}{\Delta^2 x_i} \leqq 1\,, \tag{3.4}$$

$$\frac{2\,D\,\Delta t}{\Delta^2 x_i} \leqq 1 \,. \tag{3.5}$$

Conditions (3.3) to (3.5) can impose a very severe restriction on the time step. For instance, with a grid distance of Δx_1 = 1 km and a maximum depth of 200 m which may be typical of a lake, $\Delta t \leqq 20$ s is obtained. This means a large computational effort.

In numerical computations one finds frequently that condition (3.2) is sufficient for stability. This occurs when an automatically-acting dissipative mechanism, the *numerical diffusion* is applied. In some cases, such *dissipative interfaces* (Abbott, 1974) are especially introduced in order to achieve stability. A dissipative interface acts as a filter, which replaces q_i by a quantity $q_i^*(q_i)$ such that

$$\| q_i^{*n} \| < \| q_i^{n} \| \,;$$

$\|\cdot\|$ is commonly chosen as the energy norm. A popular scheme is the so-called α - algorithm:

$$q_i^{*n} = \alpha q_i^{n+1} + (1-\alpha) q_i^{n}\,, \qquad 0 \leqq \alpha \leqq 1\,. \tag{3.6}$$

The dissipative operator (3.6) requires for $\alpha = 0$ (fully explicit) that the Courant condition (3.3) be satisfied; for $0 < \alpha < 1/2$ the procedure is conditionally, and for $1/2 \leqq \alpha < 1$ it is unconditionally stable. It was shown by Crank-Nicholson (1947) that the best accuracy is obtained with $\alpha = 1/2$. For increasing α improved stability is obtained at the cost of a loss in accuracy. Therefore, Courant numbers of about 10 should not, in general, be exceeded.

There remains a further type of numerical instability to be mentioned, which is caused by the non-linear inertia terms and is therefore called *non-linear instability*. It is generated by an energy transfer from the lower to the higher modes in the wave spectrum. It results, asymptotically in a concentration of wave energy in the smallest resolvable mode, that is a wave length of twice the grid distance. The solution may then become unstable, or it may oscillate unphysically. Non-linear instabilities can be filtered out by weighted dissipative interfaces which taper especially the higher wave modes, see Abbott (1974).

Numerical diffusion improves the stability behaviour of the method, but at the same time it makes the convergence worse. In the momentum equation (2.1) this error remains in the order of the truncation error. For the transport equation (2.3), however, the second law of the thermodynamics can on occasion gravely be violated.

3.2 Space integration

The spatial integration is mostly done by finite difference methods (FDM), finite element methods (FEM), or spectral methods (SM). In the first two cases the space domain is represented by a finite number of computational points in a 3D-Eulerian computational grid, in the latter case the unknown function is approximated continuously by a finite number of spectral modes. The methods have specific advantages and disadvantages.

In the *FDM* mostly regular grids are used, e.g. the staggered grid of Hansen (1956), a so-called Arakawa - C type grid. In order to discuss the numerical problems one encounters with spatial differencing, it suffices to consider the balance of acceleration terms and horizontal gradients in the x-direction of the system (2.14), (2.15). Thus we study the matrix equation

$$\frac{\partial}{\partial t} q_i(x,t) = B_{ij} \frac{\partial}{\partial x_j} q_j(x,t) , \tag{3.7}$$

in which B_{ij} is a quasi-linear differential operator. The straight-forward difference respresentation of (3.7) is given by the explicit scheme of Hansen (1956)

$$\frac{1}{\Delta t}\left[(q_i)_m^{n+1} - (q_i)_m^n\right] - \frac{B_{ij}}{\Delta x}\left[(q_j)^n_{m+1/2} - (q_j)^n_{m-1/2}\right] = 0. \tag{3.8}$$

The indices n, m denote grid points in time and space, respectively. Numerical stability requires that (3.8) satisfies the Courant condition (3.3). By introducing the α - algorithm

$$(q_i{}^*)_m = \frac{\alpha}{2}(q_i)_{m+1/2} + (1-\alpha)(q_i)_m + \frac{\alpha}{2}(q_i)_{m-1/2} , \qquad 0 \leqq \alpha \leqq 1 \tag{3.9}$$

the algorithm (3.8) becomes unconditionally stable. For instance, $\alpha = 1$ yields the wellknown stable scheme of Lax (1954):

$$\frac{1}{\Delta t}\left[(q_i)_m^{n+1} - \frac{1}{2}\left((q_i)^n_{m+1/2} + (q_i)^n_{m-1/2}\right)\right]$$

$$- \frac{B_{ij}}{\Delta x}\left[(q_j)^n_{m+1/2} - (q_j)^n_{m-1/2}\right] = 0 .$$

Application of (3.9) physically acts as a horizontal viscosity with an eddy coefficient of

$$A = \frac{\alpha}{2} \frac{\Delta^2 x}{\Delta t} .$$

In case of an improper relation between Δx and Δt this coefficient may become unphysically large and thus cause very strong numerical damping.

The FDM is very well investigated, it is an extremely efficient algorithm and easy to translate into a computer program.

In the *FEM* the given water body is subdivided into a finite number of finite elements of mostly irregular shape. The unknown quantities q_i^k are calculated in the nodal points k, and the distribution of the field variables within the elements is described by certain shape functions ϕ^k:

$$q_i(x_j,t) = \phi^k(x_j,t)\, q_i^k , \qquad k = 1, \ldots, K .$$

The finite system of algebraic equations for the determination of q_i^k is usually obtained from the Euler-Lagrange equations of a variational formulation. The necessary functional is found either from the differential equations (2.1), (2.2), (2.3) by an error minimization principle or directly from the original physical conservation laws. In the first case, often the method of weighted residuals is used, it contains as a special case the well-konwn Galerkin procedure.

For (3.7) this leads to the system

$$\int_d \int_{\Delta t} \left(\frac{\partial \phi^k}{\partial t} q_i^k - B_{ij} \frac{\partial \phi^k}{\partial x} q_j^k \right) \phi^\ell \, dt\, dx = 0 , \qquad (3.10)$$

with $k, \ell = 1, \ldots, K$ and d length in x-direction.

For the FEM the same stability restrictions are valid as for the FDM. It can be proved that the Galerkin method (3.10) works with an intrinsic α-algorithm of $\alpha = 2/3$. Its solutions are stable but strongly damped by numerical diffusion. The main advantage of the FEM is the very well adaption of the computational grid to the natural topography.

The *SM* is described in detail by Heaps (this volume). It is based on analytic expansions of the unknown variables with respect to certain basis functions ϕ^k. The numerical procedure consists of the determination of the unknown coefficients. Mostly, the vertical dependency is expressed by a superposition of spectral modes (i = 1, 2):

$$v_i(x_j, x_3, t) = \sum_{k=1}^{K} a_k(x_j)\, \phi^k(x_3) .$$

It turns out that in many cases a small number of modes is sufficient to fairly approximate observed vertical structures. Then, a spectral method might be more economical than a multi-level difference scheme.

For long-shaped lakes also the horizontal cross dimension may be represented by spectral modes.

4. ACTUAL METHODS AT THE INSTITUTE OF OCEANOGRAPHY, HAMBURG

4.1 3D - barotropic dynamics

Most recently, a very economic semi-implicit FD - algorithm has been developed by J. Backhaus, K. Duwe and R. Hewer (Backhaus, 1983; Duwe and Hewer, 1982). The motivation for its deduction was the obvious mismatch of the local and the global time step in the classical explicit FDM.

In coastal sea areas and in lakes the depth h may range from 5 to 500 m. With a typical grid distance of Δx = 1000 m the Courant criterion (3.3) then results in the following restrictions for the time step, see table 1:

h [m]	$\Delta t \leqq$ [s]
5	100
500	10

Table 1

Critical time step as a function of depth from Courant's criterion (3.3).

The global time step is governed by the maximum depth in the whole region, this is in our example 10 s. In a local area of only 5 m depth the critical time step is 100 s. The computational grid has a finite resolution down to a wave length of $2\Delta x$ = 2000 m. This corresponds to a numerical cut-off period of $T = 2\Delta x/\sqrt{2gh}$ = 200 s. In view of this a global Courant time step of 10 s means a waste of computer time.

In the following the horizontal semi-implicit procedure is described. It will be formulated for the simplified equations

$$\frac{\partial U}{\partial t} + gH \frac{\partial \zeta}{\partial x} = X - \tau_b^x , \tag{4.1}$$

$$\frac{\partial V}{\partial t} + gH \frac{\partial \zeta}{\partial y} = Y - \tau_b^y , \tag{4.2}$$

$$\frac{\partial \zeta}{\partial t} + \frac{\partial U}{\partial x} + \frac{\partial V}{\partial y} = 0 , \tag{4.3}$$

with the notations:

$U = H\overline{v_1}$ East component of the total transport,

$V = H\overline{v_2}$ North component of the total transport,

$H = h + \zeta$ actual total water depth,

X, Y remaining (optional) terms in the U, V equations of motion.

Equations (4.1), (4.2), (4.3) are simply obtained from (2.14), (2.15), (2.16).

Using again the staggered Arakawa-C grid the following difference equations can be deduced:

$$U^{n+1} = F^x \left(U^n + \Delta t \left(X^n - \frac{g\bar{H}^x}{2\,\Delta x}(\zeta_E^{n+1} - \zeta_W^{n+1} + \zeta_E^n - \zeta_W^n)\right)\right), \tag{4.4}$$

$$V^{n+1} = F^y \left(V^n + \Delta t \left(Y^n - \frac{g\bar{H}^y}{2\,\Delta x}(\zeta_N^{n+1} - \zeta_S^{n+1} + \zeta_N^n - \zeta_S^n)\right)\right), \tag{4.5}$$

$$\zeta^{n+1} = \zeta^n - \frac{\Delta t}{2\,\Delta x}(U_E^{n+1} - U_W^{n+1} + U_E^n - U_W^n + V_N^{n+1} - V_S^{n+1} + V_N^n - V_S^n), \tag{4.6}$$

where the indices E, W, N, S denote relative space coordinates. Furthermore,

$$F^x = \left(1 + \left(\frac{r\Delta t}{\bar{H}^{x2}}\sqrt{\bar{U}^2 + \bar{V}^2}\right)^n\right)^{-1},$$

$$F^y = \left(1 + \left(\frac{r\Delta t}{\bar{H}^{y2}}\sqrt{\bar{U}^2 + \bar{V}^2}\right)^n\right)^{-1}$$

are nondimensional friction functions. The overbar indicates a centered quantity. Next, equations (4.4) ans (4.5) are introduced into the divergence terms of (4.6). This yields a linear system of equations for ζ in which (n+1) level velocities are eliminated:

$$\begin{aligned} &- C_W^x \zeta_W^{n+1} - C_N^y \zeta_N^{n+1} + (1 + C_N^y + C_E^x + C_S^y + C_W^x)\zeta^{n+1} \\ &- C_E^x \zeta_E^{n+1} - C_S^y \zeta_S^{n+1} = B_1^n + B_2^n, \end{aligned} \tag{4.7}$$

with

$$B_1 = C_W^x \zeta_W^n + C_N^y \zeta_N^n + \left(1 - (C_N^y + C_E^x + C_S^y + C_W^x)\right) \zeta^n + C_E^x \zeta_E^n + C_S^y \zeta_S^n ,$$

$$B_2 = \frac{\Delta t}{2\,\Delta x} (\tilde{U}_W - \tilde{U}_E + \tilde{V}_S - \tilde{V}_N)^n ,$$

$$\tilde{U} = U^n + F^x (U^n + \Delta t\, X^n) ,$$

$$\tilde{V} = V^n + F^y (V^n + \Delta t\, Y^n) ,$$

$$C^x = g \left(\frac{\Delta t}{2\,\Delta x} \right)^2 (F^x \bar{H}^x)^n ,$$

$$C^y = g \left(\frac{\Delta t}{2\,\Delta x} \right)^2 (F^y \bar{H}^y)^n .$$

The system (4.7) is solved by successive over-relaxation. Test runs proved that, for a relaxation coefficient of $\omega = 1.2$, 5 to 10 iterations were sufficient for convergence. Once ζ is found, using equations (4.4) and (4.5), the horizontal transports are explicitely calculated.

This scheme has been tested for tidal processes in coastal areas. It was found that with Courant numbers up to 15 results hardly differed from those obtained with explicit models.

For a 3D model the implicit treatment of the vertical dimension for a barotropic flow differs from that of a baroclinic flow. In the barotropic case the above procedure is applied to the vertically integrated equations, thereby determining the water elevation. Then, in a second step, the depth variations of the horizontal velocities are determined using a semi-implicit formulation of vertical eddy viscosity (Sündermann, 1971) (i=1, 2)

$$\begin{aligned} v_i^{n+1} = v_i^n + \frac{A\,\Delta t}{2\,\Delta^2 z} (v_{iU}^{n+1} - 2\, v_i^{n+1} + v_{iD}^{n+1} \\ + v_{iU}^n - 2\, v_i^n + v_{iD}^n) + F^n , \end{aligned} \qquad (4.8)$$

where the indices U, D denote relative vertical positions (up, down) and F represents all remaining terms of equation (2.1). Equation (4.8) can be solved by a successive elimination procedure. In the baroclinic case, one first uses the continuity equation and the kinematic boundary condition at the surface, and derives a linear system of equations for the water elevation, which is solved by successive over-relaxation. Then, the velocities are calculated explicitely (Duwe, Hewer and Backhaus, 1983).

These 3D numerical FD algorithms are unconditionally stable. Some applications are discussed in section 5.

4.2 Tracer convection and diffusion

The numerical treatment of the convection-diffusion equations (2.3) and (2.16) involves specific difficulties (Richtmyer and Morton, 1967; Grotjahn and O'Brien, 1976) which mainly originate from the different scales of the convective and turbulent motion. The high discretization errors of FDM techniques lead often to physically unrealistic results. Apart from numerical instabilities, fundamental mechanical or thermodynamical principles can be violated. Fronts will not be sufficiently resolved due to numerical diffusion (Maier-Reimer, 1973).

Most of these difficulties can be by-passed by the application of tracer techniques (TM). In these, the water body is interpreted as a finite set of water particles carrying fixed physical properties. The state of a physical quantity within an Eulerian grid element of given discretization is then defined as the mean value of the states of all particles present in the cell. This viewpoint can be taken, in principle, for all physical properties such as velocity, density, temperature or dissolved substance (called "tracer"). The transport model based on this tracer concept consists of (1) the numerical simulation of the particle paths in a turbulent current field and (2) the computation of an Eulerian balance at certain time sections. In so doing one must distinguish between passive tracers such as contaminant or heat and active tracers, such as density (Maier-Reimer and Sündermann, 1982).

Compared with the established finite techniques tracer methods have the great advantage of nearly completely avoiding the undesirable numerical diffusion. They are capable, therefore, of simulating sharp gradients in natural waters. Furthermore, because of their intrinsic structure they afford computation only at those places, where transport processes really occur (according to the time depending particle coordinates). Thus, tracer methods are economically adapted to these concrete natural situations, where the propagation of particles is concentrated to small restricted areas.

In order to avoid in the numerical solution of (2.3) physically unrealistic results, such as the violation of the second law of thermodynamics, a critical Reynolds number of 2 should not be exceeded (Gorenflo, 1968):

$$Re = \frac{u\Delta x}{A} < 2 , \tag{4.9}$$

in which Δx denotes the horizontal grid size. Using the empirical rule that the diffusion coefficient A can be taken as proportional to velocity times water depth (Bowden, 1963), the rule (4.9) implies that the horizontal grid size must be of the order of the depth. This is practically unachievable. As a way out of this difficulty often "upwind" differences are used. This scheme of first order contains such a high artificial diffusivity that the condition (4.9) is automatically fulfilled. Different authors (Boris and Book, 1973; Egan and Mahoney, 1972; Duwe, 1982) have proposed improved upwind schemes with a diminished artificial diffusivity.

The outlined difficulties do not appear in the tracer methods described in the following. For passive tracers a Monte Carlo method is used.

According to the deduction of the diffusion equation in statistical mechanics and in turbulence theory, the property c is represented at a finite number of particles which move with a velocity typical of their instantaneous position. It is assumed that this velocity can be separated into a large scale organized flow, characterized by a mean velocity $\overline{v_i}$ and a smaller scale random fluctuation v_i' such that $v_i = \overline{v_i} + v_i'$. This means that each particle at any time step Δt suffers a displacement x_i which can be chosen randomly from a statistical distribution $f(x_i)$ (Bugliarello and Jackson, 1964)

$$x_i = v_i' \, \Delta t \, .$$

Thus, turbulence is described as the probability that a particle moves a certain distance during one time step.

The relation between such a large scale Brownian motion and an isotropic diffusion coefficient A is given by the expression (Einstein, 1905):

$$A = \frac{1}{2\Delta t} \int_{-\infty}^{+\infty} x_i^2 \, f(x_i)\,dx_i \Big/ \int_{-\infty}^{+\infty} f(x_i)\,dx_i \, . \tag{4.10}$$

It only involves the variance of f, but not its specific form. For the simulation of a constant, isotropic diffusion coefficient, therefore, the simplest distribution, a tophat distribution, is sufficient. Knowing a rough estimate of the diffusion coefficient A, equation (4.10) can be used to estimate the band width V' of the turbulent fluctuations v_i', $v_i' \in (-V', V')$. For a tophat distribution one obtains (Bork and Maier-Reimer, 1978):

$$A = \frac{1}{6} V'^2 \, \Delta t \, .$$

When applying the Monte Carlo technique, instead of equation (2.3), one must only solve the advection equation

$$\frac{\partial c}{\partial t} + v_j \frac{\partial c}{\partial x_j} = S \, . \tag{4.11}$$

If c is a dynamically active quantity, such as temperature or salinity, tracer methods of the kind described above will not suffice: the interaction of the quantity c with the field of advecting currents is then so strong that an unrealistically high number of particles would be needed to ensure sufficient accuracy. For lake problems, even small differences in salinity or temperature cause strong currents. A difference of 1 %, for instance, must be regarded as strong; for a satisfactory representation, at least 10'000 particles would be needed in each cell. In order to overcome this problem, an extension of the interpretation of the particles is needed. This has been done by Maier-Reimer and Sündermann (1982) with the so-called water blob tracing technique. In addition to the statistical displacement of the particle coordinates a buoyancy flux and an irreversible mixing are considered. This method will not further be explained here because it is important only for baroclinic processes.

5. APPLICATIONS

In the following, four examples of current investigations which were studied at the Institute of Oceanography Hamburg are briefly discussed.

5.1 Wind-driven circulation in the Belt Sea

This work was carried out by S. Müller-Navarra (1983). He used the semi-implicit FDM technique developed by J. Backhaus (see section 4.1) to predict the barotropic wind-driven circulation in the transition zone between North Sea and the Baltic. Figures 1 and 2 show the (nearly steady state) mean flow fields, 1.5 days after the onset of a 10 m/s wind from East, an implicit and an explicit method were employed. The time steps were 600 s and 93 s, respectively, and the horizontal grid distance was approximately 5000 m. The implicit procedure worked about 2.5 times faster than the explicit one. Almost no difference in the results can be detected.

5.2 Tidal flow in the Elbe estuary

These computations were performed by K. Duwe (1982). He used the semi-implicit 3D FDM technique described in section 4.1 with a time step of 300 s (Courant time step 21 s) and a horizontal grid distance of 500 m; in the vertical 6 computational planes were used.

Figure 3 shows the horizontal velocity distribution at the surface for the flood current. Grid cells without an arrow are representing dry tidal flats. For the same tidal phase isotaches were drawn; for a representative cross section they are shown in figure 4.

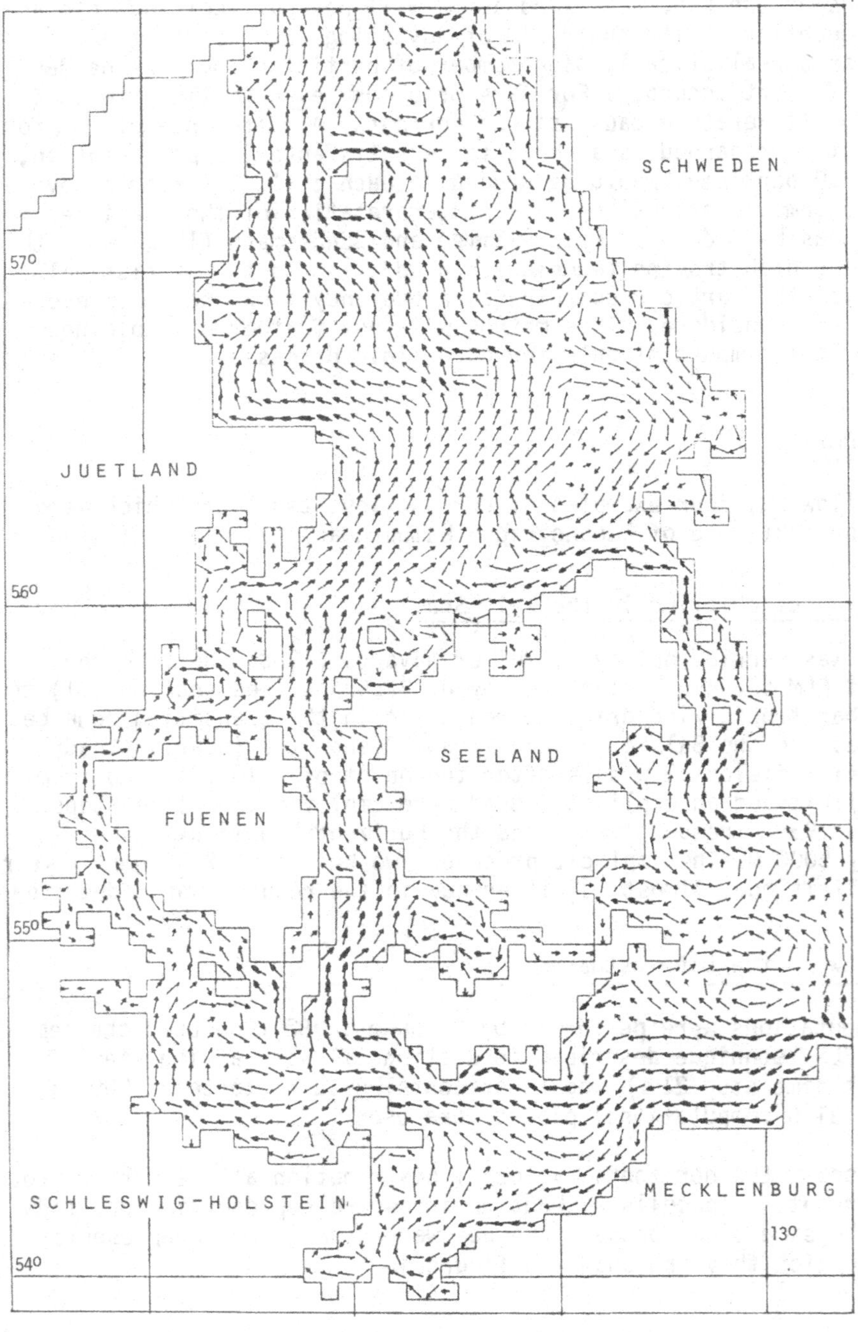

Figure 1 Mean flow field in the Belt Sea for and East wind of 10 m/s, 1.5 days after onset (semi-implicit calculation).

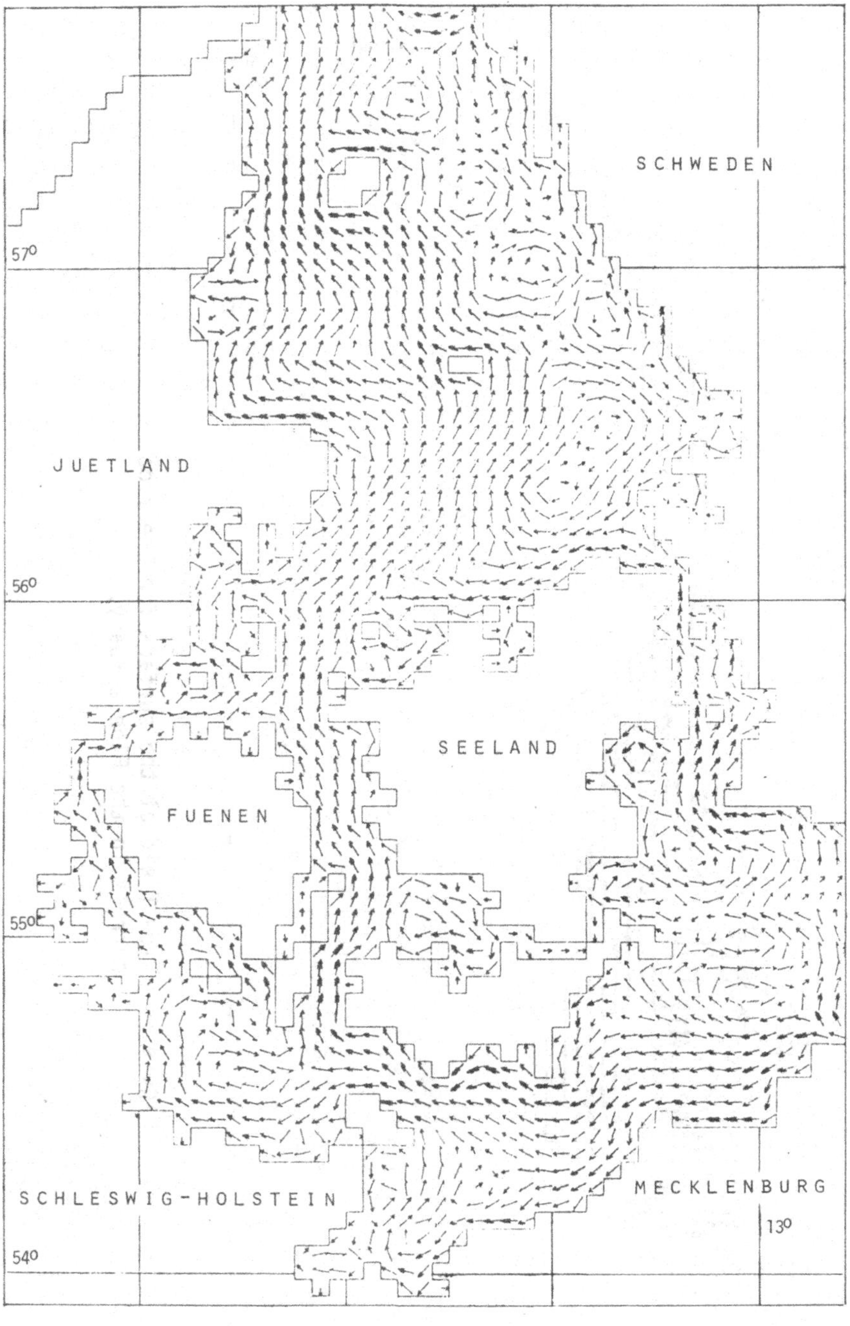

Figure 2 Mean flow field in the Belt Sea for an East wind of 10 m/s, 1.5 days after onset (explicit calculation).

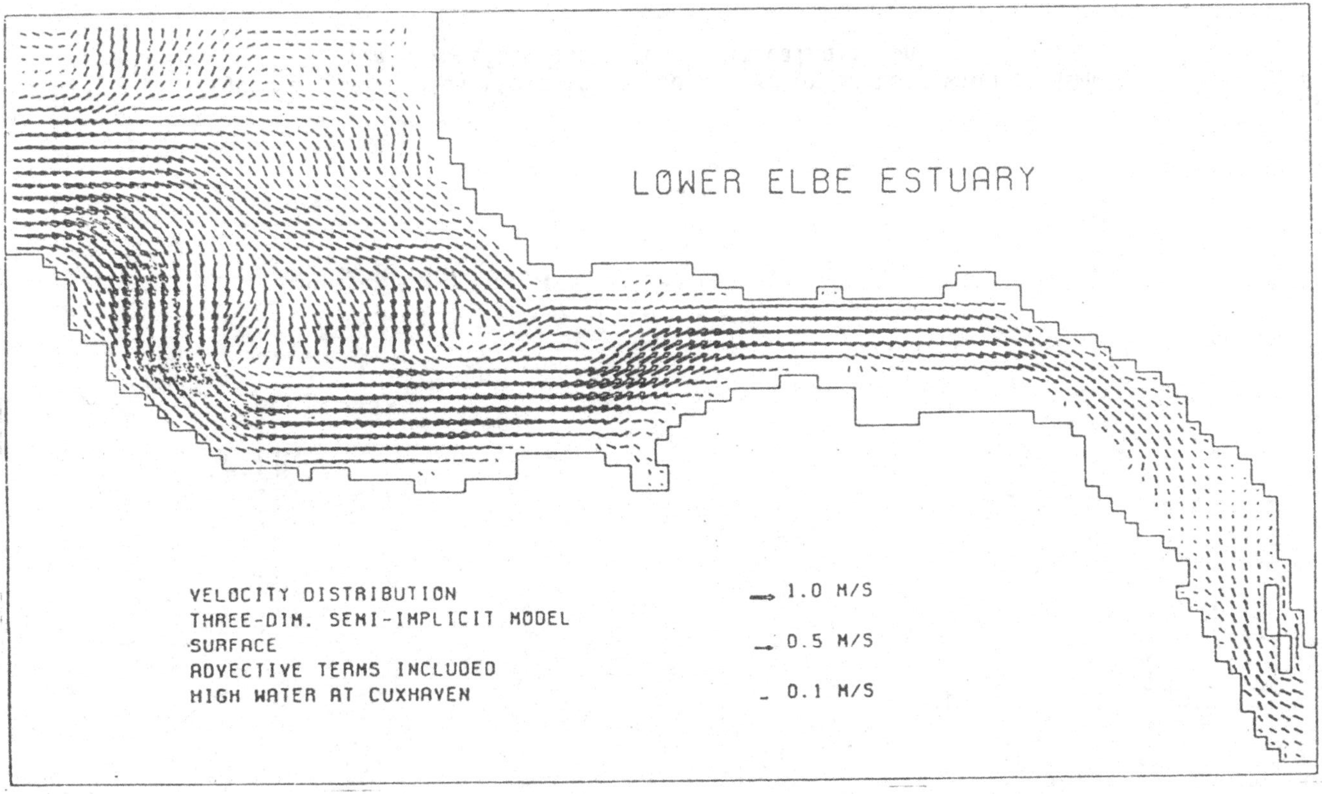

Figure 3 Flow field at the surface for the flood current in the Elbe estuary.

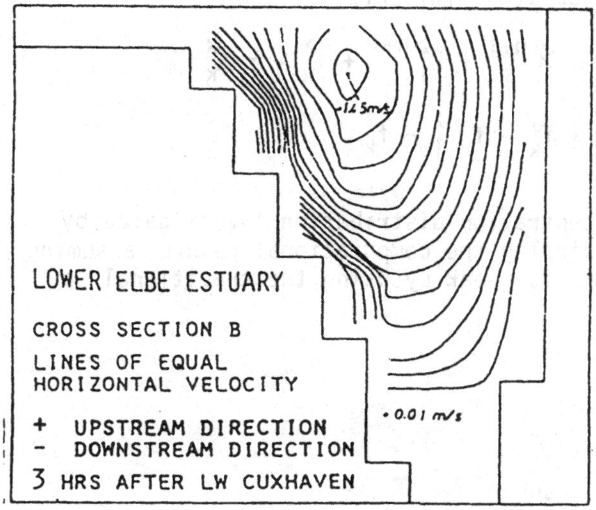

Figure 4

Isotaches for the flood current in a cross section of the Elbe estuary.

5.3 Flux-corrected-transport (FCT) methods for the solution of the convection equation

This investigation, started by K. Duwe (1982), is still under investigation. The 3D convection equation

$$\frac{\partial c}{\partial t} + v_j \frac{\partial c}{\partial x_j} = 0,$$

is solved. For each coordinate direction it is treated separately and then solved numerically in three steps (with $i = 1$ and $x_i = x$, $v_i = u$):

1) Upstream differencing

$$\overset{x}{C}{}_k^{n+1} = C_k^n - \frac{\Delta t}{\Delta x}\left(u_{k-1}^n \,\Delta C_{k-1} + u_k^n \,\Delta C_k \right),$$

with $\Delta C_{k-1} = C_k - C_{k-1}$, and $\Delta C_k = 0$ if $u_k \lessgtr 0$; $k-1$, k denote the upstream and the downstream directions respectively.

2) Antidiffusive flux in the downstream direction:

$$f_k^x = \mathrm{sign}(\Delta C_k^{n+1}) \cdot \max\Big(0, \min\big(\mathrm{sign}(\Delta C_k^{n+1})\,\Delta C_{k-1}^{n+1},$$
$$\varepsilon_k\,|\Delta C_k^{n+1}|,\ \mathrm{sign}(\Delta C_k^{n+1})\, C_{k+1}^{n+1}\big)\Big),$$

$$\varepsilon_k \doteq \frac{\Delta t}{\Delta x}\,|u_k^n|\,.$$

3) Total flux: (C^y and C^z are calculated similarly)

$$\Delta C_i^{n \to n+1} = C_i^{x\,n+1} + C_i^{y\,n+1} + C_i^{z\,n+1} + f_{k-1}^x - f_k^x$$

$$+ f_{k-1}^y - f_k^y + f_{k-1}^z - f_k^z .$$

Figure 5 shows an initial concentration distribution investigated by Zalesak (1979). Figures 6 and 7 display the computational results assuming one full rotation of the concentration peak by using the traditional upstream and Duwe's FCT algorithms.

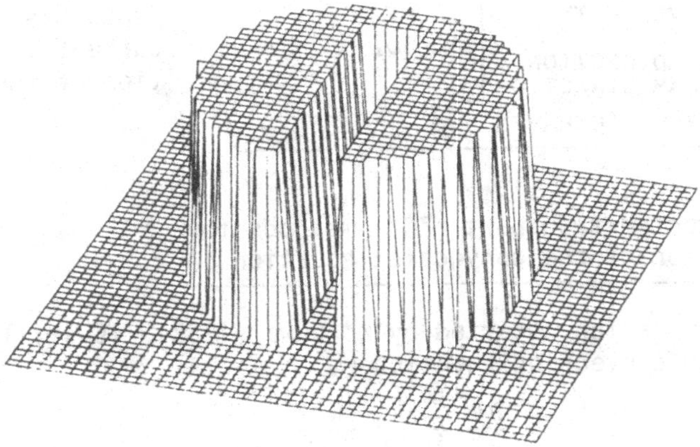

Figure 5 Initial concentration distribution of a two-dimensional test case of a convection computation on a rotating plate (Zalesak, 1979).

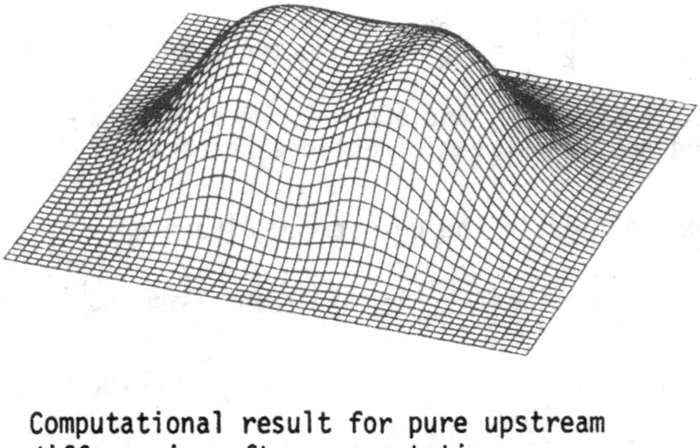

Figure 6 Computational result for pure upstream differencing after one rotation.

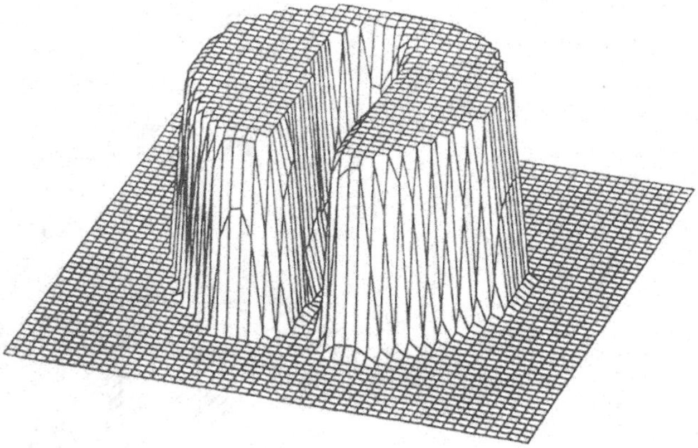

Figure 7 Computational result for FCT with limitation of gradient transport.

5.4 Propagation of radioactive Cs^{137} in the North Sea

This study was carried out by Maier-Reimer (1975) using the passive tracer technique (see section 4.2). It treats the propagation of radioactive Cs^{137} in the North Sea penetrating through the Straits of Dover. In this case field observations are available (Kautsky, 1973); they are representative for a state two years after release and show higher concentrations in a narrow band of about 100 km along the Dutch-German-Danish coast up to the Kattegat, see Figure 8.

The main circulation is due to tides and wind. Computations were carried out with the M_2-tide and a homogeneous mean wind field of WSW 3 m/s. The governing advection velocity in (2.16) is determined in two steps. First, the quasistationary current field $\overline{v_i}$ $(i = 1, 2)$ is calculated for a periodical tidal cycle. Because of the presence of gravity waves, stability is controlled by the Courant criterion (3.3) requiring a time step of $\Delta t = 90\,s$ for a horizontal grid size of $\Delta x = 10$ km. For reasons of computational economy this time step cannot be maintained for the two years simulation of a propagation process.

In a second step, therefore, the displacement of water masses within one tidal cycle is computed. This is done by a numerical integration of the ordinary differential equation

$$\frac{dx_i}{dt} = \overline{v_i}\left(x_i(t), t\right).$$

Thus the Lagrangean field of water mass displacements with respect to

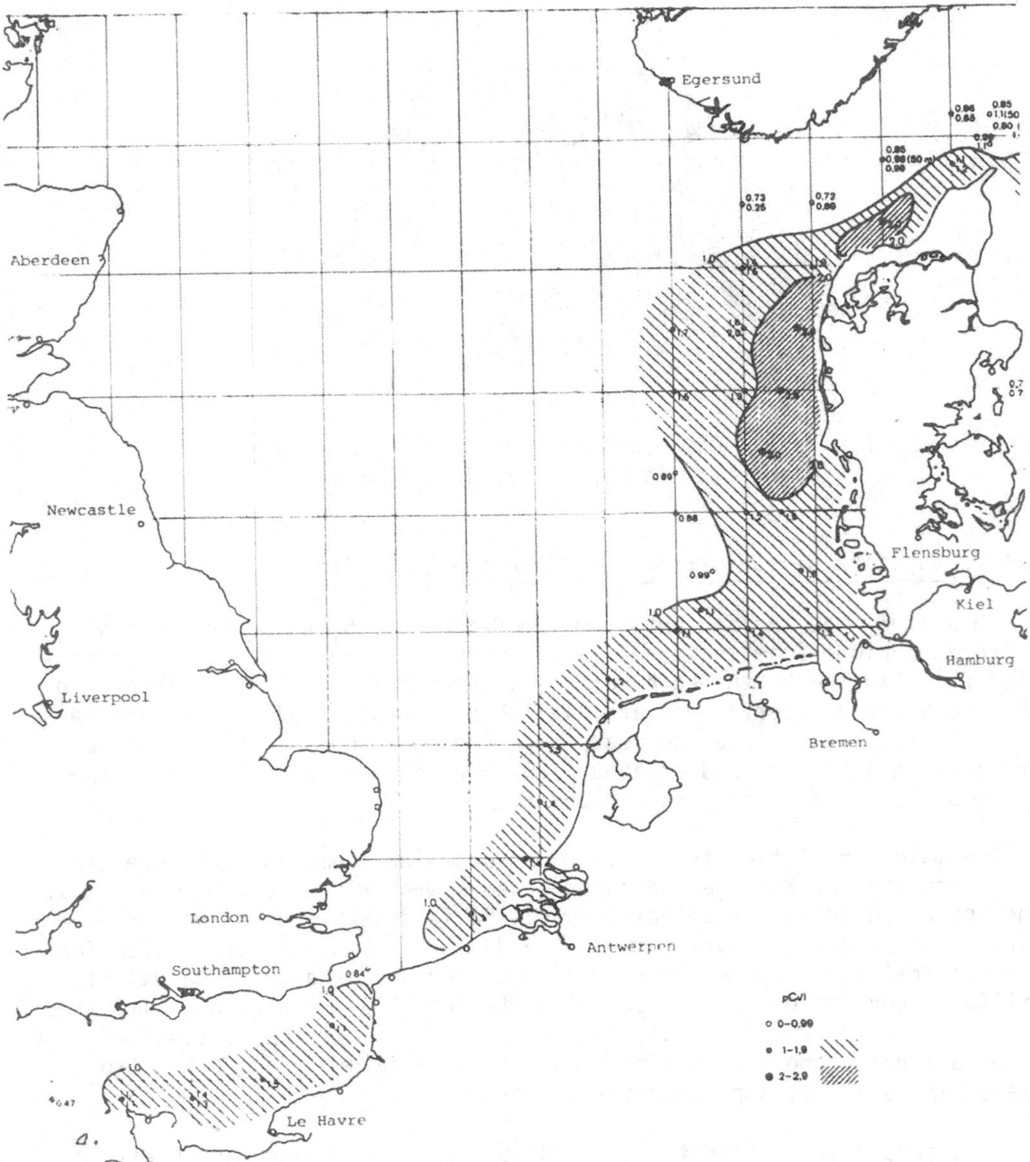

Figure 8 Observed concentration of radioactive Cs^{137} in the Southern North Sea (in p Ci/l) two years after the release in the English Channel (Kautsky, 1973).

tides is obtained. Knowing this field, the long term integration of the governing equations can be carried out now with a time step of one tidal period which is approximately 12 hours.

The Monte Carlo computation was performed with a diffusion coefficient A = 25 m^2/s; it determined the band width V' of the turbulent fluctuations. The continuous source in the Straits of Dover was simulated by increasing the number of particles involved in the computation by a factor of 5 at each time step. Figure 9 shows the result of the Monte Carlo simulation after two years. Each particle represents a certain mass. The corresponding concentration is obtained from the two-dimensional density by dividing it by the water depth. This picture is in a good agreement with observations, see Figure 8.

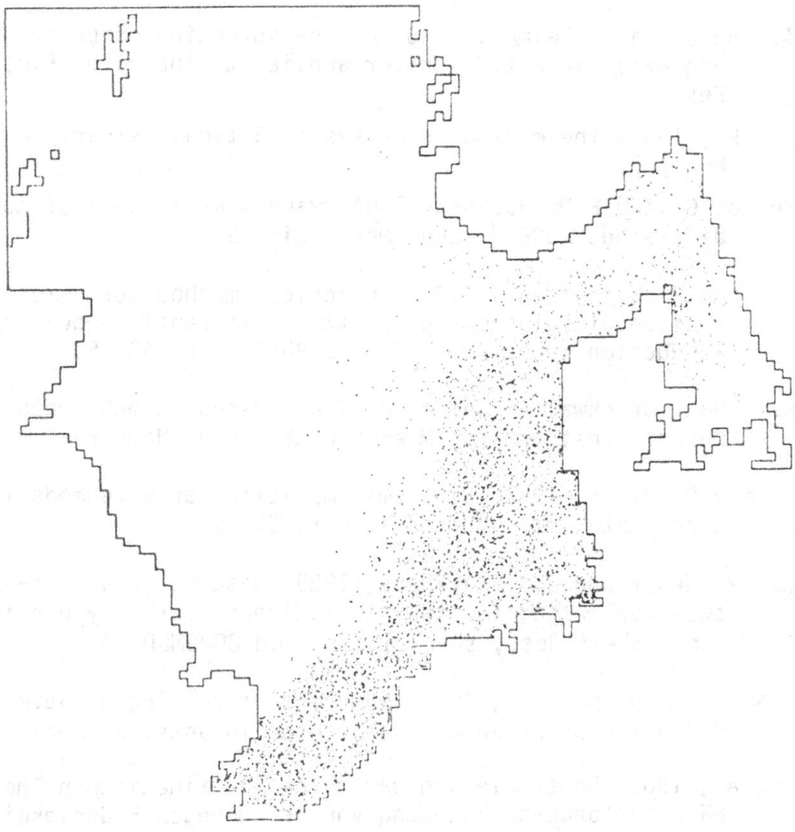

Figure 9 Computed propagation of a tracer within two years caused by the M_2 - tide and a mean annual wind assuming a continuous source in the English Channel. The wind was taken as WSW 3 m/s (Maier-Reimer, 1975).

REFERENCES

Abbott, M.B., 1974. Continuous flows, discontinuous flows, and numerical analysis. J. Hydr. Res. 12, 4.

Backhaus, J.O., 1983. A semi-implicit primitive equation model for shelf seas, outline and practical aspects of the scheme. Cont. Shelf Res., special issue on JONSMOD '82.

Backhaus, J.O., 1983. A semi-implicit scheme for the shallow water equations for application to shelf sea modelling. In press.

Boris, J.P. and D.L. Book, 1973. Flux corrected transport. I. Shasta. A fluid transport algorithm that works. J. Comp. Phys. 10.

Bork, I. and E. Maier-Reimer, 1978. On the spreading of power plant cooling water in a tidal river applied to the river Elbe. Adv. Wat. Res. 1.

Bowden, K.F., 1963. The mixing processes in a tidal estuary. J. Air Wat. Poll. 7.

Bugliarello, G. and E.D. Jackson, 1964. Random walk study of convective diffusion. ASCE J. Eng. Mech. Div. 94.

Crank, J. and P. Nicholson, 1947. A practical method for numerical integration of solutions of partial differential equations of heat conduction type. Proc. Cambr. Phil. Soc. 43, 50.

Duwe, K., 1982. Brackwasserzone eines Tideflusses. Unpublished internal report, Institut für Meereskunde, Univ. Hamburg.

Duwe, K. and R. Hewer, 1982. Ein semi-implizites Gezeitenmodell für Wattengebiete. Dt. Hydr. Zeitschr. 35.

Duwe, K., R. Hewer and J.O. Backhaus, 1983. Results of a semi-implicit two-step method for the simulation of markedly nonlinear flow. Cont. Shelf Res., special issue on JONSMOD '82.

Egan, B.A. and J.R. Mahoney, 1972. Numerical modelling of advection and diffusion of urban area source pollutants. J. Appl. Met. 11.

Einstein, A., 1905. Ueber die von der molekular-kinetischen Theorie der Wärme geforderte Bewegung von in ruhenden Flüssigkeiten suspendierten Teilchen. Ann. Phys. 4.

Gorenflo, R., 1968. Differenzenschemata vom Irrfahrttypus für die Differentialgleichung von Fokker-Planck-Kolmogorov. Zeitschr. ang. Math. Mech. 69.

Grotjahn, R. and J.J. O'Brien, 1976. Some inaccuracies in finite differencing hyperbolic equations. Monthly Weath. Rev. 104.

Hansen, W., 1956. Theorie zur Errechnung des Wasserstandes und der Strömungen in Randmeeren nebst Anwendungen. Tellus 8.

Kautsky, H., 1973. The distribution of the radio nuclide Caesium 137 as an indicator for North Sea water mass transport. Dt. Hydrogr. Zeitschr. 26.

Lax, P.D., 1954. Weak solutions of nonlinear hyperbolic equations and their numerical applications. Comm. Pure App. Math. 7.

Maier-Reimer, E., 1973. Hydrodynamisch-numerische Untersuchungen zu horizontalen Ausbreitungs- und Transportvorgängen in der Nordsee. Mitt. Inst. Meereskd. Univ. Hamburg 21.

Maier-Reimer, E., 1975. Zum Einfluss eines mittleren Windschubes auf die Restströme der Nordsee. Dt. Hydrogr. Zeitschr. 28.

Maier-Reimer, E. and J. Sündermann, 1982. On tracer methods in computational hydrodynamics. In: Abbott, M.B. and J. Cunge (eds.) Engineering applications of computational hydraulics. Pitman Boston-London-Melbourne.

Müller-Navarra, S., 1983. Simulation von Bewegungsvorgängen im Uebergangsgebiet zwischen Nordsee und Ostsee. Diploma thesis, Univ. of Hamburg.

Richtmyer, R.D. and K.W. Morton, 1967. Difference methods for initial value problems. Interscience Publishers New York.

Simons, T.J., 1980. Circulation models of lakes and inland seas. Can. Bull. Fish. Aquat. SCI 203.

Sündermann, J., 1971. Die hydrodynamisch-numerische Berechnung der Vertikalstruktur von Bewegungsvorgängen in Kanälen und Becken. Mitt. Inst. Meereskd. Univ. Hamburg 19.

Sündermann, J., 1979. Numerical modelling of circulation in lakes. In: Graf, W.H. and C.H. Mortimer (eds.) Hydrodynamics of lakes. Elsevier Amsterdam-Oxford-New York.

Zalesak, S.T., 1979. Fully multidimensional flux corrected transport algorithms for fluids. J. Comp. Phys. 31.

Hydrodynamics of Lakes: CISM Lectures
edited by K. Hutter, 1984
Springer Verlag Wien-New York

SOME ASPECTS OF BAROCLINIC CIRCULATION MODELS

J. Kielmann[1] and T.J. Simons[2]

[1]Institut für Meereskunde an der Universität Kiel, W. Germany

[2]Canada Centre for Inland Waters, Burlington, Canada

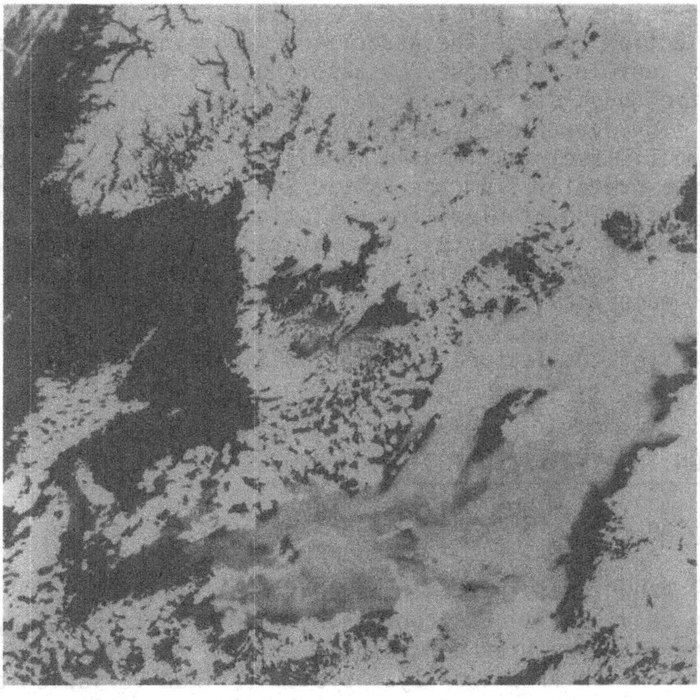

Infrared satellite picture of the Baltic Sea. The visible area in the West is the Bornholm Basin. Observe the beautiful vortices.

1. INTRODUCTION

These lectures deal with problems of baroclinic circulation of lakes and inland seas. The term "circulation" needs further explanation: What is meant by circulation? Generally, we tend to define circulation as the rectified effects of all possible thermo-hydrodynamic processes in a lake or sea. Rectification is a consequence of averaging the possibly non-linear processes over some predefined space and time scales. Thus, we speak of the "world ocean circulation" when time scales of some hundred years and space scales of some hundred kilometers are involved, and we talk about the "summer circulation of a lake" and may mean by it the average of several years of stratified circulation patterns in summer. However, even the response of a lake to a few days of wind forcing is often referred to as "circulation"; in this case it means that also wave processes of this time scale have to be included.

In the following we are mostly concerned with this latter definition. We will ask the question what happens in a stratified lake after a few days of constant wind forcing. We are less interested in the generation of barotropic seiches but rather in the quasi-static response. This is a matter of selection because the associated wave generation is already covered by different lectures (see Mysak or Hutter, this volume). Principal problems on the quasi-static response are dealt with in section 2, where an idealized two-layer flow is discussed. In section 3 the numerical formulation of multi-level/multi layer models is presented with some quasi-static response examples from the Baltic Sea. In view of the importance of advection-diffusion problems for pollution, for transport simulation of suspended or dissolved matter, for water quality models and for advective processes in gneral, section 4 deals with some aspects of solving such problems numerically by finite difference methods.

A great part of the lecture will follow Simons (1980).

2. CIRCULATION IN STRATIFIED TWO-LAYER MODELS

2.1 Introduction

As is well known, natural basins are often stratified in such a manner, that they may be visualized as containing mainly horizontal layers of fluid of different densities separated by moving material interfaces. Figure 1 shows a sketch of a two-layer system (primed variables belong to the lower layer). As the interfaces of multi-layer models may intersect the bottom, the number of layers is variable in the horizontal direction. For the two-layer case, therefore, it is usually assumed that the basin

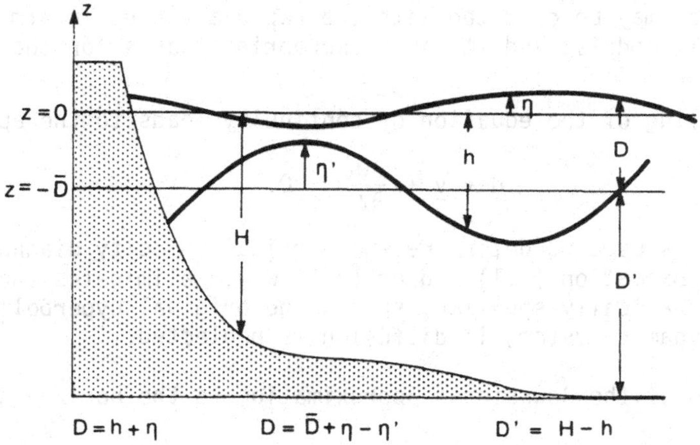

Figure 1 Sketch of a two-layer system.
Definition of variables.

has vertical walls at the intersection of the average interface with the bottom, in order to keep the number of layers constant (Hutter, p. 28).

Multi-layer models with impermeable moving interfaces are seldom used in circulation studies, because the intersection of interfaces with the free surface and the bottom leads to complications. A four-layer model was used in upwelling studies by McNider and O'Brien (1973), but the most common version of the class of moving interface models is the two-layer configuration. Although the latter has been applied to basins with realistic topography, its main area of application has been in theoretical studies of idealized basins. It is for this reason that the pertinent equations for the most conventional two-layer model, which consists of two layers of fluid of constant density, are listed here.

2.2 Two-layer equations

In the following we briefly review the derivation of the governing equations for the two-layer model (see also Hutter, p. 28). The vector notation used refers to the horizontal coordinates only.

Under the assumption of hydrostatic balance the vertical momentum equation reduces to

$$\frac{\partial p}{\partial z} = - \rho g, \quad \text{(z upward)}. \tag{2.1}$$

The assumption eliminates vertical accelerations due to gravity effects and, therefore, precludes explicit treatment of free convection associated with unstable stratifications. However, it is this principle, more than any other, that may be credited with the rapid advances in atmospheric as well as oceanic models, and it is an appropriate basis for the present discussion.

Proper scaling of the equation of continuity leads to the approximation

$$\text{div } \underline{v} + \frac{\partial w}{\partial z} = 0. \tag{2.2}$$

This equation is used to determine the vertical velocity diagnostically. It is the approximation (2.1) and/or (2.2) which eliminates the sound waves, or, mathematically speaking, spoils the original hyperbolicity of the thermo-hydrodynamic system, if diffusion is neglected.

Application of the Boussinesq approximation to the horizontal momentum balance leads to

$$\frac{\partial \underline{v}}{\partial t} + \nabla\cdot(\underline{v}\underline{v}) + \frac{\partial (w\underline{v})}{\partial z} - \underline{f}\times\underline{v} = -\frac{1}{\rho_0}\nabla p + \frac{1}{\rho_0}\frac{\partial \underline{\tau}}{\partial z}, \tag{2.3}$$

where ρ_0 is the average density, and $\underline{\tau}$ the Reynolds stress; horizontal friction is neglected.

The internal pressure distribution in the upper layer follows immediately from the dynamic boundary condition at the water surface, which requires the pressure to be continuous. With this boundary condition, (2.1) may be integrated to give

$$p = p_s + \rho g(\eta - z), \tag{2.4}$$

where p_s is the atmospheric pressure at the air-sea interface, η is the elevation of the free surface above the mean water level, and ρ is the upper layer density. Similarly, since the pressure must be continuous at the interface $z = -h(t,x,y)$, the lower layer presure is

$$p' = p_s + \rho g(\eta + h) - \rho' g(h + z).$$

The horizontal pressure gradients are uniform in the vertical within each layer and given by

$$\begin{aligned} \nabla p &= \nabla p_s + \rho g \nabla\eta, \\ \nabla p' &= \nabla p_s + \rho g \nabla\eta - (\rho' - \rho) g \nabla h, \end{aligned} \tag{2.5}$$

where the prime refers to the lower layer and h is the depth of the interface measured from the equilibrium surface level.

The continuity equation and horizontal momentum equations are obtained in vertically integrated form by applying the rules for interchanging differentiation and integration, for example,

$$\frac{\partial}{\partial x}\int_{-h}^{\eta} u\,dz = \int_{-h}^{\eta}\frac{\partial u}{\partial x}\,dz + u(\eta)\,\frac{\partial \eta}{\partial x} - u(-h)\,\frac{\partial(-h)}{\partial x}. \qquad (2.6)$$

The primary boundary condition to be imposed is that the normal component of the velocity vector must vanish at any solid boundary, such as the bottom of the basin and the lateral walls. On the other hand, at a moving material surface such as the air-water interface, a water particle must follow the motion of that surface. This is known as the kinematic boundary condition as contrasted to the dynamic boundary condition used to integrate the hydrostatic equation (2.1), (2.4). Specifically then, the velocity components at the bottom and free surfaces must satisfy the following conditions

$$z = \eta: \quad w = \frac{\partial \eta}{\partial t} + \underline{v}\cdot\nabla\eta\,,$$

$$z = -h: \quad w = -\frac{\partial h}{\partial t} - \underline{v}\cdot\nabla h\,, \qquad (2.7)$$

$$z = -H: \quad w = -\,\underline{v}\cdot\nabla H\,,$$

where H measures the depth of the basin relative to the mean water level. We define the layer volume transports

$$\underline{V} = \int_{-h}^{\eta}\underline{v}\,dz\,, \qquad \underline{V}' = \int_{-H}^{-h}\underline{v}'\,dz\,, \qquad (2.8)$$

and write the layer thickness and layer transports (see also Fig. 1):

$$D = \eta + h, \quad D' = H - h, \quad \underline{V} = D\bar{\underline{v}}, \quad \underline{V}' = D'\,\bar{\underline{v}}' \qquad (2.9)$$

where h is the distance between the interface and the equilibrium surface, hence h decreases if the interface moves upward. $\underline{v}$, $\bar{\underline{v}}'$ are the vectors of average velocity.

The equations of continuity now become

$$\frac{\partial D}{\partial t} + \nabla\cdot\underline{V} = 0, \qquad \frac{\partial D'}{\partial t} + \nabla\cdot\underline{V}' = 0, \qquad (2.10)$$

and the equations of motion are

$$\frac{\partial \underline{V}}{\partial t} + \nabla\cdot\underline{V}\bar{\underline{v}} - f\underline{V}\times\underline{k} = -\frac{gD}{\rho_o}\nabla\left(\frac{p_s}{g} + \rho\eta\right) + \frac{\tau_s - \tau_i}{\rho_o}\,,$$

$$\frac{\partial \underline{V}'}{\partial t} + \underline{\nabla}\cdot\underline{V}'\bar{\underline{v}}' - f\underline{V}'\times\underline{k} = -\frac{gD'}{\rho_o}\nabla\left(\frac{p_s}{g} + \rho\eta - \rho'\varepsilon h\right) + \frac{\tau_i - \tau_b}{\rho_o}\,, \qquad (2.11)$$

where $\underline{\tau}_i$ represents the momentum transfer across the interface, that is, an interface stress analogous to surface and bottom stresses $\underline{\tau}_s$, $\underline{\tau}_b$, $\varepsilon = (\rho' - \rho)/\rho'$. The two layers are coupled through the interface displacements and interface stresses.

It may be stated that each of the system (2.10), (2.11) constitutes a strict hyperbolic system with 3 real eigenvalues, each corresponding to a possible wave (two are gravity waves, one is a Rossby wave). In the following discussions, the non-linear volume transport terms in (2.11) will be neglected.

2.3 Normal modes of a two-layer system

If friction and non-linear terms are neglected, it can be shown that the vertical differential equations of a continuously stratified system possess an infinite number of eigensolutions which determine the shape of the current profile in the vertical (e.g. Simons (1980), see Hutter on internal waves, pp. 39-80). These eigensolutions are called modes. If non-linear terms of the two-layer system (2.10), (2.11) are neglected and if the deviations of the interfaces from their equilibrium values are small, it can be shown that the two-layer model of a rectangular flat basin has exactly 2 vertical modes (e.g. Simons, 1980).

If the internal interface η' is defined as the deviation from some equilibrium position, positive upward, the upper and lower layer interfaces and transports may be written as linear combinations of these modes:

$$\eta = \eta_0 + \varepsilon\left(\frac{D'}{D+D'}\right)^2 \eta_1 , \qquad \underline{V} = \frac{D\underline{V}_0 + D'\underline{V}_1}{D+D'} , \tag{2.12}$$

$$\eta' = -\frac{D'}{D+D'}\eta_1 , \qquad \underline{V}' = \frac{D'(\underline{V}_0 - \underline{V}_1)}{D+D'} , \tag{2.13}$$

where η_n, $\underline{V}_n$ $(n = 0,1)$ refer to the different modes which are solutions to the differential equations (e.g. Simons, 1980)

$$\frac{\partial \underline{V}_n}{\partial t} - f\,\underline{V}_n \times \underline{k} = -g\lambda_n \nabla\eta_n + \frac{\underline{\tau}_s}{\rho_0} , \qquad \frac{\partial \eta_n}{\partial t} + \nabla\cdot\underline{V}_n = 0 , \tag{2.14}$$

in which

$$\lambda_0 = D + D' , \qquad \lambda_1 = \frac{\varepsilon D D'}{D+D'} .$$

The relations to η, η' are approximately given by

$$\eta_0 = \eta + \frac{\varepsilon D'}{D+D'}\,\eta' , \qquad \underline{V}_0 = \underline{V} + \underline{V}' , \tag{2.15}$$

$$\eta_1 = -\frac{D+D'}{D'}\,\eta' , \qquad \underline{V}_1 = \underline{V} - \frac{D}{D'}\underline{V}' . \tag{2.16}$$

Obviously $\underline{V}_0$, the zero mode transport, is the total transport of the water column. Inspection of (2.14) shows that the differential equation for $n=0$ has the same form as that for an unstratified lake (see e.g. Sündermann, pp. 209-233), with the exception that the free surface η_0 in (2.15) is corrected due to the elevation of the internal interface η'. In a homogeneous ocean the pressure surfaces cannot be inclined against the density surfaces; therefore, we call $\underline{V}_0$ the barotropic volume transport. The first order mode is due to stratification and is called first baroclinic mode. Subtraction of $\underline{V}_0$ from the layer transport (2.12), (2.13) results in $\underline{V} = -\underline{V}'$, i.e. the layer transports of the baroclinic mode are of opposite sign. The baroclinic mode has zero net transport. Furthermore, the sign of the interface perturbation η' is opposite to that of the surface η as can be seen from (2.12), (2.13).

It should be pointed out that barotropic and baroclinic modes cannot be separated for arbitrary topography. We will reconsider this problem later.

2.4 External and internal modes

In view of (2.15), (2.16) we define the vertically integrated transport and shear between the two layers as

$$\tilde{\underline{V}} \equiv \underline{V} + \underline{V}' , \qquad \underline{V}^* \equiv \frac{D'\underline{V} - D\underline{V}'}{D+D'} = \frac{D'D}{D+D'}(\bar{\underline{v}} - \bar{\underline{v}}'), \tag{2.17}$$

where $\underline{V}$, $\underline{V}'$ can be recovered by

$$\underline{V} = \frac{D}{H}\tilde{\underline{V}} + \underline{V}^* , \qquad \underline{V}' = \frac{D'}{H}\tilde{\underline{V}} - \underline{V}^* \tag{2.18}$$

The definition of $\underline{V}^*$ was obtained by multiplying the baroclinic part of equation (2.14) by $D'/(D+D')$. With these definitions the equations of continuity, (2.10), become

$$\frac{\partial \eta}{\partial t} + \nabla\cdot\tilde{\underline{V}} = 0, \qquad \frac{\partial D}{\partial t} + \nabla\cdot\left(\underline{V}^* + \frac{D\tilde{\underline{V}}}{D+D'}\right) = 0. \tag{2.19}$$

The equations of motion (2.11) may be linearized: for $\eta \ll H$, $D + D' \approx H$, $\eta' \gg \eta$ they become

$$\frac{\partial \tilde{\underline{V}}}{\partial t} - f \tilde{\underline{V}} \times k = - gH \nabla (\frac{p_s}{\rho_o g} + \eta) + g \varepsilon D' \nabla D + \frac{\underline{\tau}_s - \underline{\tau}_b}{\rho_o},$$

$$\frac{\partial \underline{V}^*}{\partial t} - f \underline{V}^* \times \underline{k} = - g\varepsilon \frac{DD'}{H} \nabla D + \frac{D'}{\rho_o H} \underline{\tau}_s - \frac{1}{\rho_o} \underline{\tau}_i + \frac{D}{\rho_o H} \underline{\tau}_b , \qquad (2.20)$$

where $\rho\eta - \rho'\varepsilon h = \rho'\eta - \rho'\varepsilon D$ has been used. From the second equation (2.20) we can see that the effective wind forcing of the shear $\underline{V}^*$ is proportional to the ratio of lower to total depth. Further, from the last section, we also know that, if friction is neglected, the quantities $\eta, \tilde{\underline{V}}$ and D, $\underline{V}^*$ arising in (2.19), (2.20) describe the barotropic and baroclinic response respectively. However, if friction is involved, the vertical current structure is determined by frictional as well as baroclinic effects (see also Heaps, pp. 155-201): the term "barotropic/baroclinic mode" is no longer applicable.

The solutions to (2.19), (2.20) are called external $(\eta, \tilde{\underline{V}})$ and internal $(D, \underline{V}^*)$ modes. The external mode refers to the vertically averaged current $\tilde{V}/H$, and the internal mode represents the deviation of the current from this mean (shear). If the rigid lid approximation (see Hutter, p. 28) is applied to $(2.19)_1$, $\partial\eta/\partial t \approx 0$, surface waves are filtered out. If furthermore,

$$\tilde{\underline{V}} \cdot \nabla \frac{D}{H} \approx - \frac{D}{H^2} \tilde{\underline{V}} \cdot \nabla H = 0,$$

i.e., if the flow is parallel to the depth contours, (2.19) simplifies to

$$\nabla \cdot \tilde{\underline{V}} = 0, \qquad \frac{\partial D}{\partial t} + \nabla \cdot \underline{V}^* = 0 . \qquad (2.21)$$

The second of these and $(2.20)_2$ describe the internal solution, which is independent of the barotropic current in this case.

We shall see later, in connection with the numerical multi-level models, that it is favourable to separate the flow into external and internal motions. The reason for this is that the phase speed of internal waves is very much smaller than that of external motions (e.g. surface waves). This means that the time step in an explicit numerical model may be by O(10) larger for the internal shear equations than for the total volume transport equations.

2.5 Steady-state two-layer circulation

Let us return to the introductory remarks on definition of the "circulation" and ask for steady-state solutions of a two-layer strafified model with friction. Even though steady state solutions for the fully nonlinear model must not exist in the limit $t \to \infty$ (only statistically) and it is also rather unlikely that steady-state circulations develop in a real lake, because of the relatively short period of summer circulation, simplified steady-state two-layer models have been considered for lakes [e.g. Lee and Liggett (1970), Liggett and Lee (1971) and on oceanic scales Nowlin (1967), Welander (1968)]. A short review of these papers is given in Simons (1980). It is difficult to obtain analytical solutions without further simplifications, even for the linearized and stationary version of (2.10), (2.11).

For deep basins the *hypothesis of quasi-compensation* is often applied. In this approximation, the surface gradients are assumed to be compensated by interface gradients to such an extent that the pressure gradients in deep water tend to vanish. If we set $\nabla p' = 0$ in (2.5) we obtain

$$\nabla \eta = \frac{\rho' - \rho}{\rho} \nabla h = \varepsilon \nabla D \quad \text{(compensation condition).}$$

By this assumption the lower layer geostrophic current vanishes, whereas the geostrophic upper layer velocity is a function of the upper layer thickness D only (Welander, 1968). Because of the vanishing geostrophic current in the lower layer the influence of bottom topography on the upper layer velocity appears to be negligible; the lower layer currents are driven by the interfacial stresses $\underline{\tau}_i$ only. For shallow seas and lakes the quasi-compensation condition is hardy ever justified. This is confirmed by the computations of Lee and Liggett (1970). Moreover, it is rather unlikely for steady-state circulation to occur in natural lakes. Thus we stop the discussion here and direct the reader to the literature cited. In the following we shall deal with time dependent two-layer circulation.

2.6 Time dependent two-layer circulation

Time dependent solutions of a two-layer system naturally include wave phenomena which are discussed in more detail by Hutter (pp. 39-80) and Mysak (pp. 81-150). As is well known, the shore zones are dominated by internal Kelvin waves and the offshore areas by Poincaré waves which often have near-inertial frequencies. From the viewpoint of water circulation the time scales of surface seiches or Poincaré waves are relatively short. Here, we are rather concerned with time scales of internal Kelvin waves, of topographic waves or longer time scales.

a) *Infinite channel/cross section solutions*

We start by discussing analytical solutions of (2.19), (2.20) for an infinitely long channel, and select the longshore axis parallel to the x-axis. We assume that the flow is independent of x and apply the rigid lid approximation. From (2.19) it follows then that $\partial\tilde{V}/\partial y = 0$, and consequently, because of vanishing transports at the shores, $\tilde{V} \equiv 0$. From the equation of motion for $\tilde{V}$, because the flow is independent of x, $\partial^2\eta/\partial x\,\partial y = 0$ can be deduced. Therefore, the pressure gradient $\partial\eta/\partial x$ is independent of y. Furthermore, we assume that the internal longshore pressure gradient is negligible as long as the pulses emitted by the end walls or shores have not reached the center of the lake. Because the phase speed of an internal gravity wave is $c = (\varepsilon g D D'/H)^{1/2} = Rf$, in which $R = c/f$ is the internal Rossby radius of deformation (see Hutter, p. 54), which is of the order O(5 km) for most lakes, it takes at least a few inertial periods to signal the shores to the lake center (Csanady, 1973).

With this latter assumption and for a constant longshore wind the equation of longshore transport $\tilde{U}$ reduces to

$$\frac{\partial\tilde{U}}{\partial t} = -\frac{gH}{\rho_o}\frac{\partial\eta}{\partial x} + \frac{1}{\rho_o}(\tau_{sx} - \tau_{bx}). \tag{2.22}$$

Averaging (2.22) over the width of the channel yields

$$g\bar{H}\frac{\partial\eta}{\partial x} = \frac{1}{\rho_o}(\tau_{sx} - \bar{\tau}_{bx}), \tag{2.23}$$

where it was assumed that the integral of the longshore transport $\tilde{U}$ over the cross section of the channel vanishes. For closed basins this is always justified. From (2.23) one may now compute $\partial\eta/\partial x$ and insert the result into (2.22). With $\tau_{bx} \sim 0$, initially, this yields

$$\frac{\partial\tilde{U}}{\partial t} = (1 - \frac{H}{\bar{H}})\frac{\tau_s}{\rho_o}. \tag{2.24}$$

This equation shows that the water will accelerate in the direction of the wind, where the depth is less than its cross sectional mean $\bar{H}$, and against the wind in deeper water. The reason is, that in the first case the wind stress overwhelms the pressure gradient, thus accelerating the water, while in the latter case the reverse happens, which causes the return flow. The width of the current band in direction of the wind depends on the shape of H (and on friction if it becomes important). In any case these considerations provide partial insight into the barotropic response.

Let us now turn to the baroclinic response, described by the second of equations (2.19), (2.20). We take the same assumptions as for the barotropic response. Since $H = H(y)$ and $\tilde{V} = 0$, the barotropic flow follows the depth contours and (2.19) may be replaced by (2.21).

Thus from (2.20)

$$\frac{\partial U^*}{\partial t} - fV^* = \frac{D'\tau_{sx}}{\rho_o H},$$

$$\frac{\partial V^*}{\partial t} + fU^* = \varepsilon g \frac{DD'}{H}\frac{\partial \eta'}{\partial y} + \frac{D'\tau_{sy}}{\rho_o H}, \qquad (2.25)$$

$$\frac{\partial \eta'}{\partial t} - \frac{\partial V^*}{\partial y} = 0 .$$

The solution of (2.25) for a channel was discussed in more detail by Csanady (1973) and Bennett (1974). Elimination of η' and U^* yields the single equation

$$\frac{\partial^2 V^*}{\partial t^2} + f^2 V^* - g\varepsilon \frac{DD'}{H}\frac{\partial^2 V^*}{\partial y^2} = \frac{D'}{\rho_o H}\left(\frac{\partial \tau_{sy}}{\partial t} - f\tau_{sx}\right) \qquad (2.26)$$

subject to the conditions

$$V^* = 0, \quad \text{for} \quad y = \pm W/2, \qquad V^* = 0, \quad \text{for} \quad t \le 0, \qquad (2.27)$$

where W is the width of the channel ($-W/2 \le y \le W/2$). For a sudden constant longshore wind and constant H, the steady-state solution to (2.26), i.e.

$$V^* - R^2 \frac{\partial^2 V^*}{\partial y^2} = -\frac{D'\tau_{sx}}{\rho_o fH} \qquad (2.28)$$

is

$$V^* = \frac{D'\tau_{sx}}{\rho_o fH}\left(\frac{\cosh(y/R)}{\cosh(W/2R)} - 1\right). \qquad (2.29)$$

This may be substituted into $(2.25)_{1,3}$, using the initial conditions $U^* \equiv \eta' \equiv 0$ at $t = 0$. Thus,

$$U^* = t\,\frac{D'\tau_{sx}}{\rho_o H}\,\frac{\cosh(y/R)}{\cosh(W/2R)}, \qquad \eta' = t\,\frac{D'\tau_{sx}}{R\rho_o fH}\,\frac{\sinh(y/R)}{\cosh(W/2R)}. \qquad (2.30)$$

This particular solution of (2.26) is called *quasi-geostrophic* or *quasi-static*, it does not fulfill the initial condition for V^*. One can see that the longshore transport is in geostrophic balance with the thermocline excursion (as is the case for internal Kelvin waves). U^* and η' and the barotropic longshore transport $\tilde{U}$ from (2.24) all increase linearly with time. The longshore baroclinic current and the thermocline excursion are confined to a narrow coastal band of width O(R) on both sides of the cross section or channel. This current is called baroclinic jet (Charney, 1955, Csanady, 1967). Figure 2 shows the principal features of the solutions, where R is again the internal Rossby radius of deformation. The

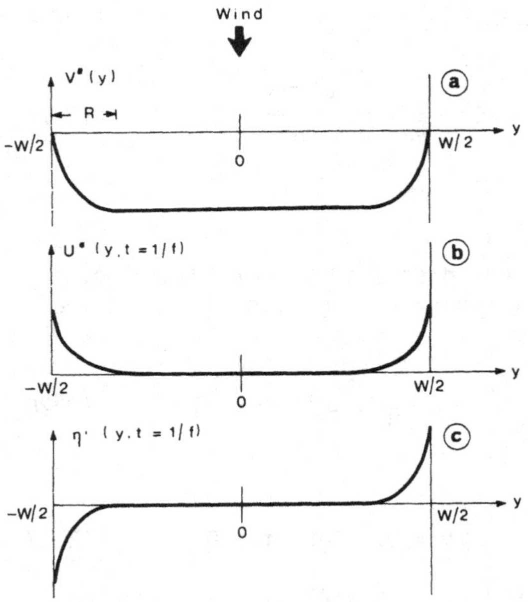

Figure 2

Internal quasi-static solution (2.29), (2.30), channel parallel winds (blowing out of the paper), t = 1/f.

a) Transverse transport

b) Longshore transport

c) Interface excursion

linear growth of the transport U* with time will be discussed later. Depending on H(y) the barotropic component of the longshore current may also be confined to the shore zones. Thus, the coastal currents appear to be a dominant feature.

To demonstrate this channel flow, we reproduce -as an example- a result of a 4-level model of Lake Ontario (Simons, 1972) shortly after the onset of a storm blowing almost parallel to the major axis of the lake. Figure 3 shows the sum of barotropic and baroclinic flow which can be interpreted by the simplified channel solutions discussed above.

The transverse component of the transport, V*, is almost constant over a cross section of the channel, pointing to the right of the wind and decreasing with a scale of O(R) at the shore lines. In order to have $V^* = 0$ at $t = 0$, we use the unforced solution of the wave equation (2.26) (H is constant):

$$\frac{\partial^2 V^*}{\partial t^2} + f^2 V^* - c^2 \frac{\partial^2 V^*}{\partial y^2} = 0, \qquad c = Rf. \tag{2.31}$$

The solutions which obey $V^* = 0$ at $y = \pm W/2$ and do not vanish for $t = 0$ could have the form

$$V^*_W \sim \cos \lambda_n y \cos \sigma_n t, \tag{2.32}$$

where

$$\lambda_n = \frac{2n-1}{W}\pi, \quad \sigma_n^2 = f^2(1 + R^2\lambda_n^2), \qquad n = 1,2,3 \ldots \tag{2.33}$$

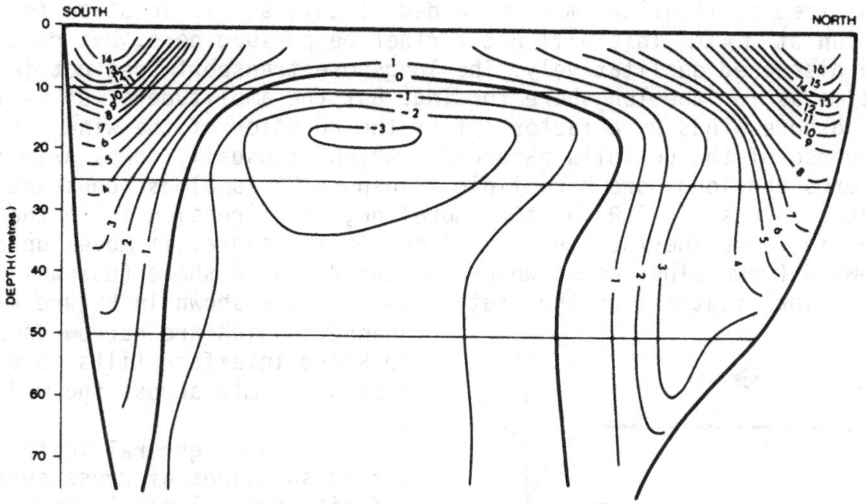

Figure 3 Eastward velocity component computed with a stratified model for a cross section of Lake Ontario which was subject to longshore winds (Simons, 1972).

The amplitudes of these Poincaré-type waves are determined by the cancelation requirement at $t=0$; this is accomplished by

$$V_W^* = \frac{4D'\tau_{sx}}{\rho_0 f\pi H} \sum_{n=1}^{\infty} \frac{(-1)^{n+1}}{(2n-1)(1+R^2\lambda_n^2)} \cos\lambda_n y \cos\sigma_n t \qquad (2.34)$$

which, for $t=0$, is simply the expansion of the negative value of (2.29), (Simons, 1980). Of course, the complete solution of the system (2.26), (2.27) could likewise have been obtained by Laplace transform techniques (Crèpon, 1969). The remaining variables U^*, η' can be computed from (2.25).

We see that the problem of quasi-static circulation is strongly connected with the wave theory of channels and enclosed basins.

The case for a sudden wind normal to the shore has $\tau_{sy} \neq 0$ but $\tau_{sx}=0$ and leads, by similar methods, to the particular solution

$$U^* = \frac{D'\tau_{sy}}{\rho fH}\left(1 - \frac{\cosh(y/R)}{\cosh(W/2R)}\right), \quad V^* = 0,$$
$$\eta' = -\frac{D'\tau_{sy}}{\rho_0 cfH}\,\frac{\sinh(y/R)}{\cosh(W/2R)}, \qquad (2.35)$$

valid for $H=$ constant. It does not vanish for $t=0$ either. Again, a series of free oscillations must be added to this static solution for cancellation at $t = 0$. This will not further be pursued here, but results may be discussed qualitatively: The longshore transport does not increase with time as for the longshore forcing. But the amplitude of U^* is larger for longshore winds by a factor $1/f$ if the duration of the wind exceeds the inverse of the Coriolis parameter, which it usually does. Away from the shores the longshore baroclinic transport U^* is almost constant and decreases on a scale $O(R)$ at the shoreline, its direction is to the right of the wind. The thermocline is confined to the shore, it moves upward on the upwind (upwelling) and downward on the downwind shore (downwelling). The principal features on the static solution are shown in Figure 4. For channels which are narrow compared to R the interface tilts more or less uniformly across the width.

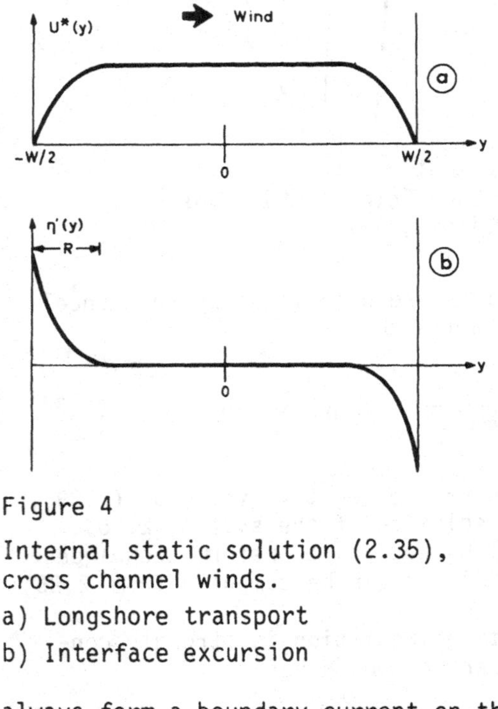

Figure 4

Internal static solution (2.35), cross channel winds.

a) Longshore transport

b) Interface excursion

So far some general features of forced solutions of cross section/infinite channel models were derived. Depth variations were only possible in cross channel direction if at all. For axi - symmetric channels this leads to axis-symmetric solutions if the non-linear terms in the momentum equation are neglected.

However, if one allows for a bottom slope in the direction of the channel axis, the currents are distributed asymmetrically along a cross section; maximum currents occur only on one side of the basin, as can be deduced from vorticity conservation arguments. In a north/south channel on the northern hemisphere with the bottom sloping down towards north, the currents will always form a boundary current on the east side of the channel. An example will be given later for the Baltic Sea.

b) Rectangular flat basin solutions

The next step in the principal investigations of the two-layer circulation is to close the channel at both ends and to seek solutions in an elongated basin of length $L > W$. We start again from $(2.19)_2$ and $(2.20)_2$, apply the rigid lid condition, neglect bottom friction and consider a wind which blows along the longer sides of the basin, but we do not neglect the internal pressure gradient in the x-direction as we did before. The equations are then

$$\frac{\partial U^*}{\partial t} - fV^* = c^2 \frac{\partial \eta'}{\partial x} + \frac{D'\tau_{sx}}{\rho H},$$

$$\frac{\partial V^*}{\partial t} + fU^* = c^2 \frac{\partial \eta'}{\partial y}, \tag{2.36}$$

$$\frac{\partial \eta'}{\partial t} = \frac{\partial U^*}{\partial x} + \frac{\partial V^*}{\partial y},$$

with c as defined before. The initial and boundary conditions are

$$U^* = 0, \quad \text{at} \quad x = \pm L/2,$$

$$V^* = 0, \quad \text{at} \quad y = \pm W/2, \tag{2.37}$$

$$U^* \equiv V^* \equiv \eta' \equiv 0, \quad \text{at} \quad t \leq 0.$$

Consider uniform wind, cross-differentiate the first two equations of (2.36), subtract and substitute the continuity equation to obtain the vorticity equation

$$\frac{\partial}{\partial t}\left(\frac{\partial V^*}{\partial x} - \frac{\partial U^*}{\partial y}\right) + f\frac{\partial \eta'}{\partial t} = 0. \tag{2.38}$$

This expresses conservation of potential vorticity for the present problem (see Hutter, p. 26). If integrated once with respect to time it states that, for a start from rest, interface perturbations must always be balanced by vorticity, thus precluding the purely static setup solution. Note that the previous channel solutions satisfy this condition.

The integrated version of (2.38) may be substituted into the divergence equation, obtained from $(2.36)_{1,2}$. The resulting wave equation is

$$\frac{\partial^2 \eta'}{\partial t^2} + f^2 \eta' - c^2 \nabla^2 \eta' = 0. \tag{2.39}$$

This equation is homogeneous, hence the forcing enters only through the boundary conditions (2.37).

Following earlier work by Taylor (1922), Lauwerier (1961) first determined a particular solution of (2.39) that satisfies the boundary conditions (2.37) at the side walls, but not at the end walls of the basin. In a second step he satisfied the remaining boundary conditions by expanding the particular solution at the end walls in terms of normal modes of the homogeneous differential equations. With regard to the time variable, the solution may also be decomposed into a quasi-static component and a series of free oscillations, in exactly the same fashion as for the channel problem. The discussion will be restricted to the quasi-static solution and will employ a graphical solution technique similar to one proposed by Csanady and Scott (1974).

Considering first equations (2.36) and (2.39) without the time derivatives, the particular static solution that satisfies the boundary conditions at the longer sides of the rectangle $(y = \pm W/2)$ is

$$\begin{aligned} \eta' &= -\frac{D'\tau_{sx}}{c^2\rho H}\,\frac{x\cosh(y/R)}{\cosh(W/2R)}, \\ U^* &= -\frac{D'\tau_{sx}}{\rho cH}\,\frac{x\sinh(y/R)}{\cosh(W/2R)}, \\ V^* &= \frac{D'\tau_{sx}}{f\rho H}\left(\frac{\cosh(y/R)}{\cosh(W/2R)} - 1\right). \end{aligned} \tag{2.40}$$

The transverse component V^* is identical to the infinite channel solution for V^*. The thermocline displacement decreases rapidly with distance from shore because the internal radius of deformation is of the order of only a few kilometers. The thermocline slope normal to the sides of the rectangle is balanced by a longshore geostrophic current that is antisymmetric with respect to both coordinates. In the interior, there is only a drift to the right of the wind.

To complete the static solution, the downwind baroclinic transport U^* must be made to satisfy $U^*(\pm L/2) = 0$ by trigonometric series expansion. However, we do not want to go into details here. We merely state that, on a scale $O(R)$, one obtains a boundary current along the end walls. The thermocline is characterized by a downwind tilt of constant slope along the $y = \pm W/2$ shores [as can already be deduced from the particular solution (2.40)]; but away from the shores, the displacements decay exponentially to the equilibrium level. The principal properties of the static solution are demonstrated in Figure 5 (after Simons, 1980).

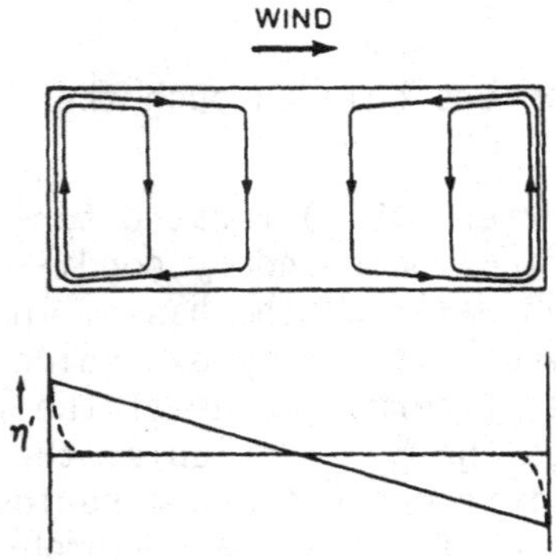

Figure 5

Stream function of static solution for the baroclinic mode of a two-layer rectangular basin and associated thermocline elevation along the boundaries (solid line) and along the major axis of the basin (broken line).

In order to complete the time dependent solution for a suddenly imposed wind, the static solution must be cancelled by a series of free oscillations of the basin at time $t=0$.

Possible waves in a rectangular basin of constant depth are Poincaré waves, which have near-inertial frequencies and dominate offshore, and the low-frequency Kelvin waves trapped close to the boundaries. We are only interested in the low-frequency behaviour of the basin and restrict the discussion to Kelvin wave solutions of (2.39). They have the form

$$\eta' = \phi(s-ct)\,e^{-n/R}, \qquad V_s = \frac{c^2}{f}\,\frac{\partial\eta'}{\partial n}, \qquad V_n = 0, \tag{2.41}$$

where s and n indicate coordinates along and normal to the boundary, respectively, and ϕ represents the longshore shape of the wave. Because the static elevation (2.40) and the wave solution (2.41) are both confined to the shore zones, as a first approximation we ignore effects of opposite shores and the rather artificial corners, und unfold the boundaries of the basin to obtain a one-dimensional display of nearshore thermocline excursions shown in Figure 6. To satisfy the initial conditions, the shape function of the Kelvin wave (2.41) must be equal, but of opposite sign, to the static solution. It is then a simple matter to graphically construct the ensuing quasi-static solution as shown by the solid line in Figure 6.

In order to compare this solution with the corresponding channel solution (2.30), we add two Kelvin waves together, one which moves downwind along the shore to the right of the wind with the simple shape function $x-ct$, the other one moving upwind along the shore to the left of the wind with the shape function $x+ct$, both with the (otherwise arbitrary) amplitude

$$A = \frac{D'\tau_{sx}}{\rho c^2 H}\,(1+e^{-W/R})^{-1}\,.$$

We obtain for the interface perturbation

$$\eta' = A\left((x+ct)\,e^{(y-W/2)/R} + (x-ct)\,e^{-(y-W/2/R}\right),$$

which can, finally, be written as

$$\eta' = \frac{D'\tau_{sx}}{\rho H}\;\frac{x\cosh(y/R) + ct\sinh(y/R)}{c^2\cosh(W/2R)}\,. \tag{2.42}$$

(2.42) cancels the static solution (2.40) at $t=0$. Away from the ends of the basin ($x\approx 0$), the quasi-static solution at the initial time is

$$\eta' = t\,\frac{D'\tau_{sx}}{c\rho H}\;\frac{\sinh(y/R)}{\cosh(W/2R)}\,. \tag{2.43}$$

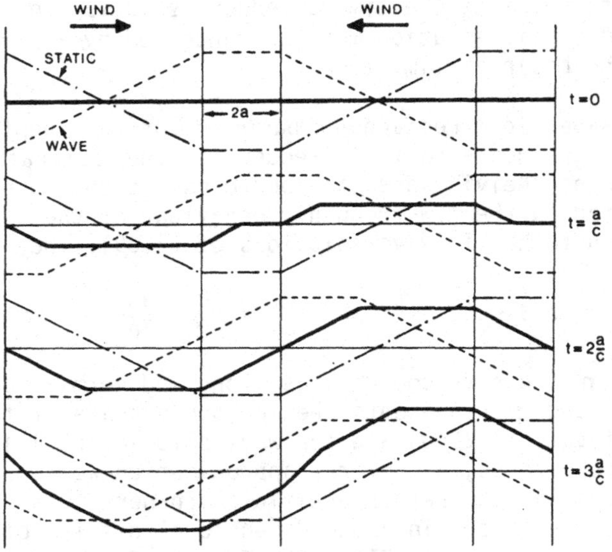

Figure 6 Thermocline excursions (solid lines) around the perimeter of a rectangular basin obtained by superposition of the static solution and a Kelvin wave with propagation speed c (after Simons, 1980). Take $a = W/2$.

(2.43) is identical to (2.30) since $c = Rf$, by definition. Thus, for the channel solution, the neglect of the internal pressure gradient does not allow the end wall effects to be signaled by the Kelvin waves: η' and U^* increase with time. In a more realistic basin, the end effects inhibit the linear increase after some initial time. This time depends on the length L of the basin and the phase speed c of the Kelvin wave, $t = L/c$.

In a study of the forced response of a continuously stratified channel of variable depth in which he used a semi-spectral model, Krauss (1979) concludes for the same reasons that in the subinertial frequency range ($\omega < f$) the longshore current component is unrealistically predicted. Apparently, the missing pressure gradient (internal or external) below the frictional Ekman layer leads to the main balance of the acceleration term and the Coriolis force, $\partial u/\partial t = fv$. For low frequencies this requires large u-values.

It is worthwhile to compare the results gained from channel or rectangular basin problems with barotropic and baroclinic wind induced processes of circular basins (e.g. Simons, 1980). In the barotropic case, where according to (2.24) and with $H \neq \bar{H}$ an initial linear increase of the longshore transport in time is obtained, the quasi-static response is mainly due to the topographic low frequency wave contribution, which for small σt grows proportionally to $\sin\sigma t \approx \sigma t$. In the baroclinic case and for constant H the quasi-static response is partly due to a Kelvin wave contribution which is again proportional to $\sin\sigma t$, where σ is the frequency of the wave. For small σt, the longshore current increases linearly in time. This Taylor expansion, after all, is the explanation for the above stated early time linear increase of the barotropic longshore transport $\tilde{U}$ (2.24) and the baroclinic transport U^* (2.30).

3. MULTI-LAYER AND MULTI-LEVEL MODELS OF STRATIFIED CIRCULATION

We have seen above that by introducing the concept of layers of a fluid the original three-dimensional problem of the stratified circulation can successfully be reduced to a number of coupled two-dimensional problems.

The extension of the two-layer formulation to a *multi-layer* is straightforward. The number of possible vertical modes will correspond exactly to the number of layers. For a continuously stratified lake, there exists an infinite number of modes. Thus, the multi-layer formulation is a *physical approximation* of the continuous problem. The distribution of variables in the layers and at the interfaces is indicated for later use in Figure 7.

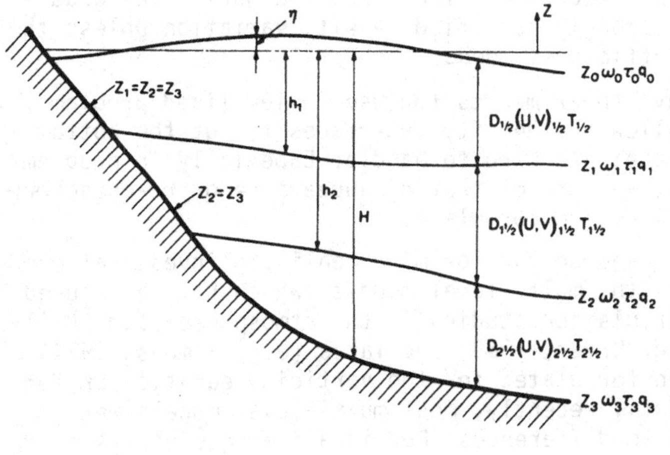

Figure 7

Vertical configuration of multilayered model.

A *multi-level* formulation of stratified circulation is distinguished from the layered model by the following facts: The interfaces Z_k are levels, i.e. planes $z = \text{const}$, which are independent of space and time and which are permeable. In other words, multi-level models are almost grid-point formulations with respect to the z-axis. Therefore, multi-level models are *mathematical approximations* of the continuous problem. Too coarse a resolution in the vertical would, therefore, falsify the solutions.

Another type of model is the so-called sigma-coordinate model where the vertical coordinate z is transformed such that the free surface and the variable bottom become coordinate surfaces (e.g. Freeman et al., 1972):

$$\hat{\sigma} = \frac{z-\eta}{D}, \qquad D = \text{total water depth.} \tag{3.1}$$

The equations have to be transformed accordingly.

If the multi-level model formulation is applied to this new system at gridpoints $\hat{\sigma}_k$, $k = 1, \ldots K$, these levels correspond to variable interfaces in the original z-dependent system. In contrast to the multi-layer models, however, the interfaces are not at all density surfaces, but fit into the general layered model formulation below, if the interfaces are defined as

$$Z_k = \hat{\sigma}_k D + \eta, \qquad k = 1, \ldots, K. \tag{3.2}$$

The σ-coordinate models have the advantage that spectral techniques can be employed in the vertical even for variable bottoms. Haidvogel (1981) conceived a model in which the vertical dependence is formulated in terms of Chebyshef polynomials. However, for steep topographies of natural basins, σ-models may cause considerable difficulties if the horizontal resolution cannot be made fine enough. If an originally horizontal density interface is considered in a σ-model, the interface is mainly deformed according to the depth variations, i.e. the interface rises in shallow areas as steeply as the depth decreases. Therefore the horizontal gradient operators may produce strongly falsified density variation unless the horizontal resolution is sufficiently dense.

The two-layer models have their merits for use in idealized problems. Multi-layer models which allow the density interfaces to cut the bottom or to crop out of the surface, are hard to handle. Especially in problems with diffusive heat fluxes, an intermittant disappearance or re-establishment of interfaces is difficult to formulate.

The most popular models used so far for more realistic investigations of circulation problems are the multi-level models, which have been used in nearly all numerical circulation studies in the atmosphere (e.g. Phillips, 1960), the ocean (e.g. Bryan, 1969) and lakes (e.g. Simons, 1974). The models have either been formulated for the vorticity equation or for the primitive equations. Until recently, most multi-level models were formulated in terms of finite differences. For idealized geometries or proper mapping functions, spectral models seem to be preferable in terms of efficiency and accuracy. A comparison for the barotropic vorticity equation which includes also the finite element approach, can be found in Haidvogel et al., (1980). Most circulation models used for irregularly shaped basins, however, are finite difference multi-level models.

We start with the formulation of layered model equations, but will specialize these equations for constant levels on need.

3.1 Layered model variables and equations

Let K denote the number of layers and let h_k, $k = 1,2,\ldots, K-1$, be the distance between the k-th interface and the equilibrium free surface ($z = 0$). The character of the interfaces is not yet specified and the layers can be separated by moving material surfaces, $h_k(t,x,y)$, fixed permeable interfaces, $h_k(x,y)$ or fixed levels where $h_k = \text{constant}$. Thus, it is convenient to denote the surfaces and the interfaces entering into the model by the general equation $z = Z_k(t,x,y)$, $k = 0,1,\ldots, K$. Recalling the notation for the surface deviation and the basin depth, the various interfaces are defined

$$
\begin{aligned}
&\text{Free surface:} \quad && Z_0 = \eta(t,x,y), && \\
&\text{Interface:} && Z_k = -h_k(t,x,y), && k = 1,2,\ldots, K-1. \\
&\text{Bottom:} && Z_K = -H(x,y). &&
\end{aligned}
\tag{3.3}
$$

The dependent variables are averaged over the depth of a layer bounded by two adjacent interfaces of any arbitrary shape. If the layer thickness is introduced, then the following layered model variables, all functions of t, x, and y, are obtained

$$
\begin{aligned}
D_{k-1/2} &\equiv Z_{k-1} - Z_k, \\
\bar{\Phi}_{k-1/2} &\equiv \frac{1}{D_{k-1/2}} \int_{Z_k}^{Z_{k-1}} \Phi \, dz, \qquad k = 1,2,\ldots, K(x,y),
\end{aligned}
\tag{3.4}
$$

where Φ is an arbitrary variable. Henceforth, the bar and the half-integer subscript refer to a layer average, whereas an integer subscript indicates the value of a variable at an interface. Furthermore, repetition of subscripts will be suppressed wherever possible. For example, by definition

$$
(D\bar{\Phi})_{k-1/2} = D_{k-1/2}\, \bar{\Phi}_{k-1/2} \,. \tag{3.5}
$$

It is convenient to define two new variables at the interfaces, namely,

$$
\begin{aligned}
\omega_k &\equiv w_k - \underline{v}_k \cdot \nabla Z_k - \frac{\partial Z_k}{\partial t}, \\
&\qquad\qquad\qquad\qquad k = 0,1,\ldots, K(x,y). \\
\chi(\Phi)_k &\equiv \gamma(\Phi)_k - \underline{\Gamma}(\Phi)_k \cdot \nabla Z_k \,,
\end{aligned}
\tag{3.6}
$$

Here, ω_k is the component of the velocity vector normal to the interface Z_k, e.g. for $k = K$, $Z_K = H(x,y)$, ω_K is the velocity component perpendicular to the bottom H, which determines the number of layers $K(x,y)$. $\underline{\Gamma}$ and γ are the components of the flux vector defined by

$$\underline{\Gamma}(\Phi) = -A_\Phi \nabla\Phi , \qquad \gamma(\Phi) = -\nu_\Phi \frac{\partial\Phi}{\partial z}, \tag{3.7}$$

in which A_Φ and ν_Φ are horizontal and vertical diffusive eddy coefficients, respectively.

The formulation (3.7) is widely used in circulation models, although it is not always possible to parameterize the diffusive fluxes in this form. For oceanic circulation models which do not resolve the observed eddy scale O(100 km), (3.7) would not work properly (Harrison, 1978). Neither is (3.7) a consistent formulation from a more formal point of view (see e.g. Hutter, p. 23). However, within the frame of our lecture we cannot go into more detail on this subject.

According to (3.7) then, if Φ is identified by momentum and temperature, respectively, $\chi(\Phi)$ represents the stress and heat flux between the layers. The prognostic equations for momentum and temperature belong to the class of conservation equations of the form

$$L(\Phi) = \delta(\Phi) + s(\Phi) , \tag{3.8}$$

where s represents the net source per unit volume and the other terms are defined by

$$L(\Phi) = \frac{\partial\Phi}{\partial t} + \nabla\cdot(\underline{v}\,\Phi) + \frac{\partial}{\partial z}(\omega\,\Phi), \qquad \delta(\Phi) = -\nabla\cdot\underline{\Gamma}(\Phi) - \frac{\partial}{\partial z}\gamma(\Phi). \tag{3.9}$$

Thus, vertical integration leads to the following layered equivalents of the advection and diffusion operators

$$L(\Phi)_{k-1/2} = \int_{z_k}^{z_{k-1}} L(\Phi)\,dz, \qquad \Delta(\Phi)_k \equiv \int_{z_{k-1/2}}^{z_{k-1}} \delta(\Phi)\,dz . \tag{3.10}$$

With the rules for interchanging differentiation and integration and with definitions (3.4) - (3.6), these operators may be written as

$$L(\Phi)_{k-1/2} = \frac{\partial}{\partial t}(D\bar{\Phi})_{k-1/2} + \nabla\cdot(D\,\overline{\underline{v}\Phi})_{k-1/2} + (\omega\,\Phi)_{k-1} - (\omega\,\Phi)_k , \tag{3.11}$$

$$\Delta(\Phi)_{k-1/2} = -\nabla\cdot\left[D\overline{\underline{\Gamma}(\Phi)}\right]_{k-1/2} - \chi(\Phi)_{k-1} + \chi(\Phi)_k . \tag{3.12}$$

Quantities operated by the divergence operators are the advective and diffusive transports through a vertical plane of unit width and bounded at top and bottom by the two interfaces under consideration.

The operators (3.11) are yet to be expressed in terms of the layered variables (3.4). To this end we use the simplest possible closure formulation and assume that the vertical average of the product of two variables may be approximated by

$$\overline{(\underline{v}\Phi)}_{k-1/2} \approx (\underline{\bar{v}}\bar{\Phi})_{k-1/2} \,. \tag{3.12}$$

A consistent interpolation formula for the value of a variable at an interface is of the type

$$\Phi_k \approx \frac{1}{2}(\bar{\Phi}_{k-1/2} + \bar{\Phi}_{k+1/2}) \,. \tag{3.13}$$

Finally, considering flux formulations of the form (3.7), the vertical average of the horizontal flux vector can be written as

$$\overline{\underline{\Gamma}(\Phi)}_{k-1/2} \approx -(A\nabla\bar{\Phi})_{k-1/2} \tag{3.14}$$

and the vertical fluxes at internal interfaces as

$$\chi(\Phi)_k \approx -2C \,\frac{\bar{\Phi}_{k-1/2} - \bar{\Phi}_{k+1/2}}{D_{k-1/2} + D_{k+1/2}} \,, \tag{3.15}$$

where A and C represent local horizontal and vertical diffusion coefficients, respectively.

It remains to integrate the source term of (3.8) and to express the result in terms of the layered variables (3.4). Integration of the hydrostatic equation (2.1) yields

$$p = p_e + p_i, \quad p_e = p_s + g\rho_o(\eta - z), \quad p_i = \int_z^\eta g(-\rho_o)\,dz \,, \tag{3.16}$$

where p_e, p_i are called external and internal pressure. In particular, for the baroclinic pressure term in the equations of motion, we obtain

$$\int_{z_k}^{z_{k-1}} \nabla p_i \,dz = D_{k-1/2}\left[\nabla S_{k-1/2} + \bar{\sigma}_{k-1/2}\,\nabla Z_{k-1/2}\right], \tag{3.17}$$

where

$$Z_{k-1/2} \equiv \frac{1}{2}(Z_{k-1} + Z_k), \qquad S_{k-1/2} \equiv \int_{Z_{k-1/2}}^{\eta} \sigma\,dz \,.$$

Here σ is proportional to the density anomaly $\sigma = g(\rho - \rho_o)$ and is related to the temperature T through an equation of state. For a multi-level model the second term in (3.17) vanishes.

Setting $\Phi = 1$ in $(3.11)_1$, the continuity equation is obtained as

$$\frac{\partial D_{k-1/2}}{\partial t} + \nabla\cdot(D\bar{\underline{v}})_{k-1/2} + \omega_{k-1} - \omega_k = 0\,. \tag{3.18}$$

This equation permits computation of either the displacement of a material surface ($\partial Z_k/\partial t$) or the apparent vertical motion through a rigid interface (ω). Computation starts with the following conditions, which apply at the free surface, the bottom, and the two types of interfaces, respectively:

$$\begin{aligned}
&\text{Free surface (impermeable)} && Z_0(t,x,y): && \omega_0 = 0,\\
&\text{Material interface (impermeable)} && Z_k(t,x,y): && \omega_k = 0,\\
&\text{Fixed interface (permeable)} && Z_k(x,y): && \frac{\partial Z_k}{\partial t} = 0,\\
&\text{Bottom (rigid, impermeable)} && Z_K(x,y): && \omega_K \equiv \frac{\partial Z_K}{\partial t} \equiv 0.
\end{aligned} \tag{3.19}$$

Relations (3.19) follow from kinematic boundary conditions. The second condition is discarded for multi-level models.

Proper specification of s in (3.8) when $\Phi = u$ or $\Phi = v$ and using (3.11) yields the equations of motion

$$\begin{aligned}
L(u)_{k-1/2} &= f(D\bar{v})_{k-1/2} - \frac{D_{k-1/2}}{\rho_0}\left[\frac{\partial}{\partial x}(p_e + s_{k-1/2}) + \bar{\sigma}_{k-1/2}\frac{\partial Z_{k-1/2}}{\partial x}\right]\\
&\qquad + \Delta(u)_{k-1/2}\\
L(v)_{k-1/2} &= -f(D\bar{u})_{k-1/2} - \frac{D_{k-1/2}}{\rho_0}\left[\frac{\partial}{\partial y}(p_e + s_{k-1/2}) + \bar{\sigma}_{k-1/2}\frac{\partial Z_{k-1/2}}{\partial y}\right]\\
&\qquad + \Delta(v)_{k-1/2}\,.
\end{aligned} \tag{3.20}$$

Again, for fixed interfaces the ∇Z-terms vanish. Finally, the conservation equation for heat is expressed as

$$L(T)_{k-1/2} = \Delta(T)_{k-1/2}. \tag{3.21}$$

Specification of the downward heat flux q_s through the free surface and of the geothermal heat ($q = 0$) at the bottom and prescription of the momentum transfers at the free surface and the bottom, respectively, yields

$$\begin{aligned}
\chi(u)_0 &= -\frac{\tau_{sx}}{\rho_0}, \quad \chi(v)_0 = -\frac{\tau_{sy}}{\rho_0}, \quad \chi(T)_0 = q_s\,,\\
\chi(u)_K &= -\frac{\tau_{bx}}{\rho_0}, \quad \chi(v)_K = -\frac{\tau_{by}}{\rho_0}, \quad \chi(T)_K = 0\,,
\end{aligned} \tag{3.22}$$

where the wind stress and temperature flux at the surface are assumed to be known, and the stress at the bottom of the basin must be expressed in terms of the flow field, for example by the relationship

$$\tau_b = \rho_o \hat{k} \, |\underline{v}_b| \, \underline{v}_b \, ; \qquad (3.23)$$

$\hat{k}$ is a nondimensional skin friction coefficient of the order $2.5 \cdot 10^{-3}$ and the bottom velocities may be approximated by the average velocities in the lowest layer.

The consistency of the above equations in terms of variance conservation is discussed in detail in Simons (1980).

Through the above formalism the original three-dimensional problem has become two-dimensional in each layer. For numerical purposes, therefore, the grid point formulation in horizontal space coordinates as well as the time stepping procedure may be copied from two-dimensional numerical models and will not be dealt with here (see Sündermann, pp. 209-233). For convenience, however, an example of a staggered grid version as used by Simons(1980) is presented in Figure 8a for the horizontal and in Figure 8b for the vertical distribution of variables.

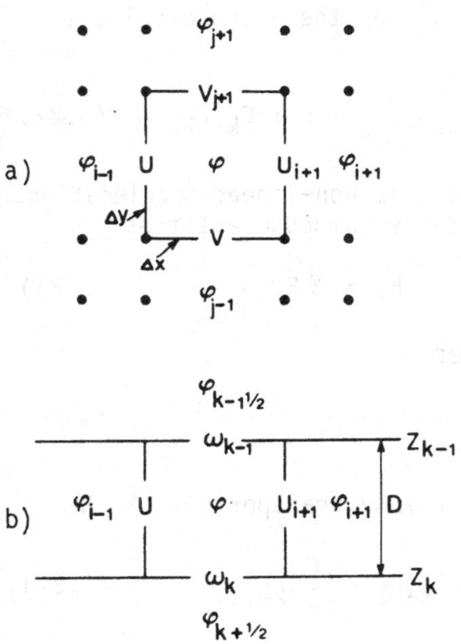

Figure 8

a) Horizontal and b) vertical staggering of variables in a layered model.

In order to advect the scalar Φ by U at the face of a grid box, ϕ has to be interpolated at this face (see Figure 8). Bryan (1966) shows that variance conservation requirements are satisfied for arbitrary grid point distributions if values of the advected variable Φ on the surfaces of the box surrounding the point (i, j, k-1/2) are approximated by simple linear interpolation.

Furthermore, it should be noted that temperature fluxes through the lateral boundaries have to be specified for each layer (may be to zero). The same is true for momentum, where frequently a no-slip condition is used.

The specific problems one is faced within the integration of the advection-diffusion equations (3.11) will be discussed in more detail below.

3.2 External and internal flow

As discussed in connection with the two-layer model, it is advantageous to reformulate the layered system of equations so that the internal structure of the flow is computed more or less separately from the external, i.e. vertically averaged flow. This allows for a larger time step in the integration of the internal solution because its time step is mainly limited by the phase speed of the internal gravity waves. In a model with a freely moving surface, the surface waves demand a relatively short time step.

Despite the fact that the gain in computer time (using different time steps) is of order O(10), the separation of external and internal flow is also required when the rigid lid approximation is applied.

The separation is done as follows: to eliminate the vertical average from the equations of motion, we subtract the equations of motion of adjacent layers. From K layer equations we thus obtain K - 1 "shear" equations. We add an equation for the vertical average by summing the equations of motion over K layers. With this equation of the external flow, we arrive again at K equations.

In a simplified form the equation of motion for the average velocity component $\bar{u}$ in the fixed layer (Z_{k-1}, Z_k) is

$$\frac{\partial}{\partial t}(D\underline{u})_{k-1/2} = f(D\bar{v})_{k-1/2} - g D_{k-1/2} \frac{\partial}{\partial x}\left(\frac{p_s}{\rho_o g} + \eta\right) + F_{k-1/2} \qquad (3.24)$$

where $F_{k-1/2}$ includes the baroclinic pressure, the non-linear accelerations and the diffusive terms. Defining the vertically summed quantities

$$D = \Sigma D_{k-1/2}, \qquad \underline{V} = \Sigma (D\underline{\bar{v}})_{k-1/2}, \qquad F_x = \Sigma F_{k-1/2} \qquad (3.25)$$

the sum of the layered equations (3.24) gives

$$\frac{\partial U}{\partial t} - fV = - gD \frac{\partial}{\partial x}\left(\frac{p_s}{\rho_o g} + \eta\right) + F_x , \qquad (3.26)$$

where U, V are components of the integrated volume transport. With the definitions

$$\underline{v}_k^{\star} = \underline{\bar{v}}_{k+1/2} - \underline{\bar{v}}_{k-1/2} , \qquad F_k^{\star} = \left(\frac{F}{D}\right)_{k-1/2} - \left(\frac{F}{D}\right)_{k+1/2} \qquad (3.27)$$

and by subtracting the adjacent layer equations (3.24), one finds

$$\frac{\partial u^{\star}}{\partial t} - f v_k^{\star} = F_k^{\star} , \qquad k = 1, \ldots, K-1 . \qquad (3.28)$$

(3.28) and (3.26) are coupled through F.

Equation (3.26) and its analogue for the V component may then be integrated together with the equation of continuity

$$\frac{\partial \eta}{\partial t} + \nabla \cdot \underline{V} = 0 \qquad (3.29)$$

while the term F is held constant for some period of time (e.g. the internal time step).

3.3 The combined effect of topography and baroclinicity

From the considerations of the channel model we know that the baroclinic response of a channel is separable from the barotropic response, even if the depth is variable in cross direction, provided that bottom friction or non-linear effects are not so important. For natural basins this separation is generally not possible.

It is quite instructive to have a look at the transport vorticity equation, which may be derived from $(2.11)_1$ by taking the curl:

$$\frac{\partial}{\partial t} \operatorname{curl} \underline{V} = -\operatorname{div}(f\underline{V}) - \frac{1}{\rho_0} J(H, p_e + p_{ib}) + \operatorname{curl}(\underline{\tau}_s - \underline{\tau}_b) \,, \qquad (3.30)$$

where p_e is the external pressure [see (3.16)] and p_{ib} the internal pressure at the bottom. One can see immediately that the Jacobian term vanishes either if $H = \text{const}$ or if the external pressure is balanced by the internal pressure in deep water.

A more suitable approach to discuss the circulation in shallow basins is to consider the relative vorticity of the mean flow, defined by

$$\zeta = \frac{\partial \bar{v}}{\partial x} - \frac{\partial \bar{u}}{\partial y} \,, \qquad \underline{\bar{v}} = \underline{V}/H \,. \qquad (3.31)$$

Dividing (2.11) by the water depth and taking the curl yields

$$\frac{\partial \zeta}{\partial t} = -\operatorname{div}(f\underline{\bar{v}}) - \operatorname{curl} \underline{\bar{P}}_i + \operatorname{curl} \frac{\underline{\tau}_s - \underline{\tau}_b}{\rho_0 H} \,, \qquad (3.32)$$

where $\underline{\bar{P}}_i$ is the vertical average of the baroclinic pressure gradient ∇p_i:

$$\underline{\bar{P}}_i = \frac{1}{\rho_0 H} \int_{-H}^{\eta} \nabla p_i \, dz \,, \qquad \left(p_i(z) = \int_z^{\eta} \sigma \, dz^*\right) \qquad (3.33)$$

Using the continuity equation (2.10), the first term on the right of (3.32) can be expressed as

$$- \operatorname{div}(f\bar{\underline{v}}) = -\beta\bar{v} + \frac{f}{H}\,\bar{v}\cdot\nabla H + \frac{f}{H}\,\frac{\partial\eta}{\partial t} \qquad (3.34)$$

where $\beta = df/dy$ is negligible against $(f/H)\partial H/\partial y$, see second term on the right. This holds because in small natural basins $\beta \ll (f/H)\partial H/\partial y$ [in the Baltic Sea by an order O(100)]. If the rigid lid approximation is applied, the third term on the right side of (3.34) is also negligible. The flow follows almost the depth contours.

In summary, (3.32) states that mean flow vorticity is generated by a mean current crossing depth contours [see (3.34)], by the combined effect of baroclinic pressure gradients and topography [see (3.33)] and by the curl of the stress divided by H; this means that in a basin with variable bathymetry, even for spatially constant winds, relative vorticity is generated. A constant wind stress is a good approximation for lakes because the meteorological scale is usually much larger than the lake scale. Moving fronts are certainly an exception to this.

The crossing of depth contours by the mean flow causes vertical velocities which couple the external and internal flow as shown for the Baltic Sea by Simons (1978a) and Kielmann (1981). Examples will be given below. For time scales of a few days the topographic/baroclinic effect in the Baltic Sea is about half of the wind effect, as computed from a numerical multi-level model of the Baltic Sea (Kielmann, 1981).

3.4 Three-dimensional steady-state diagnostic models

As has already been discussed for the two-layer models, steady-state conditions will hardly exist for stratified lake circulation. However, for the ocean circulation, it is possible to compute currents and surface elevations using the observed density field as a diagnostic variable. For details of the procedure the reader is referred to Sarkisyan (1977). In a simplified manner this method was applied to the Baltic Sea by Kowalik and Staśkiewiez (1976). One of their interesting results was that on a climatological time scale the combined effect of baroclinity and topography exceeds the pure wind effect (by about a factor of 2).

3.5 Examples of time-dependent circulation in a multi-level model of the Baltic Sea

In order to demonstrate some of the theoretically predicted principal features of baroclinic circulation for an inland sea we present examples from multi-level computations of the Baltic Sea. Figure 9 shows a map of the Baltic Sea together with depth contours. It consists of a number of partially deep basins separated by sills and channels. The mean depth is

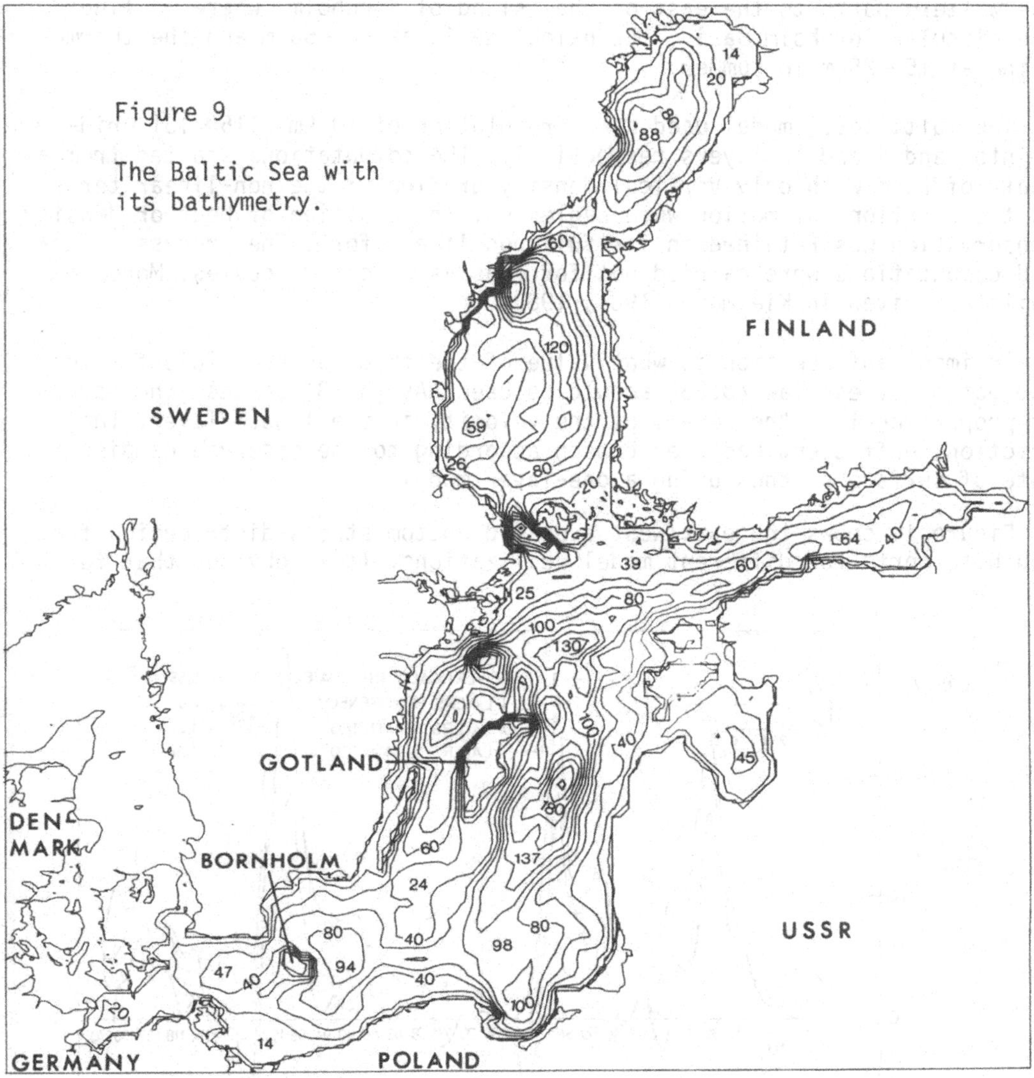

Figure 9

The Baltic Sea with its bathymetry.

50 m, whereas maximum depths down to some hundred meters are possible in the center of the basins.

The stratification, which is determined by the temperature and salinity distribution, results from the climatological surplus of fresh river-waters in the north and east, and from salt exchange in the west through the Danish Belts. This leads to a two-layer stratification of less salty water on top of heavier salty water, and the interface slopes down towards north. In summer a thermocline builds up and produces a three-layer structure. In the western part, to the east of the island of Bornholm, where we find the circular Bornholm Basin, the halocline is at 50 - 60 m and the thermocline at 15 - 25 m in summer.

The multi-level model used had a resolution of 10 km (115 × 131 gridpoints) and 4 and 10 layers respectively. The computations started from a state of rest with only vertical density gradients; the non-linear terms in the equations of motion were neglected. The equation of heat or density conservation was retained in its full non-linear form. The process oriented computations were carried out for quasi-static time scales. More details are given in Kielmann (1981, 1983).

An important question is what value of the friction coefficient k for the bottom stress law (3.23) should be used. As (3.23) states, the stress is proportional to the square of the velocity in the lowest layer. The friction coefficient had been chosen according to the observed damping rate of surface seiches using a one-layer model.

Figure 10 shows the east/west averaged bottom stress distribution from south to north for different model realizations. It is obvious that for

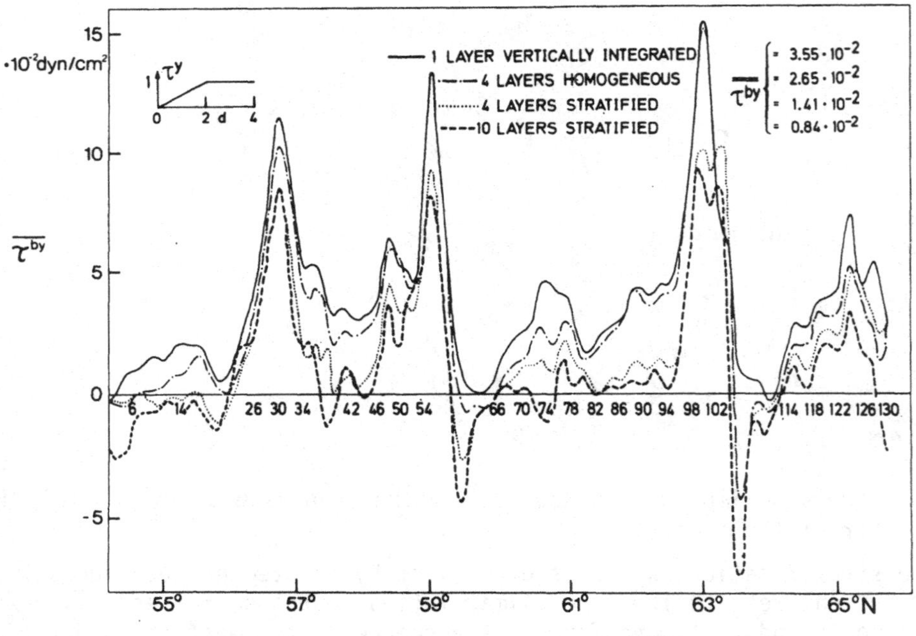

Figure 10 East/West average of the bottom stress $\overline{\tau_b{}^y}$ of the stratified and unstratified Baltic Sea after 4 days of south wind. The overall average is $\overline{\overline{\tau_b{}^y}}$.

experiments in the stratified Baltic with $\hat{k}$ fixed the baroclinic model computations show much less frictional influence, e.g. on average, one-layer friction is 4 times as large as the ten-layer friction.

Examples of the horizontal current structure for winds which are parallel to the main axis are presented in Figure 11 and Figure 12. They show the average current in 5 - 15 m (Ekman layer) and 15 - 25 m (below Ekman layer). In both layers we find the coastal boundary currents on the north/

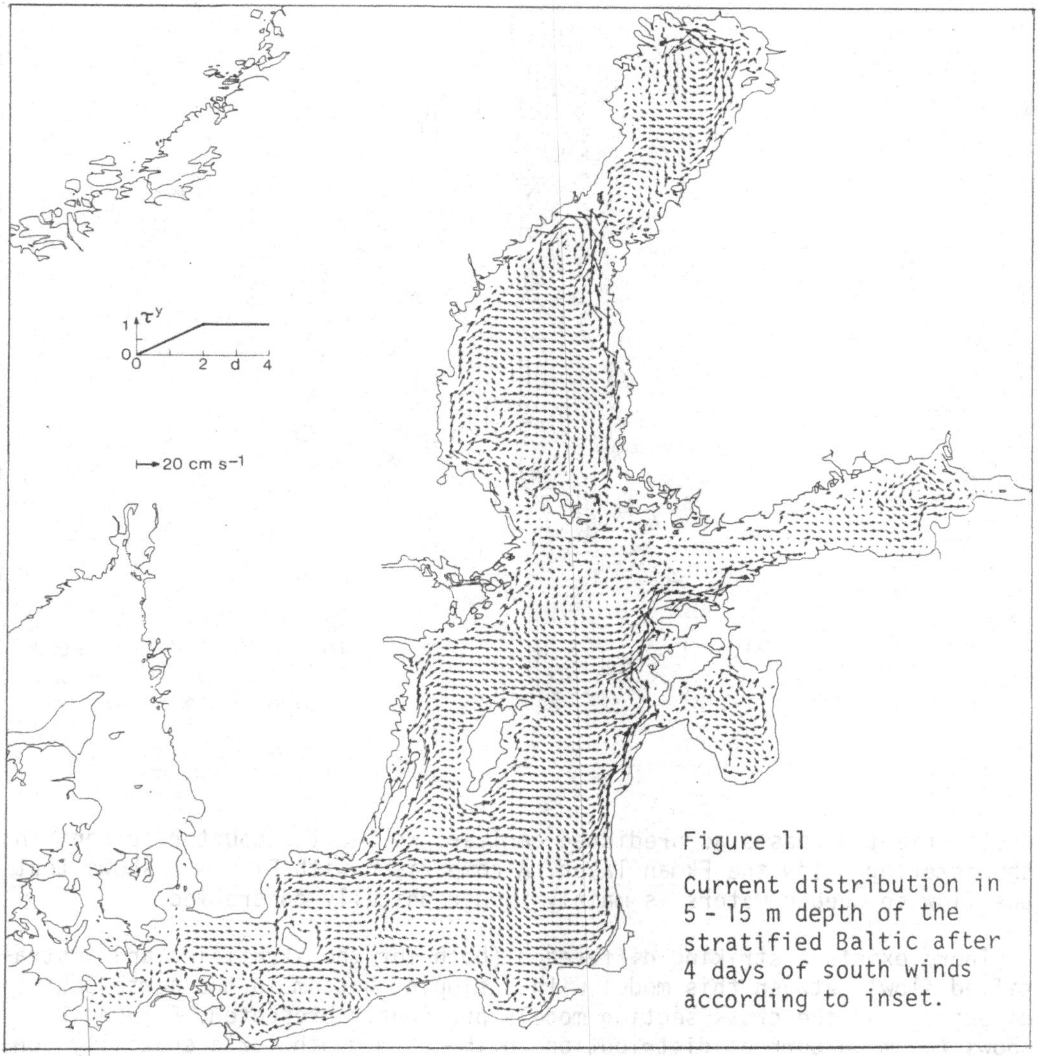

Figure 11

Current distribution in 5 - 15 m depth of the stratified Baltic after 4 days of south winds according to inset.

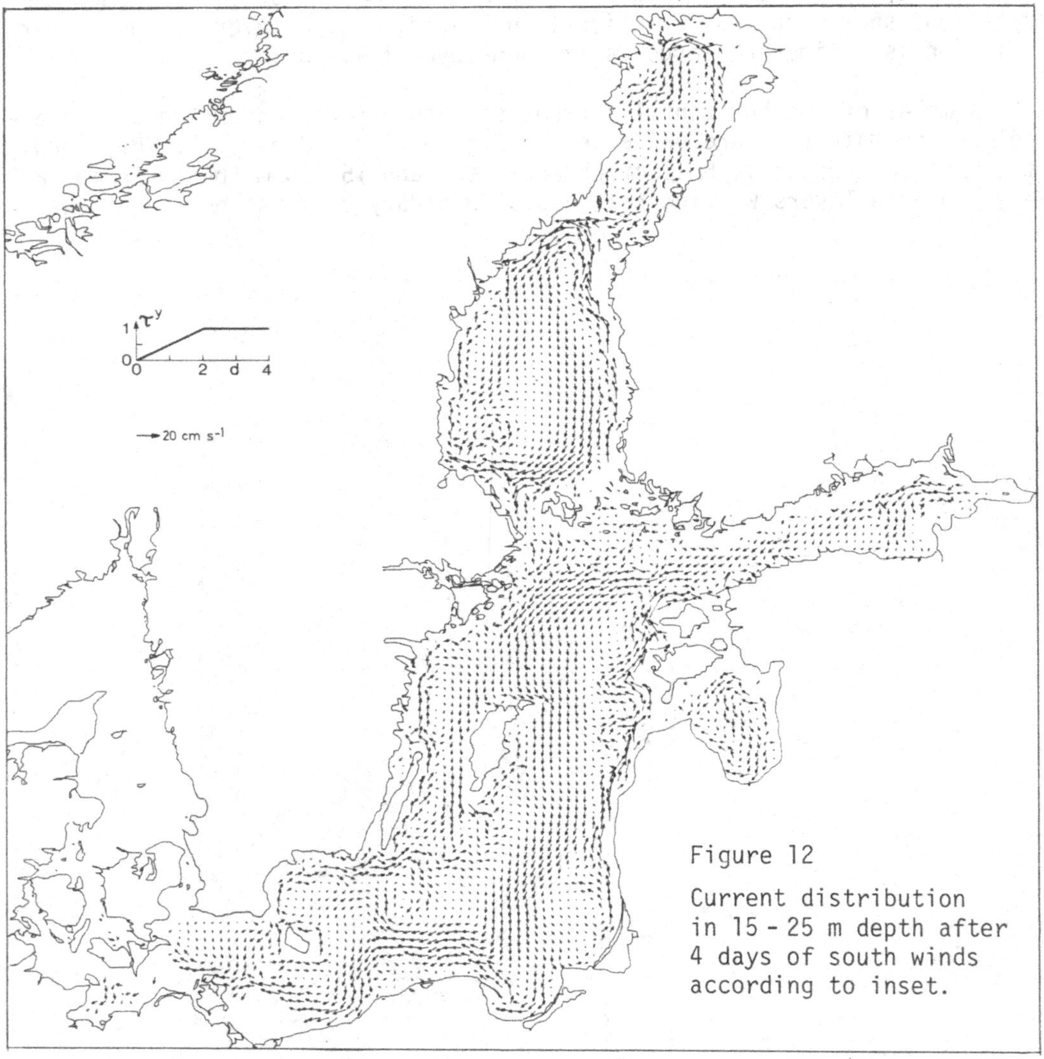

Figure 12

Current distribution in 15 - 25 m depth after 4 days of south winds according to inset.

south oriented coasts as predicted earlier. We see the counter current in the interior below the Ekman layer. A comparison with Figure 9 shows that the flow in deeper waters is mainly topographically controlled.

There exists a striking difference (both for the stratified and unstratified flow) between this model with a sloping bottom in the north/south direction and the cross section models previously presented. Figure 13 shows the mean current distribution in 15 - 25 m depth for a spatially constant west wind (quasi-static solution of the multi-level model). On the east side the southern basin of the Gulf of Bothnia shows a strong asymmetric current band. The asymmetry may be explained by looking at the east/west averaged depth $\bar{H}(y)$ which increases as one moves northward.

From oceanic circulation models it is known that the western boundary currents exist because of the β-term in (3.32), (3.34). For small basins with variable topography the term $-(f/H)\partial H/\partial y$ replaces β (3.34). Therefo-

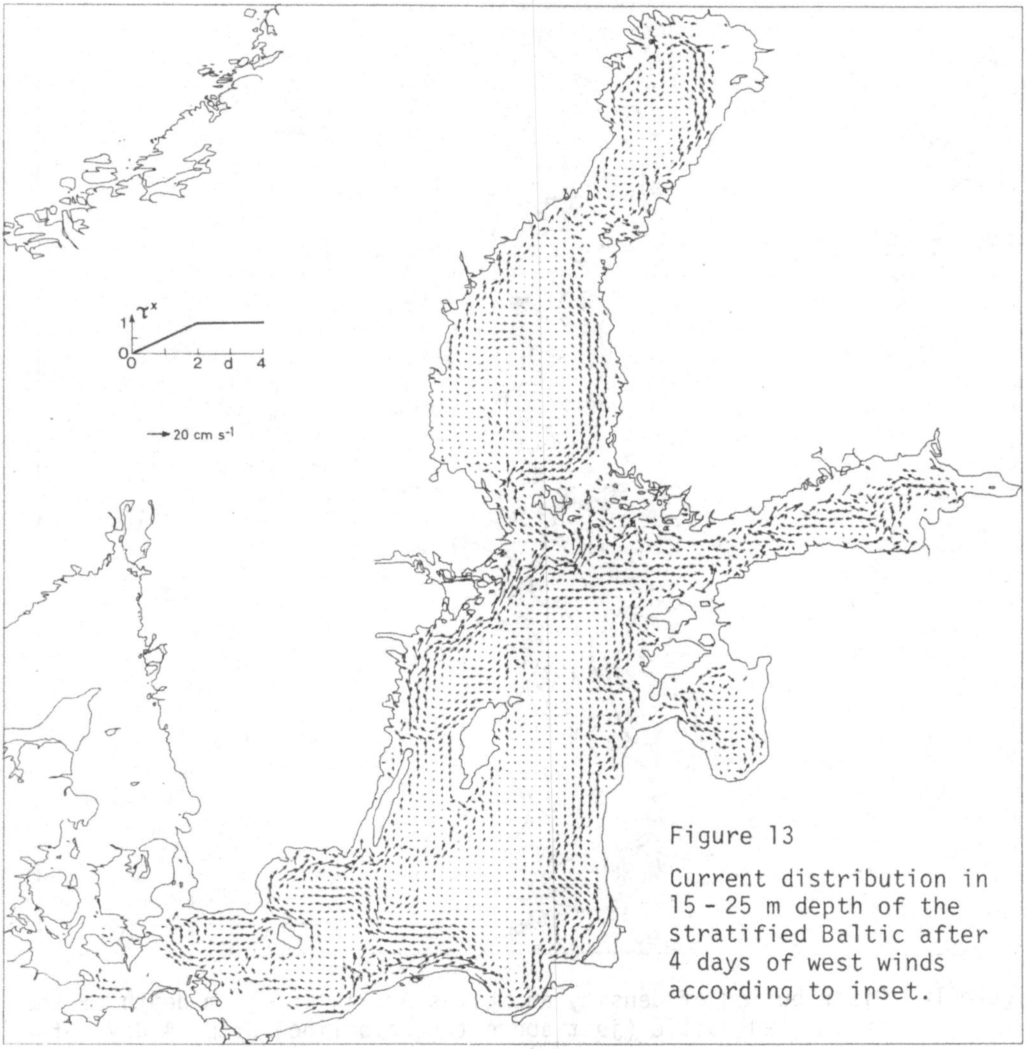

Figure 13

Current distribution in 15 - 25 m depth of the stratified Baltic after 4 days of west winds according to inset.

re, if $H(y)$ becomes shallower as one moves northward in a basin with westerly or easterly winds, one will also see stronger currents along the western boundary. However, if $H(y)$ is sloping downward toward north, one obtains an eastern boundary current. This had already been stated by Weenink (1958). The sign of $\partial/\partial y(\tau_{sx}/H)$ determines the direction of the flow as may be derived from (3.32). From this we may conclude that for spatially constant westwinds and with $\partial\bar{H}/\partial y > 0$ an anticlockwise circulation

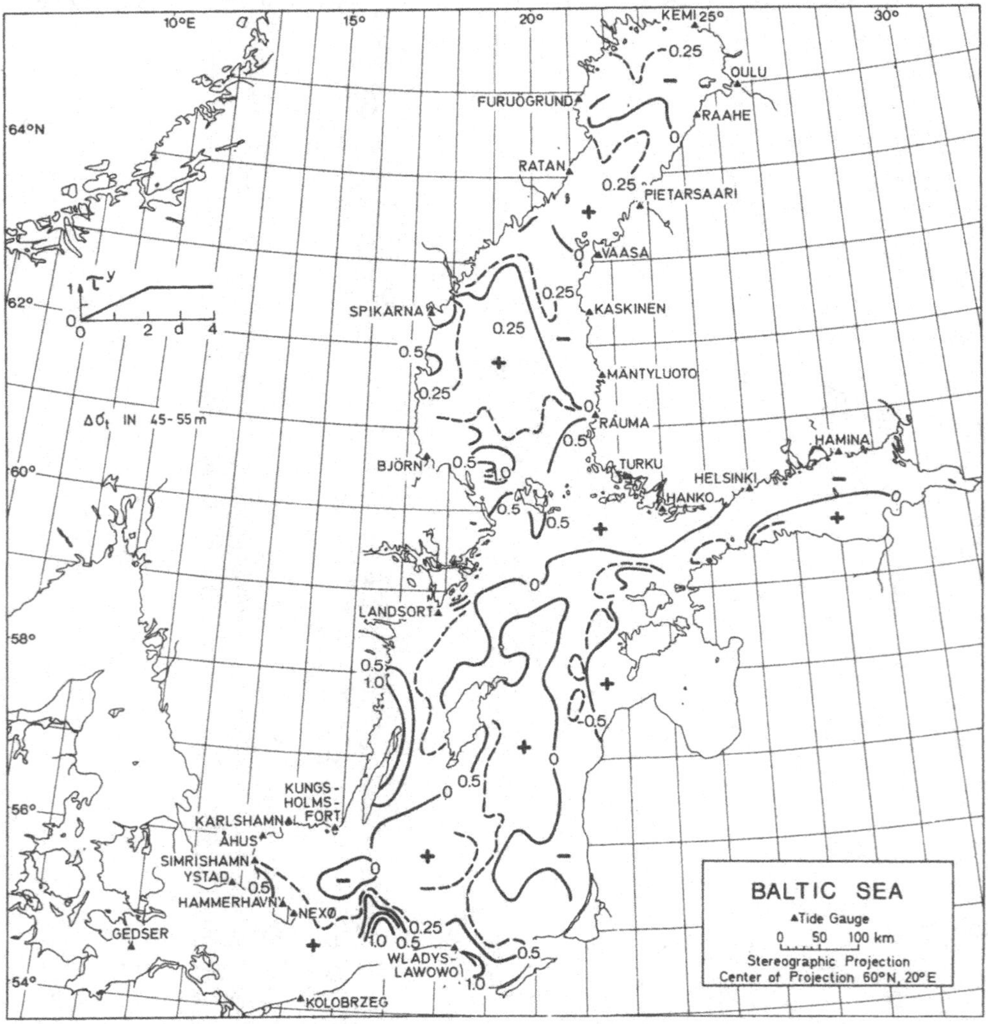

Figure 14 Distribution of density anomalies $\Delta\sigma_T$ in 45 - 55 m depth of the stratified Baltic (just above the halocline) after 4 days of south winds according to inset.

will be generated with concentrated flow on the eastern side.

It has been discovered for the Baltic that considerable differences exist locally between homogeneous and stratified flows. For example, a number of topographically induced vortex patterns disappear in the stratified flow, which are present under homogeneous conditions. For details see Kielmann (1981).

There is a tendency for a mixture of quasi-compensation and combined topography/stratification effects in the Baltic Sea. Figure 14 shows the density anomalies above the halocline (45 - 55 m) for unit stress wind from the south. It can be seen that for most part of the basin the interfaces slope locally down wind as would be expected in case of compensation. However, in shallower waters like the Bornholm Basin or the Gotland Deep, the compensation is not so dominant.

Figure 15, on the other hand, indicates that the strongest density anomalies are close to the bottom where the largest density gradients occur. The anomalies are well correlated with the topographically induced vertical velocities.

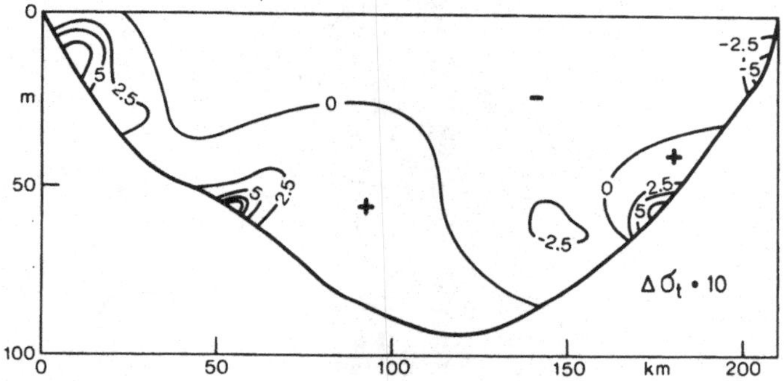

Figure 15 Density anomalies $\Delta\sigma_T \cdot 10$ across a south/north section through the center of the Bornholm Basin after 4 days of south wind according to inset in Figure 14.

Figure 16 shows the difference of the velocity in 0 - 15 m between homogeneous and stratified circulation, for a south wind stress as indicated in the inset. At many places, see e.g. the Bornholm Basin, the differences are structured as vortex patterns and are due to the homogeneous flow. Inspection of the bottom torque in (3.32) shows that this term works apparently against the curl (τ/H).

Verfication-runs that would predict the current structure of the Baltic Sea pointwise have - different from many lake current simulations - only partially been satisfactory (Simons, 1978, Kielmann, 1981). One reason is

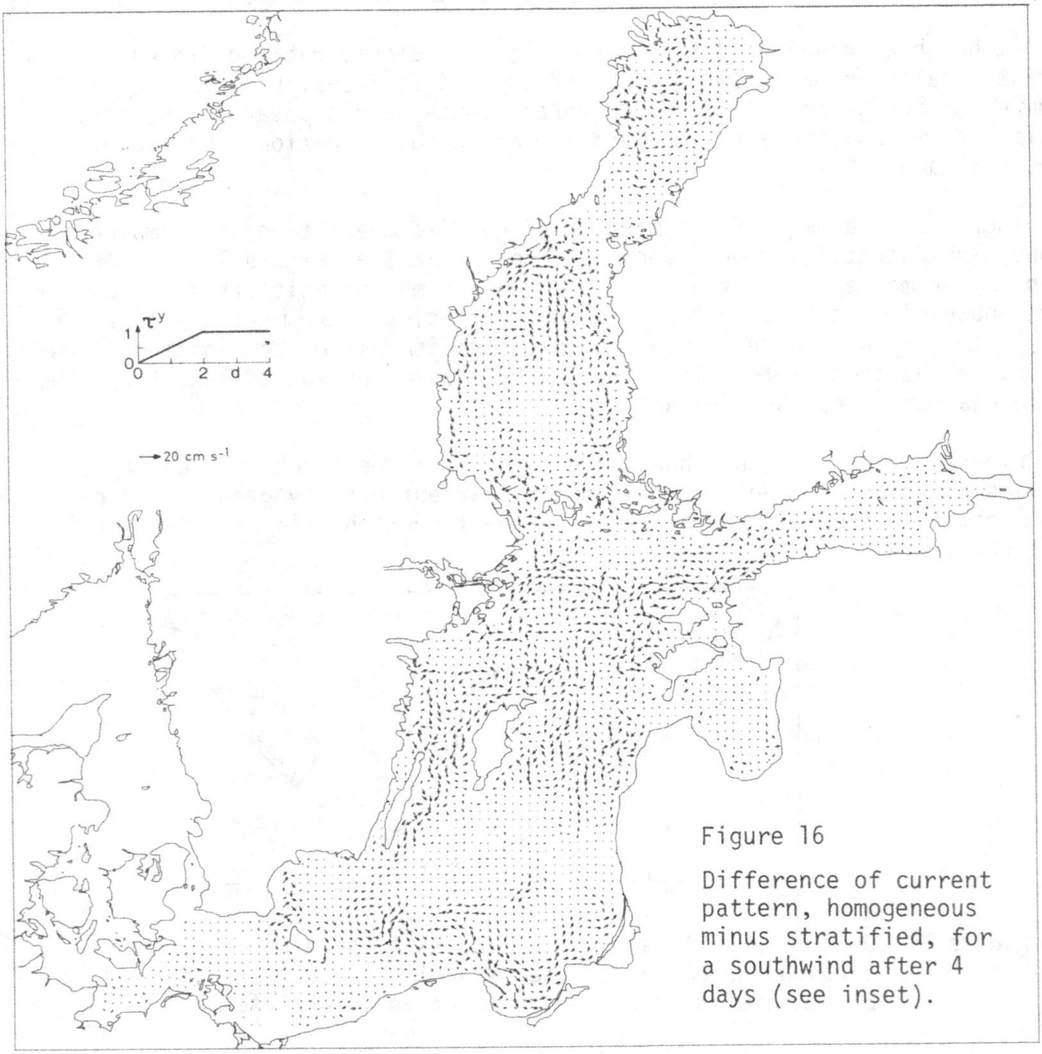

Figure 16

Difference of current pattern, homogeneous minus stratified, for a southwind after 4 days (see inset).

that the circulation for time varying winds consists of close circulation cells which migrate through the basins. The currents change then both.direction and speed accordingly. Small errors in the determination of the cell position may lead to completely different currents in the model. These cells are due to topographic waves, details of which are given by Mysak (pp. 81-128). Multi-level models are clearly able to generate these waves (Kielmann, 1983).

In terms of environmental and water quality studies there is another point which is important: Even in case the current structure is well predicted by some models it is still unknown here what problems will arise in the simulation of advection/diffusion processes of passive (or active) tracers? We deal with this question in the next section.

4. THE NUMERICAL SOLUTION OF ADVECTION-DIFFUSION EQUATIONS

With the introduction of multi-level circulation models of lakes and oceans the nonlinear advection terms take on considerabel importance; if not in the equations of motion, then certainly in the temperature and salinity equations. The recent interest in hydrodynamic modelling as a tool to study water quality problems led to the use of advection-diffusion equations and their approximate treatment to simulate transports of dissolved or suspended matter in natural basins, e.g. Lam and Simons (1976).

For simplicity we consider the one-dimensional equation

$$\frac{\partial c}{\partial t} + u \frac{\partial c}{\partial x} = \frac{\partial}{\partial x}\left(A \frac{\partial c}{\partial x}\right), \tag{4.1}$$

where u is the advection velocity, and A may be a turbulent diffusion coefficient. c may either be a passive tracer H, the temperature, salinity or a velocity component.

One problem associated with the solution of equations like (4.1) is the problem of nonlinear instability (e.g. $c = u$). This means essentially that due to aliasing errors (misinterpretation of higher harmonics by the computational grid) energy accumulates at short wavelengths (e.g. Mesinger and Arakawa, 1976). Such higher harmonics can be suppressed by eddy diffusion or smoothing terms. A particularly effective scheme is one proposed by Lax and Wendrow (1960); it is commonly applied to model strongly nonlinear wave phenomena (e.g. Simons, 1978b). Different methods must be used for problems where the built-in damping of such schemes is undesirable. In the meteorological literature considerable attention has been devoted to the design of numerical techniques that avoid the problem of nonlinear instability. These schemes belong to either of two classes: spectral techniques or conservative finite difference schemes.

Another problem with (4.1) arises in the case of advection of variables with steep gradients if conventional finite differences or even finite element techniques are used. As a typical illustration of the behaviour of such schemes, consider a multi-level model with centered differences in the vertical, as applied to a stratified basin with a sharp temperature gradient of some level. The computed deformation of the temperature profile in the presence of upwelling and downwelling, respectively, but without diffusion, is displayed in Figure 17. The gradients are smoothed out ahead of the moving frontal interface, while wave-like phenomena appear behind the front. In this example the waves cause physical instability, and are automatically removed by the mechanism for simulating convection that must be incorporated in any three-dimensional model. However, it is clear that in this kind of a model the initially steep temperature gradient will soon be dissipated. Similar problems occur in the simulation of horizontal propagation of concentration patterns, in many cases leading to locally negative concentrations or other anomalies. The stated

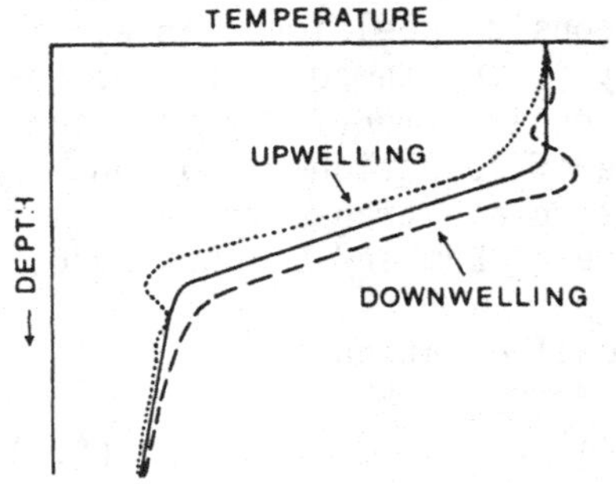

Figure 17

Typical effect of central difference scheme on computed temperature profiles in upwelling and downwelling areas, respectively (after Simons, 1980).

problem is not related to time truncation only, but rather to central spatial differences. Thus, oscillatory grid dispersion may occur even when the steady-state problem is attacked.

It is worthwhile to mention that depending on the values of u and A, (4.1) changes its character. If $u = 0$ and $A \neq 0$, (4.1) is parabolic; but it is hyperbolic when $u \neq 0$ and $A = 0$. If u, A are functions of x and t, the character may change locally and with time.

4.1 Central difference scheme (CDS)

For the moment, set $\partial c/\partial t = 0$ and assume that u and A are constants. If we use the boundary condition $c(x=0) = a_0$ and $c(x=1) = a_1$, an exact solution of (4.1) is

$$c = a_0 + (a_1 - a_0)(e^{\delta x} - 1)/(e^{\delta} - 1), \tag{4.2}$$

where $\delta = u/A$. In order to construct the finite difference solution of (4.1) with the above assumptions, we use central differences, so

$$u \frac{c_{j+1} - c_{j-1}}{2\Delta x} = A \frac{c_{j+1} - 2c_j + c_{j-1}}{(\Delta x)^2}, \tag{4.3}$$

which can be rewritten as

$$\frac{A}{2(\Delta x)^2} \Big((2-\alpha)c_{j+1} - 4c_j + (2+\alpha)c_{j-1} \Big) = 0, \tag{4.4}$$

where $\alpha = u\,\Delta x/A$ is called the cell Reynolds number and $j = 1,\ldots,N$ denumerate the grid points. Together with the boundary condition $c_0 = a_0$, $c_{N+1} = a_1$, (4.4) is a system of equations which has the solution (Peyret and Taylor, 1981)

$$c_j = a_0 + (a_1 - a_0)\,\frac{\left(\frac{2+\alpha}{2+\alpha}\right)^j - 1}{\left(\frac{2+\alpha}{2-\alpha}\right)^{N+1} - 1}. \qquad (4.5)$$

They show that (4.5) is a second-order approximation to the exact solution (4.2) in the sense that if $\Delta x \to 0$, $j\,\Delta x = x_j$ is held fixed and $\delta = O(1)$. However, if the cell Reynolds number $\alpha > 2$, the term $[(2+\alpha)/(2-\alpha)]^j$ in (4.5) changes sign at every other grid point, producing alternating behaviour of the function c.

If the same difference schem is used for the time-dependent problem (4.1),

$$\frac{d}{dt}\,c = \frac{A}{2(\Delta x)^2}\,M\underline{c}\,, \qquad (4.6)$$

is obtained in which $\underline{c}$ is the vector with components $c_1,\ldots,c_N$ and M is a tri-diagonal matrix which has the value -4 on the diagonal, $(2+\alpha)$ on the sub-diagonal and $(2-\alpha)$ on the super-diagonal according to (4.4); for the moment we shall not consider boundary conditions. Assuming that $\underline{c} = \underline{\hat{c}}\,e^{\lambda t}$ (4.6) implies the eigenvalue problem

$$\left(\frac{A}{2(\Delta x)^2}\,M - \lambda E\right)\underline{\hat{c}} = 0\,. \qquad (4.7)$$

When $\alpha < 2$ the eigenvalues take the form

$$\lambda_s = \frac{A}{(\Delta x)^2}\left(-2 + \sqrt{4-\alpha^2}\,\cos\frac{s\pi}{N+1}\right), \quad s = 1,\ldots,N, \qquad (4.8)$$

i.e. λ_s is real and negative. However, when $\alpha > 2$, (4.7) has complex eigenvalues and oscillations may occur (Price et al., 1966), which is unphysical. In order to avoid the above problems α must be made smaller by using a finer spacing. This can become very costly in terms of computer time. Another remedy is to add large artificial diffusion, but that would make the original problem unrealistic.

It should be pointed out that a central difference scheme in space and time (leap frog) for the approximation of (4.1) has no numerical built-in damping in case $A = 0$ (hyperbolic case). This can be verified, if we assume c to be proportional to $e^{i\kappa\Delta x}$ (κ = wavenumber). Then the modulus $|\lambda|$ of the complex number

$$\lambda = c_j^{\,n+1} / c_j^{\,n} = |\lambda|\,e^{i\theta} \qquad (4.9)$$

describes the change of amplitude of c during one time-step and θ the phase change. For the leap frog scheme (stable if $\varepsilon = u\,\Delta t/\Delta x \leq 1$)

$$\frac{c_j^{n+1} - c_j^{n-1}}{2\,\Delta x} = -u\,\frac{c_{j+1}^{n} - c_{j-1}^{n}}{2\,\Delta x} \tag{4.10}$$

one obtains two solutions for λ (as is well known one solution corresponds to a computational mode), but for both $|\lambda| = 1$, i.e. there is no damping (e.g. Mesinger and Arakawa, 1976).

4.2 Upstream difference scheme (UDS)

In order to avoid the above discussed problems non-centered difference schemes in space (and time) may be used. But as we shall see, this introduces alternative difficulties.

If one considers only the advective part of (4.1) the equation is the simplest first-order hyperbolic equation possible. It has the general solution $c = h(x-ut)$ where h is an arbitrary function. The lines $x-ut$ are called characteristics, and c is constant on these lines. With this in mind a difference scheme is constructed which depends on the slope of the characteristics, i.e. depends on the sign of u. The new value c_j^{n+1} is computed by tracing the characteristic passing through c_j^{n+1} back to the previous time level where the solution can be computed by linear interpolation from neighboring grid points. More details are given in Mesinger and Arakawa (1976).

The difference schemes are then (dependent on u)

$$\frac{\partial c_j}{\partial t} + u\,\frac{c_j - c_{j-1}}{\Delta x} = 0, \qquad \text{for} \quad u > 0, \tag{4.11}$$

$$\frac{\partial c_j}{\partial t} + u\,\frac{c_{j+1} - c_j}{\Delta x} = 0, \qquad \text{for} \quad u < 0. \tag{4.12}$$

In what follows we assume $u > 0$ and work only with (4.11). With the transformation $\tau = ut/\Delta x$ (4.11) has the form

$$\frac{d}{d\tau}\,c_j(\tau) = c_{j-1}(\tau) - c_j(\tau), \tag{4.13}$$

of which the solution is (Mesinger and Arakawa, 1976):

$$c_j(\tau) = \sum_{n=-\infty}^{j} c_n(0)\,\frac{e^{-\tau}\,\tau^{j-n}}{(j-n)!}, \tag{4.14}$$

this is a sum of Poisson distributions; $c_n(0)$ are the initial conditions. For strictly positive $c_n(0)$ the $c_j(\tau)$ can never become negative. However, as time goes on, a rectangular pulse would diffuse into a bell-shaped curve. This means that false diffusion is implicitly incorporated in the method (Roache, 1972).

If we would use a forward time-step in (4.11) and add physical diffusion [$A \neq 0$ in (4.1)], we could perform a similar eigenvalue analysis as for the central difference scheme, see (4.6). Lam (1975) showed that the eigenfunctions of the resulting tri-diagonal matrix have a non-oscillatory character with no restriction on the cell Reynolds number.

Figure 18 gives an impression of the different behaviour of CDS (dotted curve) and of UDS (dot-dashed curve) for the simulation of a one-dimensional plume with an almost vertical front (Lam, 1977). Whereas the CDS shows strong oscillations, UDS smoothes the front heavily. Without the applied physical diffusion ($\alpha = 200$) CDS would even produce negative concentrations. This does not happen with UDS. However, if we use forward time-stepping,

$$c_j^{n+1} = c_j^n - \varepsilon (c_j^n - c_{j-1}^n), \quad \varepsilon = u \Delta t/\Delta x \tag{4.15}$$

the damping coefficient (4.9) computed from (4.15) is

$$|\lambda|^2 = 1 - (1-\varepsilon) 4\varepsilon \sin^2(\kappa \Delta x/2). \tag{4.16}$$

Scheme (4.15) is stable for $\varepsilon \leq 1$. If $\varepsilon = 1$ (which cannot hold everywhere if u is variable) then $\lambda = 1$, i.e. neither damping nor diffusion occurs. If $\varepsilon = 1/2$ then $\lambda = 1 - \sin^2(\kappa \Delta x/2)$ and damping is very heavy. It is a desired effect that the shortest wave ($2\Delta x$-wave) is completely dissipated, but damping is also strong for longer waves.

It is interesting to note that the upstream difference scheme (4.15)

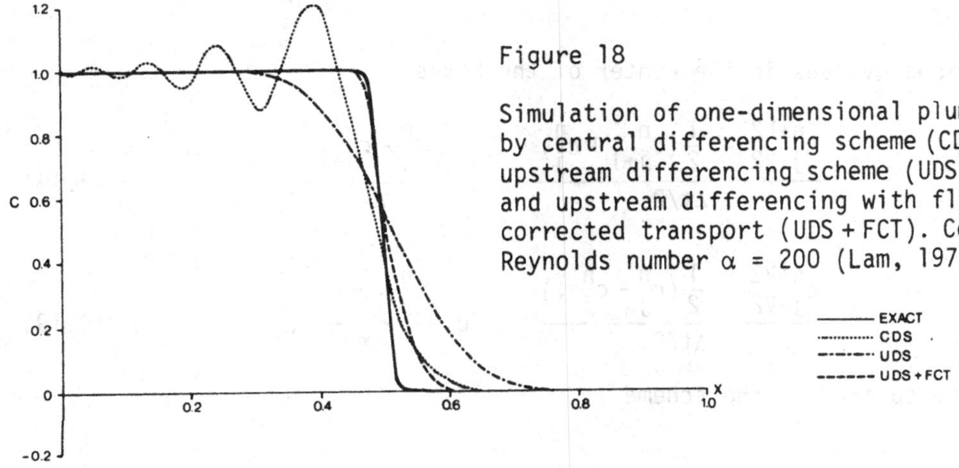

Figure 18

Simulation of one-dimensional plume by central differencing scheme (CDS), upstream differencing scheme (UDS), and upstream differencing with flux corrected transport (UDS + FCT). Cell Reynolds number α = 200 (Lam, 1977).

can be expressed in a different form by adding and subtracting the terms $-\varepsilon c_j{}^n$, $\varepsilon c_{j+1}{}^n$ to give

$$c_j{}^{n+1} = c_j{}^n - \varepsilon(c_{j+1}{}^n - c_j{}^n) + \varepsilon(c_{j-1}{}^n - 2c_j{}^n + c_{j+1}{}^n). \qquad (4.17)$$

Neglecting the third quantity on the right of (4.17), which has the form of a smoothing operation, the remainder is simply a downstream formulation of the advection equation, which is unconditionally unstable. Thus (4.17) means, that adding sufficient smoothing stabilizes an unstable scheme.

The above suggests that neither scheme, CDS or UDS, is well suited to model the advection-diffusion equation in problems with sharp gradients. For completeness we briefly discuss another scheme which has been in use for a long time, especially for shock wave problems.

4.3 The Lax-Wendrow scheme (LWS)

This scheme (Lax and Wendrow, 1960), promoted by Richtmyer and Morton (1967), may be constructed in a similar way from the characteristics as the UDS by quadratic interpolation. It may also be written as a two-step procedure, where first, provisional values are computed halfway in time and space and then used for the next half-step. Figure 19 shows a stencil of associated grid points.

Figure 19

The space time stencil used for the construction of the Lax-Wendrow scheme.

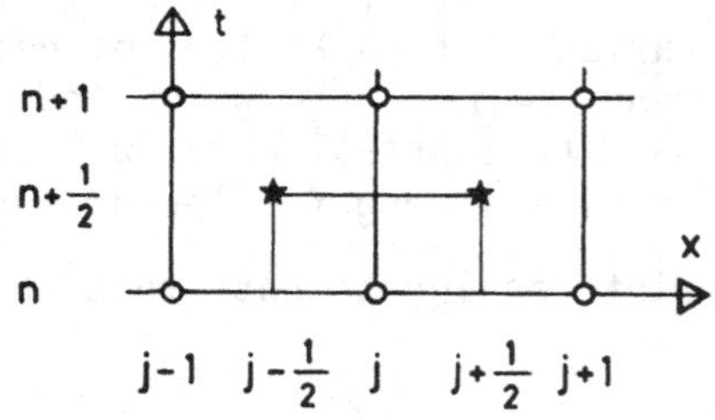

Provisional values in the center of the boxes

$$\frac{c_{j+1/2}^{n+1/2} - \frac{1}{2}(c_{j+1}^n + c_j^n)}{\Delta t/2} = u\,\frac{c_j^n - c_{j+1}^n}{\Delta x}, \qquad (4.18)$$

$$\frac{c_{j-1/2}^{n+1/2} - \frac{1}{2}(c_j^n - c_{j-1}^n)}{\Delta t/2} = u\,\frac{c_{j-1}^n - c_j^n}{\Delta x}, \qquad (4.19)$$

are used to produce the scheme

$$\frac{c_j^{n+1} - c_j^n}{\Delta t} = u \frac{c_{j-1/2}^{n+1/2} - c_{j+1/2}^{n+1/2}}{\Delta x} . \tag{4.20}$$

(4.18) - (4.20) is the Lax-Wendrow (2-step) scheme. Substitution of (4.18) and (4.19) into (4.20) yields

$$\frac{c_j^{n+1} - c_j^n}{\Delta t} = -u \frac{c_{j+1}^n - c_{j-1}^n}{2\,\Delta x} + \frac{u^2\,\Delta t}{2} \frac{c_{j+1}^n - 2c_j^n + c_{j-1}^n}{(\Delta x)^2} . \tag{4.21}$$

Evidently, the second quantity on the right side is the finite difference approximation of a diffusion term with diffusion coefficient $A^* = u^2\,\Delta t/2$, corresponding to numerical diffusion. Thus, again we have constructed a diffusive scheme. Its damping coefficient (4.9) has the form

$$|\lambda|^2 = 1 - (1 - \varepsilon^2)\,4\,\varepsilon^2 \sin^4\kappa \frac{\Delta x}{2}, \quad \varepsilon \leq 1 . \tag{4.22}$$

A comparison of the USD-damping coefficient (4.16) with (4.22) for different values of ε, κ shows that the LWS enjoys much less numerical diffusion than UDS. However, as one could guess already from the construction of LWS (central in space, non-central in time), the behaviour of LWS ranges between CDS and UDS: It may produce oscillations with negative concentration (Struve, 1978), but not as strong as CDS because the diffusion limits this process.

Certainly, a variety of other differencing schemes is possible, but they all are loaden with the inherent problem of too much diffusivity or non-positiveness of the solution for c. A completely different approach, using Langrangian techniques, can be found in Sündermann (pp. 221-231). Nevertheless, if it would be possible, somehow, to counterbalance the numerical diffusion, UDS or LWS could be preferable difference schemes.

It should be pointed out that all three difference methods can be extended to three-dimensional problems.

4.4 The flux corrected transport (FCT)

The above discussion has shown that there is no way of suppressing numerical diffusion and simultaneously having the desired accuracy. Therefore, it seems quite reasonable to try to add some "anti-diffusion" to the schemes which balance the unwanted numerical diffusion. Boris and Book (1973), Book et al. (1975, 1981) offer such a method and call it flux-corrected transport technique (FCT).

The method consists of the following steps: (i), one uses one of the schemes (CDS, UDS, LWS), and adds artificial diffusion where necessary (for CDS, LWS) so to reach positiveness and, (2), one eliminates false diffusion A_f added to CDS, LWS for example, or which was inherent in the scheme (e.g. UDS). In principle the second step is of the form

$$\frac{\partial c}{\partial t} = - A_f \, \partial^2 c/\partial x^2 \, . \tag{4.23}$$

In practice, one may first perform the anti-diffusion and then compute the transport plus false diffusion. As an example, for LWS, the total operation can be written as

$$c^{n+1} = \Big((I+F)(I+T)+D\Big)\, c^n \tag{4.24}$$

where I, F, T, D are operators with the following meaning

- I Identity operator $Ic = c$,
- F Anti-diffusion operator,
- T Transport operator $Tc = \partial u\, c/\partial x$,
- D (Artificial) diffusion operator $Dc = \eta(c^n_{j+1} - 2c^n_j + c^n_{j-1})$,

where η determines the strength of smoothing. η may be a function of u: $\eta = \eta(u)$. The operator $F = -D$ can be written as

$$Fc = -(A_{j+1/2} - A_{j-1/2}) \, , \tag{4.25}$$

where $A_j = \eta\,(c^n_{j+1/2} - c^n_{j-1/2})$. The A_j may be considerd as fluxes which are successively added and subtracted, thus satisfying conservation conditions. However, positiveness cannot be warranted.

To achieve positiveness, a limiting flux is introduced by defining the operator F as

$$Fc = L_{j-1/2} - L_{j+1/2} \, , \tag{4.26}$$

where $L_{j+1/2}$ has the complicated form

$$L_{j+1/2} = \eta S\cdot \mathrm{Max}\left[0, \mathrm{Min}\Big(S(c_j - c_{j-1}),\ |c_{j+1} - c_j|,\ S(c_{j+2} - c_{j-1})\Big)\right] \tag{4.27}$$

with $S = \mathrm{sign}(A_{j+1/2})$. We can see that this correction depends also on neighbouring values, which become important in case of steep gradients. In case the minimum is equal to $|c_j - c_{j-1}| \neq 0$, $L_{j+1/2} = A_{j+1/2}$ as before. Otherwise, this formulation does not permit that local maxima or minima are generated. For a detailed discussion of the properties of (4.27) and the associated problems, the reader is referred to the above cited literature.

For UDS no artificial diffusion needs to be added because it is already inherent in the T-operator:

$$c^{n+1} = \big((I+F)(I+T)\big)\, c^{n} .$$

An example of UDS + FCT is represented for comparison in Figure 18 (dashed curve) demonstrating a substantial improvement on the one-dimensional plume problem. An Example of a two-dimensional plume problem is presented by Sündermann, pp. 209-233).

It is possible to extend these methods to two- and three-dimensional problems. Experts claim that about 90% of the removable error is corrected by the latest FCT algorithms (Book et al., 1981).

4.5 Example for variance conservation

The other aspect of the integration of the advection-diffusion equation is the suppression of non-linear instability in case that the "concentration" c is identified with temperature or velocity. We neglected this problem because the advection velocity was assumed to be constant. However, if one deals with real non-linear problems, slight modifications of the above presented schemes are necessary. They can all be made to conserve specific quantities (like mass). Variance conserving schemes conserve also the square of a quantity, e.g. kinematic energy.

The most general rules for constructing conservative difference schemes are found in Bryan (1966). They all neglect time truncation. As an example, we present the conservative difference formulation of the layered operator $L(\Phi)$ defined in (3.11) and recall the distribution of grid points from Figure 8b (Simons, 1980).

It will be shown that conservation requirements are satisfied if values of the advected variable, Φ, on the surfaces of the box surrounding the point $(i,j,k-1/2)$ are approximated by simpler linear interpolation (index $i,j,k-1/2$ dropped). After replacing the derivatives by ordinary centered differences, the expressions $(3.11)_1$ for the operator L become

$$L(1) = \frac{\partial D}{\partial t} + \frac{1}{\Delta x}(U_{i+1} - U) + \frac{1}{\Delta y}(V_{j+1} - V) + \omega_{k-1} - \omega_k , \qquad (4.28)$$

$$\begin{aligned} L(\Phi) = \frac{\partial}{\partial t}(D\bar{\Phi}) &+ \frac{U_{i+1}}{2\,\Delta x}(\bar{\Phi} + \bar{\Phi}_{i+1}) - \frac{U}{2\,\Delta x}(\bar{\Phi} + \bar{\Phi}_{i-1}) \\ &+ \frac{V_{j+1}}{2\,\Delta y}(\bar{\Phi} + \bar{\Phi}_{j+1}) - \frac{V}{2\,\Delta y}(\bar{\Phi} + \bar{\Phi}_{j-1}) \qquad (4.29) \\ &+ \frac{1}{2}\omega_{k-1}(\bar{\Phi} + \bar{\Phi}_{k-3/2}) - \frac{1}{2}\omega_k(\bar{\Phi} + \bar{\Phi}_{k+1/2}) . \end{aligned}$$

Conservation of mass implies that $L(1) = 0$. It follows that by summing (4.28) or (4.29) over all layers and all grid points only the first terms on the right-hand sides survive. This is because of the cancellation of all other terms and the fact that U, V and ω must vanish at the appropriate boundaries. Thus, the volume integral of Φ is not affected by the non-linear advection terms. Obviously, this property of (4.29) is not lost if (4.28) is substituted in (4.29) with the result

$$L(\Phi) = \frac{\partial}{\partial t}(D\bar{\Phi}) - \frac{1}{2}\bar{\Phi}\frac{\partial D}{\partial t} + \frac{1}{2\,\Delta x}\left[(U\bar{\Phi})_{i+1} - U\bar{\Phi}_{i-1}\right] + \frac{1}{2\,\Delta y}\left[(V\bar{\Phi})_{j+1} - V\bar{\Phi}_{j-1}\right] + \frac{1}{2}(\omega_{k-1}\bar{\Phi}_{k-3/2} - \omega_k\bar{\Phi}_{k+1/2}) \, . \tag{4.30}$$

Multiplying by $\bar{\Phi}$ yields

$$\bar{\Phi}L(\Phi) = \frac{\partial}{\partial t}\left(\frac{1}{2}D\bar{\Phi}^2\right) + \frac{1}{2\,\Delta x}(U_{i+1}\bar{\Phi}_i\bar{\Phi}_{i+1} - U_i\bar{\Phi}_{i-1}\bar{\Phi}_i) + \frac{1}{2\,\Delta y}(V_{j+1}\bar{\Phi}_j\bar{\Phi}_{j+1} - V_j\bar{\Phi}_{j-1}\bar{\Phi}_j) + \frac{1}{2}(\omega_{k-1}\bar{\Phi}_{k-3/2}\bar{\Phi}_{k-1/2} - \omega_k\bar{\Phi}_{k+1/2}\bar{\Phi}_{k-1/2}) \, , \tag{4.31}$$

where some subscripts, i, j, k-1/2, have been reintroduced to emphasize the symmetry of the terms. Again the summation of (4.31) over all points reduces to the summation of the first term on the right, by virtue of the same cancellation effects and boundary conditions mentioned after (4.29). This first term is just the time rate of change of $\Phi^2/2$ for a layer, and the volume integral of Φ^2 is not affected by the non-linear advection terms. Consequently, the scheme is conservative in the sense defined above.

In summary, it is concluded that the advection of a given variable is to be computed by surrounding the grid point by a box as illustrated in Figure 8, using a linear interpolation for the advected variable on the surfaces of the box, and requiring that the advecting velocities on these surfaces be related to each other by (4.28). It is then clear that values of the advecting velocities, U, V, ω, at the surfaces of the box do not have to be grid point values, as mentioned before. Velocity components may be obtained by interpolation if, for reasons of other computations, it is more convenient to define one or more of these variables at grid points different from those shown in Figure 8.

5. SYMBOLS

A	eddy coefficient (horizontal)
α	Cell Reynolds number $u\,\Delta x/A$
β	df/dy: planetary β-parameter
c	phase speed or concentration
D, D', D_k	layer thickness
$\Delta x, \Delta t$	space/time increments
ε	$(\rho'-\rho)/\rho'$
η	free surface
η'	internal interface perturbation
f	Coriolis parameter
$\underline{\Gamma}$	flux vector
$H, \bar{H}$	water depth, averaged water depth
h	interface depth
i, j	horizontal grid indices
J	Jacobian operator $J(g,h) = \frac{\partial g}{\partial x}\frac{\partial h}{\partial y} - \frac{\partial g}{\partial y}\frac{\partial h}{\partial x}$
$\hat{k}$	friction coefficient
$\underline{k}$	unit vector in z-direction
k	vertical layer index
λ	eigenvalue
ν	eddy coefficient (vertical)
ω	generalized vertical velocity
p_e	external pressure
p_i	internal pressure
p	pressure
p_s	atmospheric pressure at the sea surface
q	temperature flux
ρ, ρ_o, ρ'	density, averaged, deviation
R	Internal Rossby radius of deformation $R = c/f = (\varepsilon g\, D\, D'/H)^{1/2}$
σ	density anomaly $\sigma = g(\rho - \rho_o)$ or frequency

σ_T	(density - 1) × 10^3
σ_n	frequency
$\hat{\sigma}$	sigma-coordinate
$\underline{\tau}$	stress vector
$\underline{\tau}_s = (\tau_{sx}, \tau_{sy})$	wind stress, surface stress
$\underline{\tau}_b$	bottom stress vector
$\underline{\tau}_i$	interfacial stress vector
u, v	velocity components in x,y-direction
$\bar{u}, \bar{v}$	averaged u, v (U/H, V/H)
U, V	transport components in x,y-direction
$\underline{V}, (\underline{\tilde{V}}, \underline{V}^*)$	volume transport vector (external, internal)
L(Φ)	differential operator on Φ
L	length of basin
W	width (of channel or basin)
x, y, z	right handed system, z upward
Z_k	interfaces

Note: Primed variables are 2nd layer variables, variables with underscores are vectors.

6. REFERENCES

Bennett, J.R., 1974. On the dynamics of wind-driven lake currents. J. Phys. Oceanogr. 4, pp. 400-414.

Book, D.L., J.P. Boris and K. Hain, 1975. Flux-corrected transport II: Generalization of the method. J. Comput. Phys. 18, pp. 248-283.

Book, D.L., J.P. Boris and S.T. Zalesak, 1981. Flux corrected transport. In: Finite difference techniques for vectorized fluid dynamics circulations. (ed. D.L. Book), pp. 29-55. Springer Verlag Berlin, 226 p.

Boris, J.P. and D.L. Book, 1973. Flux-corrected transport I: SHASTA, A fluid transport algorithm that works. J. Comput. Phys. 11, pp. 38-69.

Bryan, K., 1966. A scheme for numerical integration of the equation of motion on an irregular grid free of nonlinear instability. Mon. Weather Rev. 94, pp. 39-40.

Bryan, K., 1969. A numerical method for the study of ocean circulation. J. Comput. Phys. 4, pp. 374-376.

Charney, J.G., 1955. Generation of oceanic currents by wind. J. Mer. Res. 14, pp. 477-498.

Csanady, G.T., 1967. Large-scale motion in the Great Lakes. J. Geophys. Res. 72, pp. 4151-4162.

Csanady, G.T., 1973. Transverse internal seiches in large oblong lakes and marginal seas. J. Phys. Oceanogr. 3, pp. 439-447.

Csanady, G.T. and J.T. Scott, 1974. Baroclinic coastal jets in Lake Ontario during IFYGL. J. Phys. Oceanogr. 4, pp. 524-541.

Crêpon, M., 1967. Hydrodynamique marine en régime impulsionnel, Pt. 2. Cah. Oceanogr. 19, pp. 847-880.

Freeman, N.G., A.M. Hale and M.B. Danard, 1972. A modified sigma equation approach to the numerical modelling of Great Lakes hydrodynamics. J. Geophys. Res. 77, pp. 1050-1060.

Haidvogel, D.B., A.R. Robinson and E.E. Schulman, 1950. Review. The accuracy, efficiency, and stability of three numerical models with application to open ocean problems. J. Comput. Phys. , pp. 3411-3453.

Haidvogel, D.B., 1981. A four-dimensional primitive equation model for coupled coastal-deep ocean studies. Technical Res. Woods Hole Oceanographic Institution No. WHOI-81-90.

Harrison, D.E., 1978. On the diffusion parametrization of mesoscale eddy effects from a numerical ocean experiment. J. Phys. Oceanogr. 8, pp. 913-918.

Kielmann, J., 1981. Grundlagen und Anwendung eines numerischen Modells der geschichteten Ostsee. Ber. Inst. f. Meereskunde Kiel, No. 87a+b.

Kielmann, J., 1983. The generation of eddy-like structures in a model of the Baltic Sea by low frequency wind forcing. Submitted to Tellus.

Kowalik, Z. and A. Stáskiewicz, 1976. Diagnostic model of the circulation in the Baltic Sea. Dt. Hydrogr. Z. 29, pp. 239-250.

Krauss, W., 1979. A semi-spectral model for the computation of mesoscale processes in a stratified channel of variable depth. Dt. Hydr. Z. 32, pp. 174-189.

Lam, D.C.L., 1975. Computer modelling of pollutant transports in Lake Erie. Proc. Int. Conf. on Math. Modeling of Environmental Problems. University of Southampton, U.K., 15p.

Lam, D.C.L., 1977. Comparison of finite element and finite difference methods for nearshore advection-diffusion transport models. pp. 115-129. In: Finite Elements in water resources. (Ed. J.W. Gray, G.F. Pinder and C.A. Brebbia). Pentech Press London.

Lam, D.C.L. and T.J. Simons, 1976. Numerical computations of advective and diffusive transports of chloride in Lake Erie during 1970. J. Fish. Res. Board Can., 33, pp. 537-549.

Lauwerier, H.A., 1961. The North Sea problem: non-stationary wind effects in a rectangular bay. Proc. K. Ned. Akad. Wet., A64, pp. 104-122, pp. 418-431.

Lax, P. and B. Wendrow, 1960. System of conservation laws. Commun. Pure Appl. Math., 13, pp. 217-237.

Lee, K.K. and J.A. Liggett, 1970. Computation for circulation in stratified lakes. J. Hydr. Div. ASCE, 96, pp. 2089-2115.

Liggett, J.A. and Lee, K.K., 1971. Properties of circulation in stratified lakes. J. Hydr. Div. ASCE, 97, pp. 15-29.

McNider, R.T. and J.J. O'Brien, 1973. A multi-layer transient model of coastal upwelling. J. Phys. Oceanogr., 3, pp. 258-273.

Mesinger, F. and A. Arakawa, 1976. Numerical methods used in atmospheric models. GARP Publ. Ser. No. 17 (W.M.O.), 64 p.

Nowlin, W.D., 1967. A steady, wind-driven, frictional model of two moving layers in a rectangular ocean basin. Deep Sea Res., 14, pp. 89-110.

Peyret, R. and T.D. Taylor, 1981. Computational methods in fluid flow. Springer Verlag Berlin, 358 p.

Phillips, N.A., 1960. Numerical weather prediction. Adv. Comput., 1, pp. 43-90.

Price, H.S., R.S. Varga and J.E. Warren, 1966. Application of oscillation matrices to diffusion convection equation. J. Math. Phys. (N.Y.) 45, pp. 301-311.

Richtmyer, R.D. and K.W. Morton, 1967. Difference methods for initial value problems. John Wiley & Sons, New York, 405 p.

Roache, P.J., 1972. Computational fluid dynamics. Albquerque, Hermosa, 434 p.

Sarkisyan, A.S., 1977. The diagnostic calculation of a large-scale oceanic circulation. The Sea, Vol. 6, pp. 363-458. (Ed. E.D. Goldberg). John Wiley & Sons, Inc., New York.

Simons, T.J., 1972. Development of numerical models of Lake Ontario. Part 2. Proc. Conf. Great Lakes Res. IAGLR, 15, pp. 655-672.

Simons, T.J., 1974. Verification of numerical models of Lake Ontario. I. Circulation in spring and early summer. J. Phys. Oceanogr., , pp. 507-523.

Simons, T.J., 1978a. Wind-driven circulations in the south-west Baltic. Tellus 30, pp. 272-283.

Simons, T.J., 1978b. Generation and propagation of downwelling fronts. J. Phys. Oceanogr., 8, pp. 571-581.

Simons, T.J., 1980. Circulation models of lakes and inland seas. Can. Bull. Fisheries and Aquatic Sci., No. 203, 145 p.

Struve, S., 1978. Transport und Vermischung einer passiven Beimengung in einem Madium mit einem vorgegebenen Geschwindigkeitsfeld. Ber. Inst. f. Meereskunde Kiel, No. 57.

Taylor, G.I., 1922. Tidal oscillations in gulfs and rectangular basins. Proc. London Math. Soc. Ser., 2(20), pp. 148-181.

Weenink, M.P.H., 1958. A theory and method of calculation of wind effects on sea levels in a partly enclosed sea. R. Neth. Meteor. Inst. Med. Verh., 73, 111 p.

Welander, P., 1968. Wind-driven circulation in one- and two-layer oceans of variable depth. Tellus 20, pp. 1-15.

Hydrodynamics of Lakes: ÇISM Lectures
edited by K. Hutter, 1984
Springer Verlag Wien-New York

MEASUREMENTS AND MODELS IN PHYSICAL LIMNOLOGY

C.H. Mortimer
Center for Great Lakes Studies
University of Wisconsin-Milwaukee
P.O. Box 413, Milwaukee, WI 53201, USA

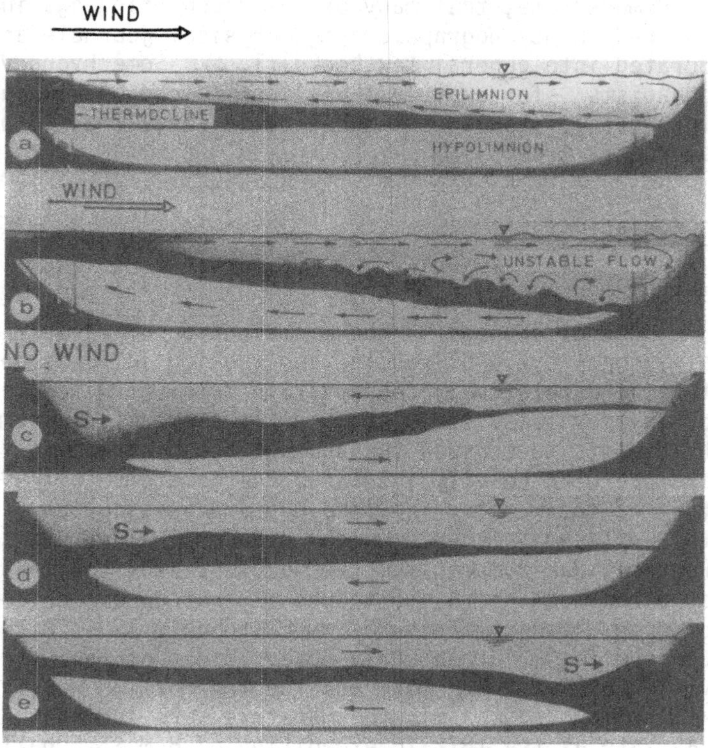

Generation and subsequent progress of an internal surge (S) superimposed on an internal seiche in a three-layered model (from (135)).

1. HISTORY: SOME STARTING-POINTS

1.1 Preamble

Within the compass of six lectures it is manifestly impossible to cover adequately the history of the interplay between field observations and models as well as recent progress and promising future trends in physical limnology. Therefore Section 1 is an annotated bibliography of starting-points or first discoveries within the field of water movements, selected (it must be admitted) on the basis of the lecturer's experience and interests. This selectivity may explain and perhaps excuse the immodest-seeming citation of his own publications. Section 2 (Recent History and Future Opportunities) will also contribute to history, but will consider in addition some promising future lines of research. The continuity between the lectures will, it is hoped, demonstrate the inseparable continuity between physical limnology and oceanography in particular and geophysical fluid dynamics in general. It is no coincidence, as the bibliography will demonstrate, that many of the recent findings in lakes have been published in oceanographic or geophysical journals and are becoming incorporated into general textbooks (1, 2). Some hydrodynamic phenomena and mechanisms -for example those constrained by boundaries- can, in fact, be more easily studied and modelled in lakes than in oceans. For others the reverse is true.

1.2 Early preoccupations

a) Thermometry

Bearing in mind that changes in the distribution of water temperature delineate the seasonal cycle of warming and cooling in lakes and also that temperature is a relatively conservative label of water movements on time scales of days or less, the history of lake hydrodynamics may be said to begin with attempts to measure the sub-surface distribution of temperature, for example with heavily insulated thermometers in 1799 (3). The subsequent story of thermometry in limnology and oceanography (4) provides examples of the profound influence which advances in instrument design exerted on progress. Maximum/minimum thermometers provided the first demonstration of a thermocline (5), although in that case (Lac Léman) the indications of deep water temperature were probably 1° - 2°C too high, because of pressure effects. Other early observations of lake stratification were reviewed by Geistbeck (6); and the thermocline was first so named by Birge (7). Improved max/min thermometers were successfully used in oceanography as late as the 1872-76 expedition of H.M.S. "Challenger", but were thereafter superseded by the reversing thermometer. Negretti's and Zambra's improved patent (4) was probably first used in a lake by Forel (8) and more systematically employed by Richter (9). With care in calibration and use, the modern standard instrument measures *in situ* temperature with an error of less than $\pm 0.01^{o}C$. For measurement near the bottom of deep lakes Strom (10) had a special thermometer constructed by Richter and Wiese (Berlin) with a range from $+2^{o}$ to $+5^{o}C$ divided in $0.01^{o}C$ intervals and with a claimed error of less than 1/5 division.

If such accuracy were needed today, it could be more conveniently achieved by electrical resistance thermometry. This method (along with the thermoclinic technique) also has a long history (11). Becquerel and Breschet (12) lowered an insulated Cu/Fe thermocouple into Lac Léman to measure the temperature profile, while keeping the reference junction at uniform air temperature in the Chateau de Chillon. Electrical resistance thermometry, introduced by Siemens to oceanography (4), was first applied in a lake by Warren and Whipple (13) in 1895. The advent of thermistors after the 2nd World War considerably simplified the technique of electrical resistance thermometry, although platinum wire coils remained in use where the highest precision was required. First described, for lake use, in 1950 (14,15), the thermistor probes are now standard equipment. The first thermistor "chain", a powerful tool for continuous simultaneous recording of temperature at selected fixed depths, was developed by the writer (16,17) to record temperatures in Windermere in 1950 and in Loch Ness two years later. The earliest device for continuous recording (but at a single depth) was Wedderburn's ingenious underwater thermograph (18), later borrowed from the Royal Scottish Museum by the writer (19) to record internal seiches in Windermere.

Much more extensive and detailed surveys in lakes, yielding quasi-synoptic pictures of temperature distribution, became possible with the invention of temperature/depth profilers deployed from moving vessels, the bathythermograph (20) and depth-undulating probes and towed thermistor chains (21). In fact, the first detailed three-dimensional study of the seasonal cycle of warming and cooling (stratification/destratification) was made (22,23) with a bathythermograph in 1942 from Lake Michigan railroad ferries. Subsequent examples of the use of thermometric time series (continuous in space on cross-basin transects, continuous in time at moored thermographs) in whole-basin dynamic studies are presented later.

b) *Oscillations in lake level (seiches)*

Readily observed, rhythmic fluctuations in lake level have long exercised a fascination and have stimulated mathematical modelling, but often with a long time gap between observation and theoretical resolution (24). The first detailed set of observations ((25), on Léman, 1730, introducing the local name "seiche") and recognition of their occurrence in many lakes (26) were, it is interesting to note, preceeded by systematic observations and conjectures by a Jesuit missionary (27) in 1671, describing the large but irregular "tides" at the head of Green Bay (a gulf which opens onto Lake Michigan) and attributing them to a combination of lunar tidal influence and to the influence of the main lake. Three centuries elapsed before those conjectures were confirmed by spectral analysis and numerical modelling (28,29,30).

With early observations and conjectures as a prelude, physical limnology was launched as a distinct branch of geophysical fluid dynamics ("L'océanographie des lacs") by Forel's lifetime study of Léman seiches and temperature regime (8). But again, in one respect priority must go to Lake Michigan, i.e. to a U.S. Army surveyor's 1872 interpretation (31) of the conspicuous 2.2 h seiche at Milwaukee as a standing wave, thereby antedating Forel's similar interpretation (32) by three years and providing yet another example of an original idea occurring to two persons at about the same time. Mathematical modelling of this seiche (as the first transverse mode (33)) confirmed the 1872 interpretation. In fact, hydrodynamic modelling may be said to have "cut its teeth" on seiches, starting with Forel's use, at the suggestion of Lord Kelvin ((8), p. 78), of Merian's equation (34) for the rectangular basin, followed by Chrystal's (35) channel equations applied to basins of various simple geometries and by Defant's (36) simple, one-dimensional iterative procedure, which calculates seiche periods and structures with remarkable success in real basins if they are not too irregular in shape. These and other computational models were later used and compared in applications to real basins (37,38,39); and these publications contain extensive bibliographies. Among the many detailed post-Forelian observations of seiches may be counted those of Bergsten (40) and Endrös (reviewed (41)).

The (Coriolis) effect of the earth's rotation upon standing waves (seiches) was first treated, from a theoretical point of view, in a rectangular gulf (42) and an elliptical basin (43) both of uniform depth. The Coriolis influence on seiche oscillations in real basins was first taken account of in the Baltic Sea (the effect on period, (44)), Lake Michigan (but with no observations to check the theory, (45)), and Lake Erie (46). The latter 1964 study was accompanied by the first spectral analysis (47) of seiche records (see also (28)). More recent numerical models, which include rotation (48,49,50,33,39), have shown satisfactory agreement with observed seiches and their spectra, thus rounding off a long chapter in lake hydrodynamics.

1.3 Internal density structure, turbulence and oscillatory responses

Stratification in lakes, i.e. the formation of a more or less distinct density interface (the thermocline), is a consequence of the seasonally-changing and storm-episodic interaction of mechanical (wind-derived) and radiative (sun-derived) fluxes. The mechanical flux generates currents and (most importantly) current shears which *promote* turbulence (51), while the positive (or negative) radiative fluxes create (or destroy) vertical density gradients and their associated buoyancy forces, which *suppress* turbulence (52). The ever-shifting balance between promotion and suppression, expressed as the Richardson Number (53), determines the short-term (storm-episodic) and long-term (seasonal) response of lakes to the forcing actions of wind and sun. To review adequately the path of discovery relating to wind and heat-flux effects and the resulting circulation patterns in lakes would far exceed the compass of these lectures. Therefore the reader is referred to reviews (54,55,56,57,58). The last review is an historical account of the pioneering work of Birge and Juday, including their study of the penetration of radiation into lakes ((59), see also (60)) and of the work of the wind in transporting heat downward (61). Another pioneer, exploring the concept and consequences of turbulence, was Schmidt (51,62).

In the remainder of this Section and in the one which follows I shall trace the history of research on oscillatory responses of stratified lakes to wind impulses and to the earth's rotation. Emphasis on this topic is justified by the fact that internal responses to wind forcing (in contrast to surface seiche responses) involve large displacements of the water layers and therefore exert a profound influence on the chemical and biological economy of the system. In the following Section some attention will also be paid to nearly-steady flows and instabilities associated with them.

When the thermal structure of lakes began to be studied systematically, large-amplitude oscillations in isotherm depth were sometimes noted (9, 63). The first interpretation of this phenomenon as a standing internal wave (64) prompted Wedderburn's thorough investigations (65,66,67,18) culminating in the first mathematical model which took basin shape and stratification into account. This, as Wedderburn's publications and the model demonstrate, was a direct development from earlier lakes in surface seiche research, which Chrystal had given to his student. Initially Wedderburn's findings were largely ignored and their interpretation even disputed (68) until the universality of their occurrence - as resonant responses of stratified lakes to wind impulses - was fully demonstrated forty years later (19,69) and the picture of layer displacement and flow during the initial wind-forcing phase had been clarified (70,71) by lake observations interpreted by physical models. The first multi-layer analytical models, approximating to real-lake stratifications but without rotation, were also developed: three homogeneous layers (19) and three layers in each of which a linear dependence on density could be selected (72). The latter model, driven by wind, was satisfactorily verified (73) using thermistor chain and wind records from Windermere (69).

The effect of earth's rotation (small but appreciable, as we have noted above, for surface seiches) is "magnified" for internal seiches, because of the small density differences between the layers. This effect, first demonstrated in Loch Ness (17), could also be displayed in a broader lake (Léman) with the help of continuous records of (i) temperature at the Geneva waterworks intake (depth 15 m), and (ii) water levels at eight stations spaced around the shore. Numerical (low-pass) filtration of the latter periods permitted the surface signature of the main internal seiche (3.5 d period) to be followed cyclonically around the basin, reproducing some features of the conceptual model in Figure 1. Publication of those results was delayed until a more comprehensive study had been made (74) of waterworks intake temperature records around Lake Michigan. That study confirmed that there was an internal wave response which followed events of strong wind-induced upwelling along one shore, accompanied by downwelling on the other. The main features of the subsequent wave response were reproduced by the well-known Kelvin*) channel-wave model which, conforming to Lake Michigan circumstances, is shore-trapped within a 10 - 20 km wide strip. The wave currents run always shore-parallel, and the wave amplitude decreases exponentially in the offshore direction - for an illustration of this model wave - running in opposing directions along the sides of a wide channel, see (2). In a narrower channel, corresponding to Léman, the behavior of the double Kelvin wave model approximates to that seen in the middle region of the conceptual model basin in Figure 1. For recall in the next lecture, it is noted here that the (linear) Kelvin model did not reproduce the step-fronted (non-linear) feature seen in the internal wave as it travelled past the water intakes in Lake Michigan (see the modified model in (79) and Figure 26 in (80)).

The Kelvin-like internal waves are shore-trapped and can be viewed as the large-basin (Coriolis-modified) modification of the longitudinal internal seiche seen in small basins. But larger stratified lakes also exhibit *cross-basin* internal seiches, which are strongly modified by earth' rotation, with features which can be modelled by internal Poincaré waves as predicted in 1963 ((74), but with some errors in the illustrated current patterns) and observed in Lake Michigan in the same year (81). Like the Kelvin-type waves, the Poincaré-type seiches also became active after large wind-induced upwelling/downwelling perturbations of the thermocline had occurred on the whole-basin scale of the example demonstrated by research vessel surveys (82). Characteristic of this response are: clockwise rotation of the current vectors; phase opposition between the currents above and below the thermocline; a cross-basin standing wave structure with one mode dominant or two or more possible modes combined; and (in large lakes) mode-dependent periodicities which approach but never

*) *The Kelvin and (later to be mentioned) Poincaré and Sverdrup waves are channel-wave models "borrowed" from tidal theory ((75, 76, 77) reviewed in (78)) illustrating, yet again, the common ground between physical oceanography and limnology.*

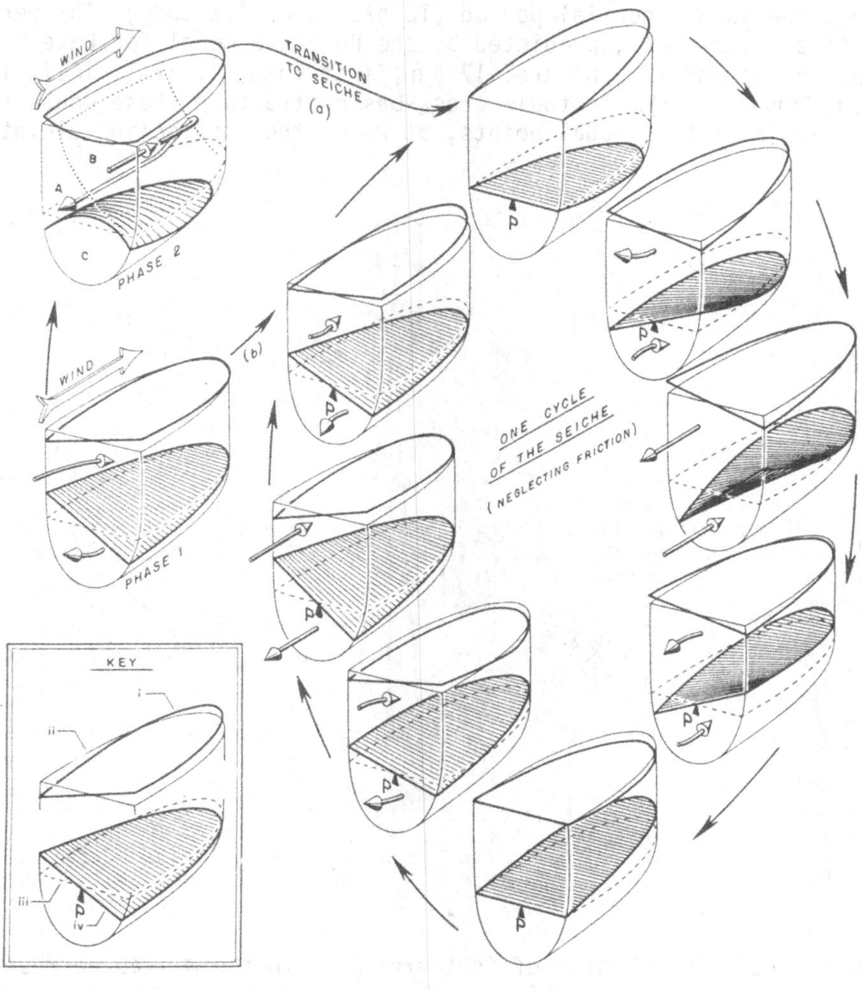

Figure 1

A conceptual model of thermocline (shaded surface) and water surface (thick line) motions during and after wind disturbance (upper left) of a two-layered, anti-clockwise rotating stratified basin (only half of which is shown) of width order 10 km. The subsequent internal seiche response is depicted at 1/8 stages of the oscillation cycle. The flow, indicated by thick arrows, is based on guesswork, particularly during the wind-forcing phase. The water surface response (thick line) represents only the (exaggerated) surface signature of the thermocline wave, i.e. the higher frequency (barotropic) surface seiche response is not illustrated. P is the nodal amphidromic point of zero elevation change, around which the internal seiche progresses cyclonically (anticlockwise). Redrawn from Mortimer, 1955.

exceed the local inertial period (12 h/sine of latitude). The periods of the first five modes, predicted by the Poincaré model for Lake Michigan (inertial period 17.5 h) are: 17.4 h (first mode)*), 17.1, 16.7, 16.1 and 15.5 h (fifth mode). In their cross-basin structure those modes respectively display 1 to 5 nodal points, at which the thermocline elevation ex-

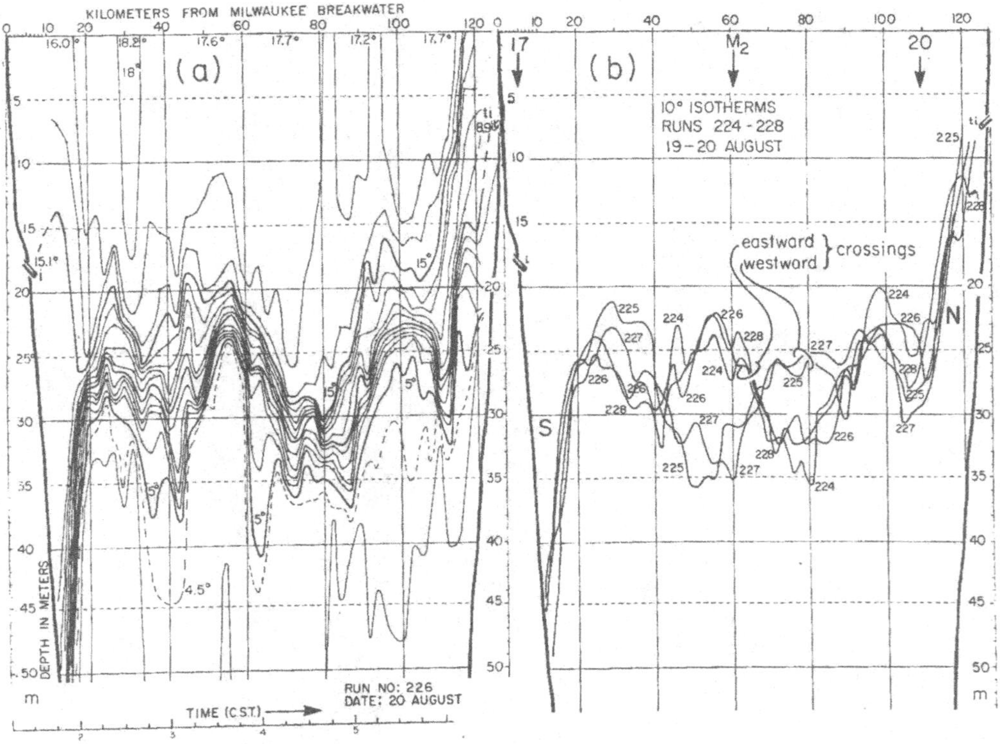

Figure 2 (a) distribution of isotherms (0°) in the Milwaukee-Muskegon section of Lake Michigan, as measured during a ferry crossing on 20 August 1963; (b) distributions of the 10°C isotherm in the same section during five consecutive ferry crossings, 19-20 August 1963. The temperatures noted near the "pipes" on either shore are the waterworks intake temperatures at the time of ferry passage. Redrawn from Mortimer, 1968 (81).

*) *Obviously, with this periodicity, the current relation of the first mode will be indistinguishable from "true" inertial motion described for oceanic examples (83) and which is also the first response to appropriately timed wind impulses in large (84) and not so large (85) lakes. The mixed inertial and Poincaré-type response is, therefore, often referred to as "near-inertial" motion.*

cursion is zero. The observed structure (81) displays, in the episode selected for Figure 2, the aftermath of two days of strong southgoing wind (blowing out of the page) which has brought about strong downwelling on the western (lefthand) shore, corresponding upwelling on the eastern shore, and a distinct cross basin thermocline structure in which several odd-numbered modes (with the fifth dominant?) appear to have been excited. The nature of this structure is revealed by the unchanging position of the mid-lake nodal point (righthand diagram, 66 km from Milwaukee) during five consecutive vessel crossings, which occupied two days. Because of the fortunate circumstance that the ferry crossed and re-crossed the mid-lake point at approximately 8 h intervals, the phase structure between 50 and 80 km is clearly displayed, leading to the inference that the average wave period was close to 16 h. This and the clockwise rotation of the current vectors (in opposition above and below the thermocline) were confirmed by repeated profilings of current and temperature from a vessel anchored in mid-lake (example illustrated in Section 2). The upwelling/downwelling region, and that of the corresponding internal Kelvin-type wave domain, it will be noted, is confined to within 20 km of the shore.

This is a convenient historical point to end the first Section. The reference list is to be found at the end of Section 2.

2. RECENT HISTORY AND FUTURE OPPORTUNITIES

Recent history may be said to begin 25 years ago (i) with development and verification of the first large-scale numerical simulation of lake motion (86) and with (ii) the first major university commitment to whole-basin investigation on the Laurentian Great Lakes, using research vessels and oceanographically-trained specialists (82). Also borrowing from oceanographic methodology, that recent era has been characterized by several intensive measurement programs involving the deployment of large arrays of moored current meters, thermographs, and wind-recording buoys. The first such campaign was mounted between March 1963 and September 1964*) in Lake Michigan (later transferred to other Great Lake basins) by the U.S. Health Service (87), aimed to provide comprehensive understanding of circulation and dispersal in various seasons and wind regimes

*) *The unprecendent scale of this pioneering effort is illustrated by the fact that 128 sets of current and temperature records, ranging from 29 to 229 days in length, were obtained at several depths at 34 moorings, in addition to vessel surveys and drifting drogue tracking to estimate horizontal diffusion rates. It is said to note, on completion of that field campaign, the team was disbanded; proper archiving was not done; and the only substantial sources of information are ((89,90) and (91)).*

and to settle, it was hoped once and for all, the question of transport and diffusion of pollutants. It was the evident over-optimism of some of the campaign planners that prompted this writer to venture then-unsupported predictions of complexities (74). Those predictions -dominance of shore-parallel to-and-fro motion near shore and near-inertial current rotation offshore - were amply confirmed by the recorders (88,89,90) and by ferry transects (see previous Figure 2); and much new light was cast on large-lake motions; but, for reasons explained in the preceding footnote, the data set was not fully exploited, nor is most of it available in archives. That crippling deficiency was avoided in the next major campaign (IFYGL 1972, described below) and in more recent surveys using arrays of modern oceanographic instruments: in the Lake of Constance (92); Lake of Zurich (93); Léman (94); Lake Huron (95); and Lake Michigan (96).

2.1 The international field year for the Great Lakes (IFYGL, Lake Ontario, 1972)

The findings and the questions raised by the 1963/64 investigations in Lake Michigan (81,89,90) stimulated interest in lake dynamics among the growing Great Lakes research community and environmental managers and laid some of the groundwork for planning the IFYGL, the most intensive lake and over-lake atmosphere experiment to date. For example, publication 90 (with M.A. Johnson's contribution to theory) was produced as a planning document for part of the Water Movements section of the IFYGL. The whole project with over 600 participants (including meteorologists, hydrologists, chemists and biologists) employed five major research vessels from Canada and the U.S.A., eighteen offshore moored stations recording meteorological parameters and water currents and temperatures at several depths, in-lake instrumented towers and land stations, three cross-lake transects traversed repetitively with depth-undulating instruments, five "coastal chains" for detailed exploration of density and flow structure within 10 km of shore; instrumented aircraft; and land stations exploring meteorology and hydrology. Waves and water levels were also recorded as were whole-basin changes in heat content (105 stations visited on 40 cruises). The unprecedentally voluminous data collections have been archived (97, 98) and a summary "IFYGL Volume" has recently appeared (99). In this the 140 references in the Water Movements chapter are a measure of the impact that the IFYGL has had (and will continue to have) on this subject. Here, in this lecture, it is possible to mention only some of the IFYGL-mediated advances in knowledge of water movements.

The IFYGL included plans to investigate more fully the mechanisms of nearshore flow; and substantial advances were made, particularly by G.T. Csanady and his co-workers manning the coastal chains. It was found that, in Lake Ontario and (by implication) in other large lakes, currents distinctly characterized as coastal are contained within 10 km of shore, attaining maximum speeds of 20 to 70 cm s^{-1} at or near the lake surface in a "current core" located between 3 and 7 km of shore, with some meandering and episodic reversals. Near-inertial motions, conforming to the above-described Poincaré wave models, are also observed near shore; but the dominant current oscillations there are of much lower frequency, appearing as nearly steady, predominantly shore-parallel flow. Mechanisms investigated included: coastal jet flow (100); intermittent upwelling (101); interaction between wind stress and the thermally-driven geostrophic current in spring (102); Kelvin wave interactions (103); and the influence of (later described) topography-related vorticity waves (104). These IFYGL results have now become incorporated into the body of knowledge referred to as coastal oceanography (1).

The combination, during the IFYGL, of the synoptic temperature surveys with simultaneous measurements of currents and meteorological forcing provided a more complete picture of the lake's dynamic responses than had previously been possible and also provided a substantial data base against which a hierarchy of numerical simulation models could be tested and evolved ((105) to (110) and a more general review of lake circulation model development (111)). Analyses of the mean current distribution patterns, displayed on a monthly basis (112,113), revealed a pattern of gyres for comparison with the models. As a consequence of this iterative evolution, a deeper understanding of the quasi-steady mean circulation patterns in large basins has been achieved.

Basin responses of higher frequency, during the stratified season and following short-lived episodes of strong wind stress, were intermittent bursts (114) of the large-amplitude inertia-gravitational waves (see spectra in later Figure 5), already described in terms of the Poincaré model for Lake Michigan. We must conclude that these are universally present in large stratified lakes. The IFYGL current and temperature records and the repeated pictures of cross-basin density structure obtained on the transects threw new light on the processes generating these waves. For example, before thermal stratification had become established basin-wide, the nearshore stratified region exhibited motions indistinguishable from the "true" inertial (Ekman) response (i.e. current rotation at the exact inertial frequency, because at that time no whole-basin Poincaré-type waves could be generated), whereas after full stratification had become established the response was either a local inertial one (if the wind impulse was strong and of short duration comparable to half the inertial period) or a basin-wide one, if the wind stress persisted long enough to set up a strong cross-basin upwelling/downwelling pattern (84, 115). In that case, resonance of the inertial motion with one of the baroclinic Poincaré modes of period distinctly less than inertial occurred. A first-mode example is illustrated in Figure 3 with a structure closely predicted by numerical simulation (117).

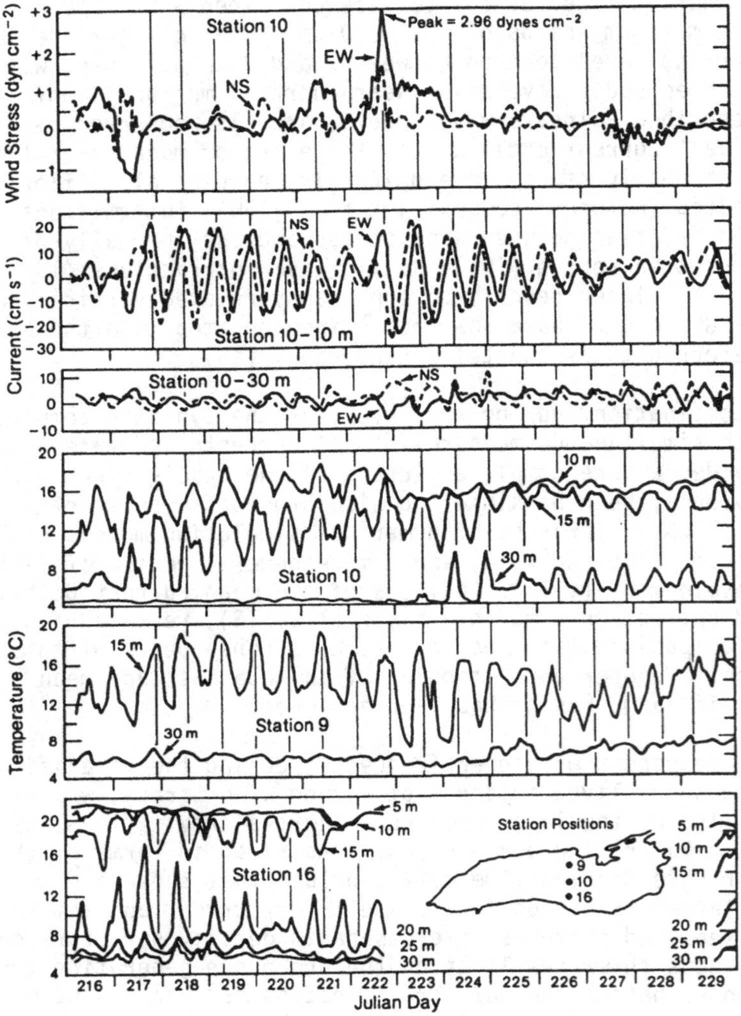

Figure 3 Lake Ontario, 3-16 August 1972: upper three panels display EW and NS components of wind speed squared (4 m above water surface, approximated measure of stress) and current speed at 10 m and 30 m depth at station 10 (see inset map); lower three panels display temperatures at varous depths at stations 9, 10 and 16. All values are hourly means (from (115)). Vertical lines at 16.9 h intervals indicate the mean periodicity of the two bursts of near-inertial waves, generated by the wind impulses on days 217 and 222. Note temperature phase relationship between stations, consistent with a first mode cross-basin seiche with a node south of station 10.

When the downwelling stroke was large, a cross-basin internal surge (interpreted and numerically simulated in (116)) was also initiated and repeatedly generated in phase with the Poincaré oscillation. Surges will be discussed, as non-linear responses, in the next section.

In general, we may conclude with (99) that "from a scientific viewpoint, the IFYGL produced an unprecedented collection of data, stimulating new insights into the physics of lakes". And, more so than in any previous endeavor, numerical simulations (tested and sharpened on solid data bases) contributed substantially to those insights. More revealing, perhaps, would be the deficiencies of the IFYGL. Some of these will be addressed in the following section.

2.2 Post IFYGL themes with potential for future orchestration

While the IFYGL and post-IFYGL analysis was engaging the attention of many North American investigators, researches on other lakes were also yielding new insights and stimulating model development. A comprehensive list would be too long for this lecture and, as the penalty of selectivity, much interesting work must pass unmentioned here. However, some topics for which further insight is clearly needed are selected for brief discussion below; and for an account of some other productive lines of study (particularly recent results) in lake physics, the following references and reviews should be consulted: water-mass exchange in river-dominated lakes (118); cooling dynamics (119,120); mixed-layer dynamics and application to a reservoir simulation model (121,122); turbulent diffusion ((123), with IFYGL results (99,124)); and reviews (57,99,125,126).

a) Evolution of models

It is now a common truism that progress in understanding and eventual management of complex environmental systems, to which lakes belong, depends upon iterative interaction between model-guided measurement campaigns and measurement-tested model development. Recent examples of this bipedal progress are emerging from a special research program directed at "fundamental problems of the water cycle in Switzerland", in which measurement campaigns were mounted in three lakes (Léman, Zurich, Lugano). Numerical simulations and analytical theory have been tested first against surface seiche observations (127,128,129); and the satisfactory agreements are support for extension of these models to more practically important baroclinic motions, for which excellent sets of observations will be used (93,94) as have been a similar set from the Lake of Constance (130). Continuing comparisons between new models and older data sets (e.g. (131,132)) are rewarding and are to be actively encouraged.

The general problem in modelling remains that of parametrization of small scale processes (133) and the influence of those processes on stratification and large-scale flow. Also, simulation of the often different nearshore and offshore regimes and of the mass and energy transfers between them -a deficiency during the IFYGL - demands further study.

b) Nonlinear (surge) features of large-amplitude internal waves

The generation and cross-basin progress of internal surges, discovered and treated (116) during the IFYGL, is one example of the breakdown of linear theory when the vertical amplitude of the motion becomes large in comparison with water depth. Earlier examples have been seen ((18,17), but little influenced by rotation); and the dynamics of the internal surge in Loch Ness has been studied in detail by means of a profiling current meter and thermograph constructed for the purpose (134). Steep fronted features of the shore-trapped internal Kelvin wave in Lake Michigan have also been analyzed in terms of non-linear theory (79); and it has recently been shown ((135), using the data set already mentioned) that large-amplitude internal seiches are accompanied by steep-fronted surges in the Lake of Zurich. Those surges are generated at and travel from either end of the basin whenever the downwelling (never the upwelling) stroke of the thermocline is large enough there. It has been suggested that such surges can be modelled as solitary waves (solitions, (136)). Because the bursts of current and the vertical motions associated with the surges probably contribute to episodic mixing as they pass (i.e. one of the small-scale processes to be parameterized in numerical models), the development and testing of a satisfactory theory is a challenge which must be taken up. The cover figure (p. 287) shows generation and progress of a surge.

c) The low-frequency end of the wave spectrum (vorticity waves)

Figure 4 displays log spectral energy, plotted against frequency, for the summer and winter wind and currents and for the summer temparature fluctuations, covering the whole interval of the IFYGL observations. The spectra permit important conclusions to be drawn: there are no dominant peaks (except perhaps for the winter storm peak) in the wind spectra; the summer current and temperature spectra show large concentrations of energy near the inertial frequency (with a smaller inertial peak in the winter current spectrum) which, in the absence of a dominant wind peak, must represent a strong resonance. (The smaller peaks near twice inertial frequency in the summer current and temperature spectra are probably evidence of non-linear behavior of the kind discussed in the previous paragraph - an hypothesis which merits testing).

Conspicuous also in the current and temperature spectra in Figure 4B is the dip in the range 0.5 to 1 cycles day^{-1} and the large rise in energy as frequency decreases below the range. Possible contributors to that

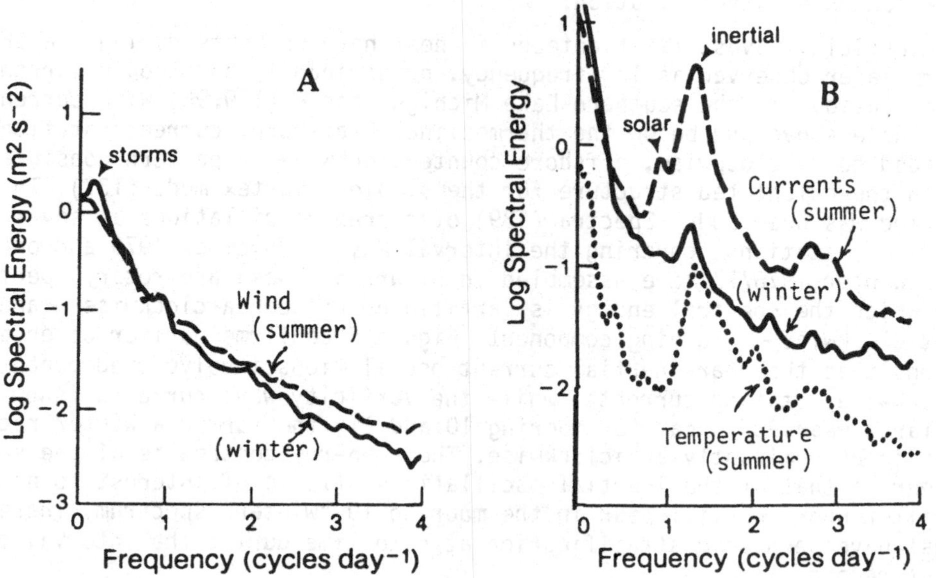

Figure 4 Lake Ontario: averaged spectra (from (113)) of EW components of (A) wind speed and (B) current speed (15 m depth) during "summer" (June-October 1972) and "winter" (November 1972 - May 1973). The summer spectrum of temperature (from (112)) is also shown (dotted line). Log spectral energy in B is in units of $cm^2 s^{-2}$ for currents and deg^2 for temperature.

low-frequency energy are the nearly steady currents of the large scale circulation and the so-called topographic waves, which influenced near-shore currents during the IFYGL (104). In this type of wave the restoring force is not gravity (as in all the wave types mentioned so far) but arises because of the prescript that absolute potential vorticity*) must be conserved. In rotating basins of variable depth that requirement produces a wave response when a fluid column is displaced (by wind, for example) over a sloping bottom. Over slopes representative of lakes the possible wave frequencies are much lower than inertial. The waves are seen as rhythmic (barotropic) perturbations of the current in the whole water column and are therefore not confined (as are internal gravity waves) to

---

*) *The varied terminology (wave of the second class, topographic Rossby wave) seems to be converging on "vorticity wave", which exhibits distinct "vortex modes" in closed basins, calculated and illustrated for a basin of simple geometric form (137). For a review of topographically trapped waves on ocean margins, see ((138) and (141)).*

the season of stratification.

Vorticity waves, first deteced in nearshore currents during the IFYGL, were later observed as low-frequency, predominantly barotropic current fluctuations in the southern Lake Michigan basin (139,96) with currents in phase above and below the thermocline. Nearshore, current rotation tended to be clockwise, offshore counterclockwise, a pattern consistent with the calculated structure for the simplest vortex mode (137). The period was near 90 h. Spectra (139) of current oscillations at several offshore stations, covering the interval May to November 1976 and one for the winter 1976/77, are assembled in Figure 5. These are rotary spectra, in which the spectral energy is partitioned between a clockwise - and an anticlockwise - turning component. Figure 5 confirms earlier observations that the near-inertial current oscillations involve predominantly clockwise rotating currents, while the vorticity wave currents (showing a large near-90 h peak for mooring 10 and 11, the former a winter record) rotate predominantly anticlockwise. Their energy content is of the same order as that of the inertial oscillations. (It is of interest to note a smaller near-inertial peak in the mooring 10 "winter" spectrum. There must have been some stratification at some time during the interval of that record.

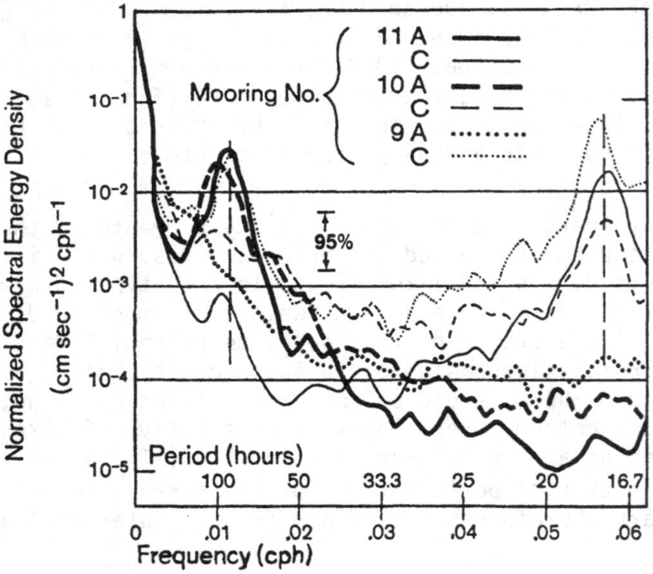

Figure 5 Lake Michigan (May-October 1976): Rotary spectra (from (139)) showing clockwise (thin lines) and anticlockwise (thick lines) components of low-frequency current oscillations at 25 m depth. The mooring No. 10 spectrum is for the 1976/77 winter.

Recalling (from Figure 4) that the wind spectrum shows no conspicuous peaks, one may ask whether the large peak near 100 h in Figure 5 represents a resonant response. In a study shortly to be published, D.J. Schwab (140) has constructed two simple numerical models (both based on the barotropic vorticity equation and with a rigid lid) to seek an answer to this question. One model used observed wind to calculate the time-dependent response of Lake Michigan for the 8 months of record in 1976. In the frequency region 1/8 to 1/3 cycles per day (cpd), corresponding to maximum energy in the forcing function and the response, the agreement with observed currents is good. Elsewhere the model underestimates the currents. The reasons for this discrepancy merit further study. They may be related (Schwab supposes) to insufficient grid resolution in the model, or to neglect of non-linear terms, or to the fact that at frequencies below 1/8 cpd stratification effects (baroclinicity, not included in the model) may be important. It has been noted, in another model/measurement comparison (132) for Lake Michigan, that barotropic oscillations appear to dominate in the 1/8 to 1/3 cpd region, whereas baroclinicity becomes important at lower frequencies.

In his second model Schwab computes the lake's response to a north-south oscillatory wind stress over the frequency range covered by Figure 5; and the average response appears to be non-resonant in this example. Over the (presumed vorticity wave) frequency range 1/8 to 1/3 cpd, the response (relatively insensitive to changes in forcing frequency) resembles a vorticity wave model (137) with two anticlockwise progressing counter-rotating gyres in the southern basin, with a more complicated pattern in the northern part.

Direct observation through current analysis, of topographic vorticity waves during the IFYGL and in Lake Michigan is a notable advance. Parallel investigations on ocean margins (reviewed in (138)) initially concentrated on the propagation of sea-level and temperature changes, in which signals from the lowest topographic wave modes predominated; but recent analysis (141) of currents and model structure has substantially improved the estimate of energy distribution between the modes (see also L.A. Mysak's lecture on lake topographic waves in this volume). An important but complex area of study promises to be the interactions between vorticity waves and inertia-gravitational waves and "steady" wind-driven or buoyancy-driven currents.

d) Frontogenesis and interactions between coastal and interior flow

The baroclinicity of nearly steady flows in Lake Michigan, and by analogy in other large lakes, has been inferred above. The most striking example is observed during the spring warming period in the Great Lakes when stratification begins in a narrow band near shore and then migrates offshore to cover the whole basin by the end of June. The offshore edge of the stratified strip forms a relatively distinct front (the "thermal bar"), on the shore side of which a steady current flows shore-parallel and cyclonically (anticlockwise) around the basin. This boundary current, in which the offshore-directed pressure gradient associated with the developing stratification is (geostrophically) balanced by the onshore-directed Coriolis force, has been modelled ((142), see Figure 6, (143)) and observed (144) by means of a satellite-tracked current-following drogue, a technique recently expanded systematically to large-lake flow studies (145). The boundary current attained a maximum daily average speed of 25 cm s^{-1} on 1 June 1976; and satellite images during the thermal bar season often show large eddies developing along the thermal front (see the example along the eastern shore in Figure 7), probably the result of the strong shear developing between the current and the offshore water mass. Questions relating to this type of instability, and the simi-

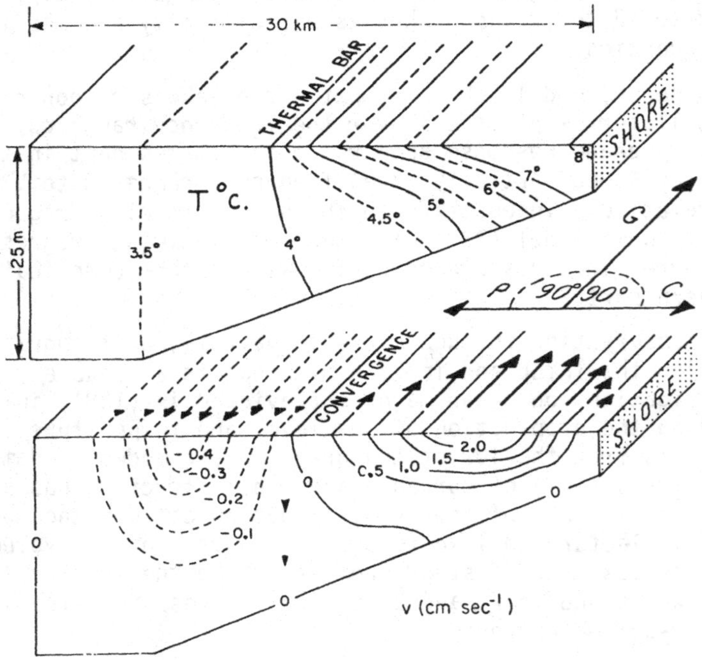

Figure 6 Numerical simulation of a thermal bar and buoyancy-driven boundary current in Lake Michigan (redrawn from (142)) showing distribution of isotherms and current after 45 days of uniform surface insolation at 400 langleys per day from an initial uniform temperature of 2°C. G indicates the shore-parallel geostrophic current, in which the Coriolis force C balances the offshore-directed buoyancy-determined pressure force P.

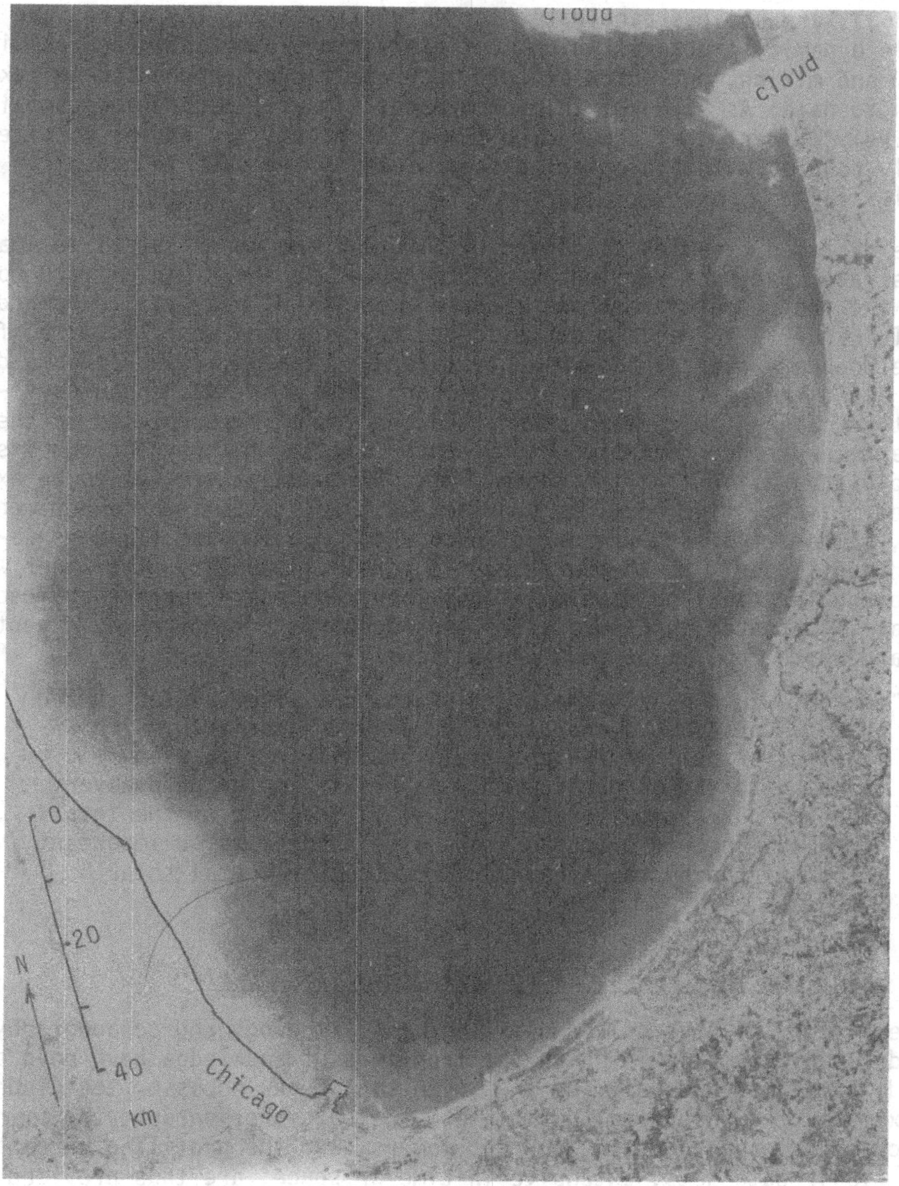

Figure 7 Landsat image (green band 4) of southern Lake Michigan, 1 May 1979. (NASA photo).

lar instability which develops at upwelling fronts in summer (modelled by (146)), are obvious candidates for future research. The eddies are mesoscale dynamic structures which must exert a profound influence an dispersal and mixing, but which are not incorporated or reproduced in lake models to date. As study of them progresses, it will probably become increasingly evident that the hydrodynamics of fronts and eddies is often closely coupled with biological dynamics, as is the case in oceanic examples (147).

The detailed dynamics disclosed by Figure 7 and other satellite images demonstrate that much remains to be discovered in the field of lake motions and that some of the most important questions have yet to be addressed. In Figure 7 there is clear evidence of cyclonic (anticlockwise) flow in the boundary current (shown up by turbidity labels, resuspended sediments along the western shore after strong N-NE wind, suspended phytoplankton ? along the eastern shore, and bending of river plumes in the SE quadrant); but eddy formation at the outer edge of the current and its possible separation from the shore in the NE quadrant are yet to be explained. A suggestive model is provided by the rotating tank experiment (148), illustrated in Figure 8, in which a colored buoyant fluid is released continuously at the surface at a vertical boundary (the inner wall of the annular tank) to develop a geostrophic boundary current analogous to that illustrated in Figure 6. As the boundary current grows in depth (h_1) and width (L), it becomes progressively

> "unstable to non-axisymmetric disturbances. The wavelength and phase velocities of the disturbances were consistent with a model of baroclinic instability of two-layer flow when frictional dissipation due to Ekman layers is included. However, when the current only occupied a small fraction of the total depth, barotropic processes were also thought to be important, with the growing waves gaining from the horizontal shear." (148).

e) *Into the future with new tools and evolving models*

The history of physical (and biological) limnology and oceanography provides many examples in which the frontiers of knowledge have been expanded, sometimes suddenly, by the invention of new tools or techniques, for example, the reversing thermometer, the bathythermograph, the thermistor chain, moored recording current meters, and the profiling current meter (134). Data sets, extensive in time or space, are thus provided for statistical analysis, separating signals from "noise" (not forgetting that one investigator's noise may be another's signal). To pursue the themes mentioned above as worthy of further intensive study - non-linear features of internal waves, parameterization of small- and meso-scale processes, interactions of gravity and vorticity waves, the shearing pro-

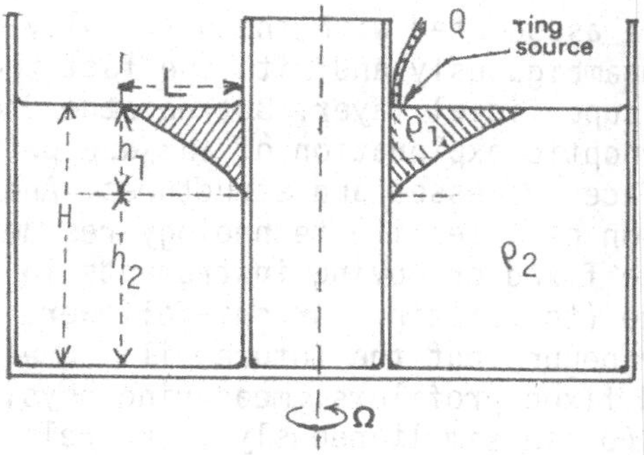

Figure 8

Development (*a* to *d*) of instability in a buoyancy-driven geostrophic boundary current, modelling a rotating annular tank (re-arranged from (148), for details see text).

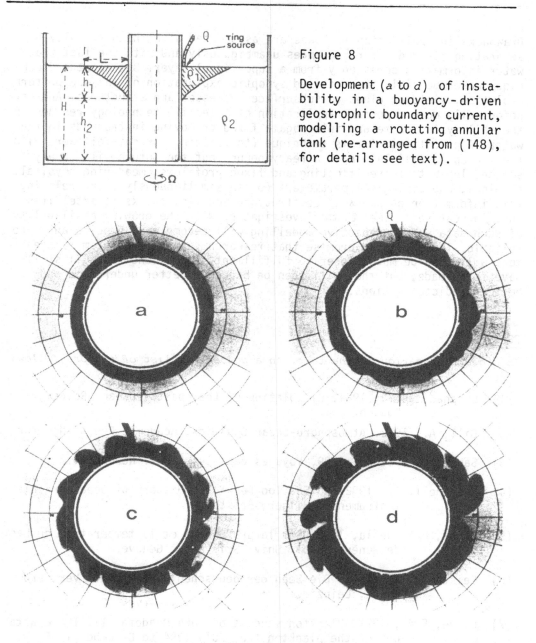

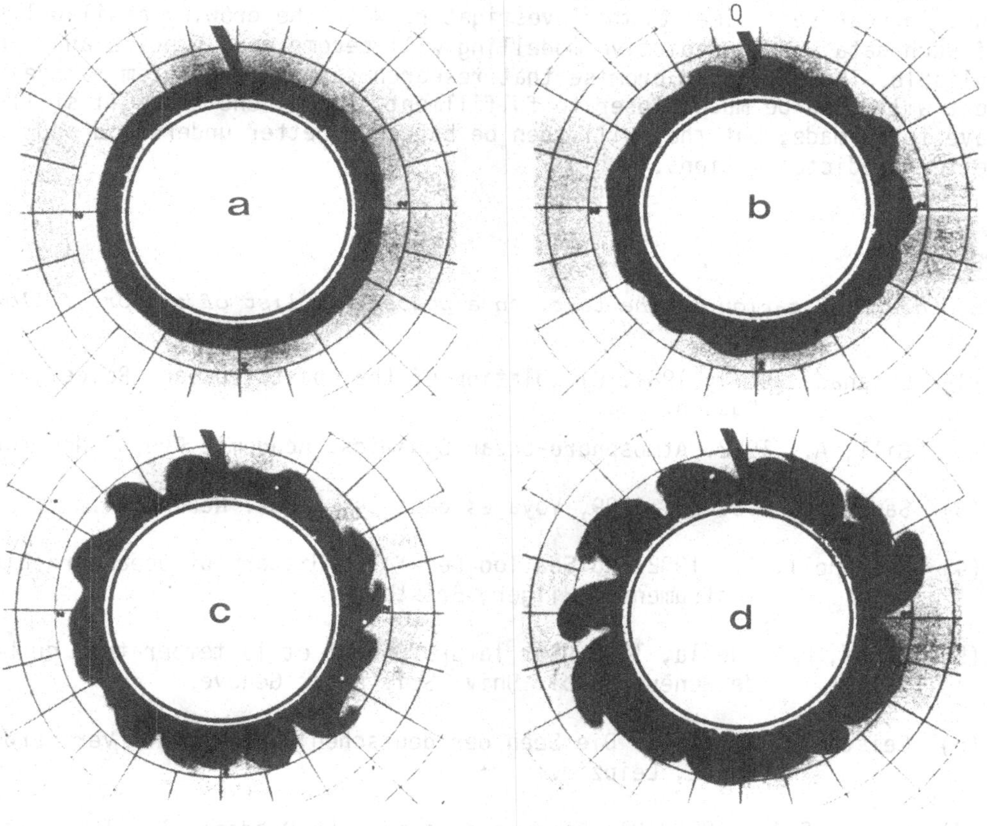

perties of nearly steady currents, and the matching of coastal and interior flow - requires the invention of improved or new tools.

For example, in recent years, synopticity on a repeat frequency of days has been rendered possible over a wide range of *horizontal* scales by satellite imagery and other forms of remote sensing (example in Figure 8).

Drawbacks in application to lakes are associated with the difficulty of separating air and water radiances unambiguously and with the fact that water information comes only from a superficial layer. But no other technique makes possible the detailed synoptic exploration of *surface* pattern evolution, yielding clues to subsurface processes and structures. Another, potentially more important application of satellite technology resides in the ability to locate and interrogate fixed or moving instruments in the water. Applications of this technique (to drifting current-followers with thermographs, (144,145) have already begun; but the future will surely see development of free-floating and fixed profilers, measuring physical, chemical, and biological parameters (often simultaneously), and relaying this information on *sub-surface* structure and flow tracks to satellites and, in near-real time, to the investigator. With the growing availability of such data sets, predictive modelling will become more precise and more reliable. The oft-heard promise that research can aid ecosystem management will then be much closer to fulfillment. Social choices will still have to be made; but they will then be based on better understood and better predicted options.

REFERENCES

In order of citation in the text. An alphabetical list of authors follows.

(1) Casanady, G.T., 1981. Circulation of the Coastal Ocean. Reidel, Boston.

(2) Gill, A., 1982. Atmosphere-ocean Dynamics. Academic Press, New York.

(3) Saussure, H.B. de, 1799. Voyages dans les Alpes. Neuchatel.

(4) McConnell, A., 1982. No Sea Too Deep: The History of Oceanographic Instruments. Hilger, Bristol.

(5) Bèche, H.T. de la, 1819. Sur la profondeur et la température du Lac de Genève. Bibl. Univ. Sci. Arts, Genève.

(6) Geistbeck, A., 1885. Die Seen der deutschen Alpen. Mitt. Ver. Erdkunde, Leipzig.

(7) Birge, E.A., 1897. Plankton studies on Lake Mendota: II, The crustacea of the plankton from July 1894 to December 1896. Trans. Wisconsin Acad. Sci. Arts Lett., II, p. 274.

(8) Forel, F.A., 1895. Lac Léman. Monographie Limnologique, 2, Lausanne.

(9) Richter, E., 1897. Seestudien. Pencks Geogr. Abh., Wien, 6, p. 121

(10) Strom, K.M., 1939. A reversing thermometer by Richter & Wiese with 1/100°C graduation. Int. Rev. Hydrol., 38, p. 259.

(11) Mortimer, C.H., 1953. A review of temperature measurement in limnology. Mitt. int. Verein. Limnol., 1, p. 25.

(12) Becquerel & Breschet, 1836. Procédé électro-chimique pour determiner la température de la terre et des lacs à diverses profondeurs. C.R. Acad. Sci., Paris, 3, p. 778.

(13) Warren, H.W. and G.C. Whipple, 1895. The thermophone, a new instrument for determining temperature. Technol. Quart., 8, p. 125.

(14) Platt, R.B. and C.S. Shoup, 1950. The use of a thermistor in a study of summer temperature conditions of Mountain Lake, Virginia. Ecology, 31, p. 484.

(15) Mortimer, C.H. and W.H. Moore, 1953. The use of thermistors for the measurement of lake temperatures. Mitt. int. Ver. Limnol., 2, p. 42.

(16) Mortimer, C.H., 1952. Water movements in stratified lakes deduced from observations in Windermere and model experiments. Un. Geod. Geophys., Bruxelles, Assn. int. Hydrol., C. rend. rapp., 3, p. 335.

(17) Mortimer, C.H., 1955. Some effects of the earth's rotation on water movement in stratified lakes. Verh. Internat. Verein. Limnol., 12, p. 66.

(18) Wedderburn, E.M. and A.W. Young, 1915. Temperature observations in Loch Earn, Part II. Trans. Roy. Soc. Edinburgh, 50, p. 741.

(19) Mortimer, C.H., 1952. Water movements in lakes during summer stratification; evidence from the distribution of temperature in Windermere. Phil. Trans. Roy. Soc. London, B, 236, p. 355.

(20) Spilhaus, A.F., 1938. A bathythermograph. J. Mar. Res. Yale, 1, p. 95.

(21) Boyce, F.M. and C.H. Mortimer, 1977. IFYGL temperature transects, Lake Ontario, 1972. Environment Canada, Tech. Bull., 100.

(22) Church, P.E., 1942. The annual temperature cycle of Lake Michigan, I. Cooling from late autumn to the terminal point, 1941-1942. Univ. Chicago, Inst. Meteorol., Misc. Rept., 4.

(23) Church, P.E., 1945. The annual temperature cycle of Lake Michigan, II. Spring warming and summer stationary periods, 1942. Univ. Chicago, Inst. Meteorol., Misc. Rept., 18.

(24) Hollan, E., D.B. Rao and E. Bäuerle, 1980. Free surface oscillations in Lake Constance with an interpretation of the "Wonder of the rising water" at Constance in 1549. Arch. Met. Geophys. Biokl., Ser. A, 29.

(25) Duillier, F. de, 1730. Remarques sur l'histoire du lac de Genève in Spon. Histoire de Genève, 2, p. 463.

(26) Vaucher, J.P.E., 1833. Mémoire sur les seiches due Lac de Genève, composé de 1803 à 1804. Mém. Soc. Phys. Genève, 6, p. 35.

(27) André, Father Louis, quoted and translated in Relation of 1671-72, and in Relation of 1676-77; Thwaites, R.G., The Jesuit Relations and Other Documents, 60, Burrow, Cleveland, Ohio, 1671, 1676.

(28) Mortimer, C.H., 1965. Spectra of long surface waves and tides in Lake Michigan and Green Bay, Wisconsin. Great Lakes Res. Div., Univ. Mich., Publ. No. 13, p. 304.

(29) Heaps, N.S., 1975. Resonant tidal co-oscillations in a narrow gulf. Arch. Met. Geophys. Bioklim., Ser. A, 24, p. 361.

(30) Heaps, N.S., C.H. Mortimer and E.J. Fee, 1982. Numerical models and observations of water motion in Green Bay, Lake Michigan. Phil. Trans. Roy. Soc., A, 306, p. 371.

(31) Comstock, C.B., 1872. Irregular oscillations in surface of Lake Michigan at Milwaukee. Ann. Rep. Survey of the Northern and Northwestern Lakes, Appendix B, 14.

(32) Forel, F.A., 1875. Les seiches, vagues d'oscillation fixe des lacs, 1er discours. Actes Soc. Helvétique Sci. Nat., Andermatt, p. 157.

(33) Rao, D.B., C.H. Mortimer and D.J. Schwab, 1976. Surface normal modes of Lake Michigan: calculations compared with spectra of observed water level fluctuations. J. Phys. Oceanogr., 6, p. 575.

(34) Merian, J.R., 1825. Ueber die Bewegungen tropfbarer Flüssigkeiten in Gefässen, Abhandl. J.R. Merian, Basel. (Attracted little attention until reproduced by Von der Mühl, Math. Ann., 28, p. 575 (1885)).

(35) Chrystal, G., 1905. On the hydrodynamical theory of seiches (with a bibliographical sketch). Trans. Roy. Soc. Edinburgh, 41, p. 599.

(36) Defant, A., 1918. Neue Methode zur Ermittlung der Eigenschwingungen (seiches) von abgeschlossenen Wassermassen (Seen, Buchten, usw.). Ann. Hydrogr., Berlin, 46, p. 78.

(37) Caloi, P., 1954. Oscillazioni libre del Lago di Garda. Arch. Met. Geophys. Bioklim., Ser. A, 7, p. 434.

(38) Servais, F., 1957. Etude théorique des oscillations libres (seiches) du Lac Tanganika. Explor. Hydrobiol. Lac Tanganika, 1946-47, Inst. Roy. Sci. Nat. Belgique, Bruxelles, 2, p. 3.

(39) Raggio, G. and K. Hutter, 1982. An extended channel model for the prediction of motion in elongated homogeneous lakes. Parts 1-3. J. Fluid Mech., 121, p. 231.

(40) Bergsten, F., 1926. The seiches of Lake Vetter. Geogr. Ann. Stockh., 1.

(41) Halbfass, W., 1923. Grundzüge einer vergleichenden Seenkunde. Bornträger, Berlin.

(42) Taylor, G.I., 1920. Tidal oscillations in gulfs and rectangular basins. Proc. Lond. Math. Soc., 2nd Ser. 20, p. 148.

(43) Jeffreys, H., 1923. The free oscillations of water in an elliptical lake. Proc. Lond. Math. Soc., 2nd Ser., p. 455.

(44) Neumann, G., 1941. Eigenschwingungen der Ostsee. Arch. Deutsch. Seewarte, Mar. Observat., 61 (Heft 4).

(45) Defant, F., 1953. Theorie der Seiches des Michigansees und ihre Abwandlung durch Wirkung der Corioliskraft. Arch. Met. Geophys. Bioklim., Wien, A, 6, p. 218.

(46) Platzman, G.W. and D.B. Rao, 1964. The free oscillations of Lake Erie. Studies in Oceanography (Hidaka Volume), pp. 359-382. Tokyo University Press.

(47) Platzman, G.W. and D.B. Rao, 1964. Spectra of Lake Erie water levels. J. Geophys. Res., 69, p. 2525.

(48) Platzman, G.W., 1972. Two-dimensional free oscillations in natural basins. J. Phys. Oceanogr., 2, p. 117.

(49) Mortimer, C.H. and E.J. Fee, 1976. Free surface oscillation and tides of Lakes Michigan and Superior. Phil Trans. Soc. London, A, 281, p. 1.

(50) Rao, D.B. and D.J. Schwab, 1976. Two-dimensional mormal modes in arbitrary enclosed basins on a rotating earth: applications to Lakes Ontario and Superior. Phil. Trans. Roy. Soc. London, A, 281, p. 63.

(51) Schmidt, W., 1917. Wirkungen der ungeordneten Bewegungen im Wasser der Meere und Seen. Ann. Hydrogr. Mar. Met., 367, p. 431.

(52) Schmidt, W., 1928. Ueber Temperatur und Stabilitätsverhältnisse von Seen. Geogr. Ann., 145.

(53) Richardson, L.F., 1925. Turbulence and vertical temperature difference near trees. Phil. Mag., 49, p. 81.

(54) Hellström, B., 1941. Wind effect on lakes and rivers. Ingen Vetensk Akad. Handl., 158.

(55) Ruttner, F., 1952. Grundriss der Limnologie. 2nd ed., Berlin, Gruyter, 232, translated by F.E.J. Fry and D.G. Frey as Fundamentals of Limnology, Toronto, University Press.

(56) Hutchinson, G.E., 1957. A Treatise on Limnology, Vol. 1. Geography, Physics and Chemistry, Wiley, New York, p. 1015.

(57) Mortimer, C.H., 1974. Lake hydrodynamics. Mitt. internat. Verein. Limnol., 20, p. 124.

(58) Mortimer, C.H., 1956. E.A. Birge - an explorer of lakes, pp. 163-211 in E.A. Birge, a memoir, by G.C. Sellery, Madison, Univ. Wisconsin Press.

(59) Birge, E.A. and Juday, C., 1929. Transmission of solar radiation by the waters of inland lakes. Trans. Wis. Acad. Sci. Arts, Lett., 24, p. 509.

(60) Sauberer, F. and F. Ruttner, 1941. Die Strahlungsverhältnisse der Binnengewässer. Becker & Erler, Leipzig, p. 240.

(61) Birge, E.A., 1916. The work of the wind in warming a lake. Trans. Wis. Acad. Sci. Arts, Lett., 18, pp. 341, 429, 495, 508.

(62) Schmidt, W., 1925. Der Massenaustausch in freier Luft und verwandte Erscheinungen. Probleme der kosmischen Physik, Vol. 7, H. Grand, Hamburg.

(63) Thoulet, J., 1894. Contribution à l'Etude des lacs des Vosges. Geography (Bull. Soc. Geogr., Paris), 15.

(64) Watson, E.R., 1904. Movements of the waters of Loch Ness as indicated by temperature observations. Geogr. J., 24, p. 430.

(65) Wedderburn, E.M., 1907. An experimental investigation of the temperature changes occurring in fresh-water lochs. Proc. Roy. Soc. Edinburgh, 28, p. 2.

(66) Wedderburn, E.M., 1911. The temperature seiche. I: Temperature observations in Madüsee Pomerania. II: Hydrodynamical theory of temperature oscillations in lakes. III: Calculation of the period of the temperature seiche in the Madüsee. Trans. Roy. Soc. Edinburgh, 47. p. 619.

(67) Wedderburn, E.M., 1912. Temperature observations in Loch Earn, with a further contribution to the hydrodynamical theory of the temperature seiche. Trans. Roy. Soc. Edinburgh, 48, pp. 629.

(68) Birge, E.A., 1910. On the evidence for temperature seiches. Trans. Wis. Acad. Sci. Arts Lett., 16, p. 1005.

(69) Mortimer, C.H., 1953. The resonant response of stratified lakes to wind. Schweiz. Z. Hydrol., 15, p. 94.

(70) Johnson, O.H., 1946. Termisk hydrologiska studier i Sjön Klämmingen. Geogr. Ann., Stockh., 1.

(71) Mortimer, C.H. 1961. Motion in thermoclines. Verh. Internat. Verein. Limnol., 14, p. 79.

(72) Heaps, N.S., 1961. Seiches in a narrow lake, uniformly stratified in three layers. Geophys. Suppl. J. Roy. Astronom. Soc., 5, p. 134.

(73) Heaps, N.S. and A.E. Ramsbottom, 1966. Wind effects on the water in a narrow two-layered lake. Phil. Trans. Roy. Soc. London, A, 259, p. 391.

(74) Mortimer, C.H., 1963. Frontiers in physical limnology with particular reference to long waves in rotating basins. Proc. 5th Conf. Great Lakes Res., Univ. Michigan, Great Lakes Div., Publ. No. 9, p. 9.

(75) Thomson, W. (Lord Kelvin), 1879. On graviational oscillations of rotating water. Proc. Roy. Soc. Edinburgh, 10, p. 92.

(76) Poincaré, H., 1910. Théorie des marées. Leçons de mécanique céleste, 3, Paris.

(77) Sverdrup, H.U., 1926. Dynamic of tides on the North Siberian shelf: result from the Maud Expedition. Geophys. Publ., 4.

(78) Platzman, G.W., 1970. Ocean tides and related waves. Amer. Math. Soc., Lectures in Appl. Math., 14, p. 239.

(79) Bennett, J., 1973. A theory of large amplitude Kelvin Waves. J. Phys. Oceanogr., 3, p. 57.

(80) Csanady, G., 1978. Water circulation and dispersal mechanisms, Chap. 2 in Lakes: Chemistry, Geology, Physics, A. Lerman, Ed., Springer-Verlag, New York.

(81) Mortimer, C.H., 1968. Internal waves and associated currents observed in Lake Michigan during the summer of 1963. Univ. Wisconsin-Milwaukee, Center for Great Lakes Studies, Spec. Report No. 1.

(82) Ayers, J.C., D.C. Chandler, G.H. Lauff, C.F. Powers and E.B. Henson, 1958. Currents and water masses of Lake Michigan. Univ. of Michigan, Great Lakes Res. Div., Publ. No. 3.

(83) Gustafson, T. and B. Kullenberg, 1936. Untersuchungen von Trägheitsströmungen in der Ostsee. Svensk, Hydrogr. biol. Komm. Skr. Ny ser. Hydrogr. No. 13.

(84) Mortimer, C.H., 1980. Inertial motion and related internal waves in Lake Michigan and Lake Ontario as responses to impulsive wind stress. I: Introduction, descriptive narrative and archive of IFYGL data. Univ. Wisconsin-Milwaukee, Center for Great Lakes Studies, Spec. Report No. 37.

(85) Bauer, S.W., W.H. Graf, C.H. Mortimer and C. Perrinjaquet. Inertial motion in Lake Geneva (Léman). Arch. Met. Geophys. Biokl. Ser. A, 30, p. 289.

(86) Platzman, G.W., 1958. A numerical computation of the surge of 26 June 1954 on Lake Michigan. Geophysica, 6, p. 407.

(87) Verber, J.L., 1964. Initial current studies in Lake Michigan. Limnol. Oceanogr., 9, p. 426.

(88) Verber, J.L., 1964. The detection of rotary currents and internal waves in Lake Michigan. Proc. 7th Conf. Great Lakes Res., Univ. Michigan, Great Lakes Res. Div., Publ. No. 11, p. 382.

(89) U.S. Department of the Interior. Lake currents (Water Quality Investigation, Lake Michigan Basin). Federal Water Pollution Control Admin., Great Lakes Region. Chicago, IL, Tech. Rep., Nov. 1967, (Principal author and editor, J.L. Verber, other contributions by C.H. Mortimer and A. Okubo).

(90) Mortimer, C.H., 1971. Large-scale oscillatory motions and seasonal temperature changes in Lake Michigan and Lake Ontario. Univ. Wisconsin-Milwaukee, Center for Great Lakes Studies, Spec. Report No. 12, Part I: Text, Part II: Illustrations.

(91) Maline, F.D., 1968. An analysis of current measurements in Lake Michigan. J. Geophys. Res., 73, p. 7065.

(92) Hollan, E., 1974. Strömungen im Bodensee. Teil 6. 6. Bericht Arbeitsgemeinschaft Wasserwerke Bodensee-Rhein (AWBR), Landesanstalt für Umweltschutz, Baden-Württemberg, Karlsruhe.

(93) Horn, W., 1981. Zürichsee 1978: Physikalisch-limnologisches Messprogramm und Datensammlung. Internal Report No. 50, Versuchsanstalt für Wasserbau, Eidg. Techn. Hochschule, Zürich.

(94) Graf, W.H., C Perrinjaquet, S.W. Bauer, J.P. Prost and H. Girod, 1979. Measuring on Lake Geneva. Hydrodynamics of Lakes, Eds. Graf, W.H. and Mortimer, C.H.; Elsevier, Amsterdam.

(95) Saylor, J.H. and G.S. Miller, 1979. Lake Huron winter currents. J. Geophys. Res., 84, p. 3237.

(96) Huang, J.C.K. and J.H. Saylor, 1982. Vorticity waves in a shallow basin. Dyn. Atmos. Oceans, 6, p. 177.

(97) Byron, J.W., Ed., 1976. International Field Year for the Great Lakes final Canadian data and information catalogue. Canadian IFYGL Data Bank, Canada Centre for Inland Waters, Department of Environment, Burlington, Ontario.

(98) Hodge, W.T., 1978. International Field Year for the Great Lakes (IFYGL) data catalogue: United States data catalogue, NOAA Technical Memorandum EDS NCC-3, National Oceanic and Atmospheric Administration, U.S. Department of Commerce, Rockwille, MD, (The archive is located at the National Climatic Center, Ashville, NC).

(99) Aubert, E.J. and T.L. Richards, Eds., 1981. IFYGL - The International Field Year for the Great Lakes, U.S. Dept. Commerce, National Oceanic and Atmospheric Admin., Great Lakes Environ. Res. Lab., Ann Arbor, MI, 48104, U.S.A.

(100) Csanady, G.T., 1977. The coastal jet conceptual model in the dynamics of shallow seas, in The Seas: Ideas and Observations on Progress in the Seas. Ed.: Goldberg, E.D.; John Wiley & Sons, New York, 117.

(101) Csanady, G.T., 1977. Intermittent "full" upwelling in Lake Ontario, J. of Geophys. Res., 82, p. 397.

(102) Csanady, G.T., 1974. Spring thermocline behavior in Lake Ontario during IFYGL. J. of Phys. Oceanogr., 4, p. 425.

(103) Csanady, G.T. and J.T, Scott, 1974. Baroclinic coastal jets in Lake Ontario during IFYGL. J. of Phys. Oceanogr., 4, p. 524.

(104) Csanady, G.T., 1976. Topographic waves in Lake Ontario. J. of Phys. Oceanogr., 6, p. 93.

(105) Simons, T.J., 1974. Verification of numerical models of Lake Ontario. Part I: circulation in spring and early summer. J. of Phys. Oceanogr., 4, p. 507.

(106) Simons, T.J., 1975. Verification of numerical models of Lake Ontario. Part II: stratified circulations and temperature changes. J. of Phys. Oceanogr., 5, p. 98.

(107) Simons, T.J., 1976. Verification of numerical models of Lake Ontario. Part III: long-term heat transport. J. of Phys. Oceanogr., 6, p. 372.

(108) Bennett, J.R., 1977. A three-dimensional model of Lake Ontario's summer circulation. Part I: comparison with observations, J. of Phys. Oceanogr., 7, p. 591.

(109) Bennett, J.R., 1978. A three-dimensional model of Lake Ontario's summer circulation. Part II: a diagnostic study. J. of Phys. Oceanogr., 8, p. 1095.

(110) Bennett, J.R. and E.J. Lindstrom, 1977. A simple model of Lake Ontario's coastal boundary layer. J. of Phys. Oceanogr., 7, p. 620.

(111) Simons, T.J., 1980. Circulation models of lakes and inland seas. Can. Bull. Fish. Aquat. Sci., No. 203.

(112) Pickett, R.L. and F.P. Richards, 1975. Lake Ontario mean temperatures and currents in July 1972. J. of Phys. Oceanogr., 5, p. 775.

(113) Pickett, R.L. and S. Bermick, 1977. Observed resultant circulation of Lake Ontario. Limnol. Oceanogr., 22, p. 1071.

(114) Marmorino, G.O. and C.H. Mortimer, 1978. Internal waves observed in Lake Ontario during the International Field Year for the Great Lakes (IFYGL) 1972. Part II: spectral analysis and model decomposition. Univ. Wisconsin-Milwaukee, Center for Great Lakes Studies. Spec. Report No. 33.

(115) Mortimer, C.H., 1977. Internal waves observed in Lake Ontario during the International Field Year for the Great Lakes (IFYGL) 1972. Part I: descriptive survey and preliminary interpretation of near-inertial oscillations in terms of linear channel-wave models. Univ. Wisconsin-Milwaukee, Center for great Lakes Studies, Spec. Report No. 32.

(116) Simons, T.J., 1978. Generation and propagation of downwelling fronts. J. of Phys. Oceanogr., 8, p. 571.

(117) Schwab, D.J., 1977. Internal free oscillations in Lake Ontario. Limnol. Oceanogr., 22, p. 700.

(118) Carmack, E.C., 1979. Combined influence of inflow and lake temperature on spring circulation in a riverine lake. J. Phys. Oceanogr., 9, p. 422.

(119) Farmer, D.M. and E.C. Carmack, 1981. Wind mixing and restrafication in a lake near the temperature of maximum density. J. Phys. Oceanogr., 11, p. 1516.

(120) Carmack, E.C. and D.M. Farmer, 1982. Cooling processes in deep, temperate lakes: a review with examples from two lakes in British Columbia. J. Mar. Res., 40 (Suppl.) p. 85.

(121) Spigel, R.H. and J. Imberger, 1980. Classification of mixed-layer dynamics in lakes of small to medium size. J. of Phys. Oceanogr., 10, p. 1104.

(122) Imberger, J. and J.C. Patterson, 1981. A dynamic reservoir simulation model - DYRESM-5. Transport Models for Inland and Coastal Waters; Ed.: H.B. Fisher, Academic Press.

(123) Jassby, A. and T. Powell, 1975. Vertical patterns of eddy diffusion during stratification in Castle Lake, California. Limnol. Oceanogr., 20, p. 530.

(124) Murthy, C.R., 1976. Horizontal diffusion characteristics in Lake Ontario. J. of Phys. Oceanogr., 6, p. 76.

(125) Edinger, J.E., D.K. Brady and J.C. Geyer, 1974. Heat exchange and transport in the environment. Electric Power Res. Inst., Publ. EPRI 74-049-00-3, The Johns Hopkins Univ., Dept. Geogr., Envir. Eng., Report No. 14.

(126) Imberger, J. and P.F. Hamblin, 1982. Dynamics of lakes, reservoirs, and cooling ponds. Ann. Rev. Fluid Mech., 14, p. 153.

(127) Hutter, K., G. Raggio, C. Bucher and G. Salvadè, 1982. The surface seiches of Lake Zurich. Schweiz. Z. Hydrol., 44, p. 423.

(128) Hutter, K., G. Raggio, C. Bucher, G. Salvadè and F. Zamboni, 1982. The surface seiches of Lake of Lugano. Schweiz. Z. Hydrol., 44, p. 455.

(129) Bauer, S.W., 1983. Simulation of Lake Geneva seiches by irregular-grid finite-difference model. Mitt. Inst. Meereskunde, Univ. Hamburg, 26, p. 199.

(130) Hollan, E., 1983. Erfahrungen mit der mathematische Modellierung grossräumiger Bewegungsvorgänge im Bodensee. Mitt. Inst. Meereskunde, Univ. Hamborg, 26, p. 154.

(131) Allender, J.H., 1977. Comparison od model and observed currents in Lake Michigan. J. of Phys. Oceanogr., 7, p. 711.

(132) Allender, J.H. and J.H. Saylor, 1979. Model and observed circulation throughout the annual temperature cycle of Lake Michigan. J. of Phys. Oceanogr., 9, p. 573.

(133) Turner, J.S., 1981. Small-scale mixing processes. Evolution of Phys. Oceanogr. Eds.: Warren, B.A. and Wunsch, C.; MIT Press, Massachusetts.

(134) Thorpe, S.A., 1977. Turbulence and mixing in a Scottish loch. Phil. Trans. Roy. Soc. London, A, 286, p. 125.

(135) Mortimer, C.H. and W. Horn, 1982. Internal wave dynamics and their implications for plankton biology in the Lake of Zurich. Vierteljahresschr. Naturforsch. Ges. Zürich, 127, p. 299.

(136) Osborne, A.R. and T.L. Burch, 1980. Internal solitons in the Andaman Sea. Science, 208, p. 451.

(137) Ball, F.K., 1965. Second-class motions of a shallow fluid. J. Fluid Mech., 23, p. 545.

(138) Mysak. L.A., 1980. Topographically trapped waves. Ann Rev. Fluid Mech., 12, p. 45.

(139) Saylor, J.H., J.C.K. Huang and R.O. Reid, 1980. Vortex modes in southern Lake Michigan. J. of Phys. Oceanogr. 10, p. 1814.

(140) Schwab, D.J., 1984. The low-frequency response of Lake Michigan currents to time-dependent and oscillatory sind forcing, (submitted to J. of Phys. Oceanogr.).

(141) Hsieh, W.W., 1982. On the detection of continental shelf waves. J. of Phys. Oceanogr., 12, p. 414.

(142) Bennett, J.R., 1971. Thermally driven lake currents during the spring and fall transition periods. Proc. 14th Conf. Great Lakes Research, Int. Assoc. for Great Lakes Research, 535.

(143) Huang, J.C.K., 1971. The thermal current in Lake Michigan. J. of Phys. Ocenaogr., 1, p. 105.

(144) Mortimer, C.H., 1977. One of Lake Michigan's responses to the sun, the wind, and the spinning earth. Forum 1977, Lectures Celebrating 20th Anniversary of College of Letters and Science, Univ. Wisconsin-Milwaukee.

(145) Pickett, R.L., R.M. Partridge, A.H. Clites and J.E. Campbell, 1983. Great Lakes satellite-tracked drifters, (in press, J. Mar. Technol. Soc.).

(146) Rao, D.B. and B.C. Doughty, 1981. Instability of coastal currents in the Great Lakes. Arch. Meteorol., Geophys., Bioklimatol., A, 30, p. 145.

(147) Pingree, R.D., 1979. Baroclinic eddies bordering the Celtic Sea in late summer. J. Mar. Biol. Ass., U.K., 59, p. 689.

(148) Griffiths, R.W. and P.F. Linden, 1981. The stability of buoyancy-driven coastal currents. Dynamics of Atmospheres and Oceans, 5, p. 281.

ALPHABETICAL LIST OF AUTHORS WITH
CORRESPONDING REFERENCE NUMBERS

AUTHOR INDEX

SUBJECT INDEX

G)

H)

I, J)

K)

L)

Printed in the United States
By Bookmasters